FRONTIERS OF STATISTICS

FRONTIERS OF STATISTICS

The book series provides an overview on the new developments in the frontiers of statistics. It aims at promoting statistical research that has high societal impacts and offers a concise overview on the recent developments on a topic in the frontiers of statistics. The books in the series intend to give graduate students and new researchers an idea where the frontiers of statistics are, to learn common techniques in use so that they can advance the fields via developing new techniques and new results. It is also intended to provide extensive references so that researchers can follow the threads to learn more comprehensively what the literature is, to conduct their own research, and to cruise and lead the tidal waves on the frontiers of statistics.

SERIES EDITORS

Jianqing Fan
Frederick L. Moore '18 Professor of Finance,
Director of Committee of Statistical Studies,
Department of Operation Research and
Financial Engineering,
Princeton University, NJ 08544, USA.

Zhiming Ma
Academy of Math and Systems Science,
Institute of Applied Mathematics,
Chinese Academy of Science,
No.55, Zhong-guan-cun East Road,
Beijing 100190, China.

EDITORIAL BOARD

High-Dimensional Data Analysis

Editors

T. Tony Cai
University of Pennsylvania, USA

Xiaotong Shen
University of Minnesota, USA

Volume 2

Frontiers of Statistics

Higher Education Press

World Scientific

NEW JERSEY · LONDON · SINGAPORE · BEIJING · SHANGHAI · HONG KONG · TAIPEI · CHENNAI

T. Tony Cai
Department of Statistics,
University of Pennsylvania,
Philadelphia, PA 19104, USA.

Xiaotong Shen
School of Statistics,
University of Minnesota,
Minneapolis, MN 55455, USA.

Copyright © 2011 by
Higher Education Press
4 Dewai Dajie, 100120, Beijing, P.R. China and
World Scientific Publishing Co Pte Ltd
5 Toh Tuch Link, Singapore 596224

ISBN 13: 978-981-4324-85-4
ISBN 10: 981-283-4324-85-x
ISSN 1793-8155

Preface

Over the last few years, significant developments have been taking place in high-dimensional data analysis, which are driven primarily by a wide range of applications in many fields, such as genomics and signal processing. In particular, substantial advances have been made in the areas of feature selection, covariance estimation, classification and regression. This book intends to examine important issues arising from high-dimensional data analysis to explore key ideas for statistical inference and prediction. The book is structured around topics on multiple hypothesis testing, feature selection, regression, classification, dimension reduction, as well as applications in survival analysis and in biomedical research.

Fundamental statistical issues underlying data have changed, when moving from low-dimensional to high-dimensional analyses. For instance, certain structures such as sparsity need to be utilized in feature selection when the number of candidate features greatly exceeds that of the sample size. As a result of high-dimensionality, traditional statistical methods designed for low dimensional problems become inadequate or break down. To meet these challenges in high-dimensional analysis, statisticians have been developing new methods and introducing new concepts, where many issues emerge with regard to how to identify or utilize certain structures for dimension reduction in inference and prediction.

There exists a vast body of literature on high-dimensional analysis, especially for prediction, classification and regression. We do not intend to give an overview of each subject but would like to mention here only a few topics of interest—feature selection, basis/grouping pursuit, multiple hypothesis testing, effective dimension reduction and projection pursuit, sparsity, high-dimensional regression and classification.

Many classification problems are often high-dimensional. In Chapter 1, Fan, Fan and Wu review contemporary classification methods, including linear discriminant analysis, naive Bayes, and loss-based methods, as well as the impact of dimensionality on classification, with a special attention towards regularization and feature selection. In Chapter 2, Liu and Wu move further to the topic of large margin classification, where they examine various state-of-art methods for various of support vector machines, and their connection with probability estimation.

In Chapter 3, Cai and Sun consider large-scale multiple testing. They begin by reviewing methods for controlling family-wise error rates and false discovery rates (FDR), as well as other pertinent issues in multiple testing. Their main focus is on optimal multiple testing procedures minimizing the false nondiscovery rate while controls the FDR. Both independent and dependent cases are considered.

The topic of high-dimensional feature (variable) selection has been a focus in recent research. In Chapter 4, Yuan reviews several popular variable selection methods, and contrast classical methods such as stepwise selection with modern methods such as regularization. In Chapter 5, Zhu, Pan and Shen examine Bayesian model selection for networks, particularly gene networks where the number of genes in a network may greatly exceed the sample size.

In Chapter 6, Li describes a number of interesting applications in genomics studies involving networks and graphical models, where the dimension under consideration is ultra-high. Various regression techniques have been reviewed, where special structures of genomic data are considered.

Survival data analysis is an important subject in biostatistics. Analysis high-dimensional survival data requires power tools. In Chapter 7, Li and Ren focus on joint modeling for censored and longitudinal data. Various models are reviewed, subject to different types of censoring. In Chapter 8, Nan reviews the recent development of feature selection in penalized regression in survival analysis, which is a marriage between high-dimensional feature selection and survival analysis. Several methods are examined, particularly for high-dimensional covariates such as gene expressions, whereas various penalties such as grouped, hierarchical penalties are discussed.

For high-dimensional data analysis, dimension reduction is essential. In Chapter 9, Yin gives a comprehensive review on sufficient dimension reduction in regression. In Chapter 10, Chen and Yang discuss combining strategies.

Finally, we sincerely hope that this book can simulate further interest from statisticians, computer scientists and engineers, and promote further collaborations among them to attack important problems in high-dimensional data analysis.

T Tony Cai, Philadelphia
Xiaotong Shen, Minneapolis
May 18, 2010

Contents

Part III Model Building with Variable Selection

Part IV High-Dimensional Statistics in Genomics

Part V   Analysis of Survival and Longitudinal Data

Chapter 8   Survival Analysis with High-Dimensional Covariates

Part VI   Sufficient Dimension Reduction in Regression

Chapter 9   Sufficient Dimension Reduction in Regression

Chapter 10   Combining Statistical Procedures

Part I

High-Dimensional Classification

Chapter 1

High-Dimensional Classification*

Jianqing Fan[†] Yingying Fan[‡] and Yichao Wu[§]

Abstract

In this chapter, we give a comprehensive overview on high-dimensional classification, which is prominently featured in many contemporary statistical problems. Emphasis is given on the impact of dimensionality on implementation and statistical performance and on the feature selection to enhance statistical performance as well as scientific understanding between collected variables and the outcome. Penalized methods and independence learning are introduced for feature selection in ultrahigh dimensional feature space. Popular methods such as the Fisher linear discriminant, Bayes classifiers, independence rules, distance-based classifiers and loss-based classification rules are introduced and their merits are critically examined. Extensions to multi-class problems are also given.

Keywords: Bayes classifier, classification error rates, distanced-based classifier, feature selection, impact of dimensionality, independence learning, independence rule, loss-based classifier, penalized methods, variable screening.

1   Introduction

Classification is a supervised learning technique. It arises frequently from bioinformatics such as disease classifications using high throughput data like micorarrays or SNPs and machine learning such as document classification and image recognition. It tries to learn a function from training data consisting of pairs of input features and categorical output. This function will be used to predict a class label of any valid input feature. Well known classification methods include (multiple) logistic regression, Fisher discriminant analysis, k-th-nearest-neighbor classifier, support vector machines, and many others. When the dimensionality of the input

---

*The authors are partly supported by NSF grants DMS-0714554, DMS-0704337, DMS-0906784, and DMS-0905561 and NIH grants R01-GM072611 and R01-CA149569.

†Department of ORFE, Princeton University, Princeton, NJ 08544, USA, E-mail: jqfan@princeton.edu

‡Information and Operations Management Department, Marshall School of Business, University of Southern California, Los Angeles, CA 90089, USA, E-mail: fanyingy@marshall.usc.edu

§Department of Statistics, North Carolina State University, Raleigh, NC 27695, USA, E-mail: wu@stat.ncsu.edu

feature space is large, things become complicated. In this chapter we will try to investigate how the dimensionality impacts classification performance. Then we propose new methods to alleviate the impact of high dimensionality and reduce dimensionality.

We present some background on classification in Section 2. Section 3 is devoted to study the impact of high dimensionality on classification. We discuss distance-based classification rules in Section 4 and feature selection by independence rule in Section 5. Another family of classification algorithms based on different loss functions is presented in Section 6. Section 7 extends the iterative sure independent screening scheme to these loss-based classification algorithms. We conclude with Section 8 which summarizes some loss-based multicategory classification methods.

2   Elements of classifications

Suppose we have some input space $\mathcal{X}$ and some output space $\mathcal{Y}$. Assume that there are independent training data $(\mathbf{X}_i, Y_i) \in \mathcal{X} \times \mathcal{Y}$, $i = 1, \ldots, n$ coming from some unknown distribution P, where Y_i is the i-th observation of the response variable and $\mathbf{X}_i$ is its associated feature or covariate vector. In classification problems, the response variable Y_i is qualitative and the set $\mathcal{Y}$ has only finite values. For example, in the cancer classification using gene expression data, each feature vector $\mathbf{X}_i$ represents the gene expression level of a patient, and the response Y_i indicates whether this patient has cancer or not. Note that the response categories can be coded by using indicator variables. Without loss of generality, we assume that there are K categories and $\mathcal{Y} = \{1, 2, \ldots, K\}$. Given a new observation $\mathbf{X}$, classification aims at finding a classification function $g : \mathcal{X} \to \mathcal{Y}$, which can predict the unknown class label Y of this new observation using available training data as accurately as possible.

To access the accuracy of classification, a loss function is needed. A commonly used loss function for classification is the *zero-one loss*:

$$L(y, g(\mathbf{x})) = \begin{cases} 0, & g(\mathbf{x}) = y, \\ 1, & g(\mathbf{x}) \neq y. \end{cases} \tag{2.1}$$

This loss function assigns a single unit to all misclassifications. Thus the risk of a classification function g, which is the expected classification error for an new observation $\mathbf{X}$, takes the following form:

$$\overline{W}(g) = E[L(Y, g(\mathbf{X}))] = E\left[\sum_{k=1}^{K} L(k, g(\mathbf{X}))P(Y = k|\mathbf{X})\right]$$
$$= 1 - P(Y = g(\mathbf{x})|\mathbf{X} = \mathbf{x}), \tag{2.2}$$

where Y is the class label of $\mathbf{X}$. Therefore, the optimal classifier in terms of minimizing the misclassification rate is

$$g^*(\mathbf{x}) = \arg\max_{k \in \mathcal{Y}} P(Y = k|\mathbf{X} = \mathbf{x}) \tag{2.3}$$

This classifier is known as the *Bayes classifier* in the literature. Intuitively, Bayes classifier assigns a new observation to the most possible class by using the posterior probability of the response. By definition, Bayes classifier achieves the minimum misclassification rate over all measurable functions:

$$\overline{W}(g^*) = \min_g \overline{W}(g). \tag{2.4}$$

This misclassification rate $\overline{W}(g^*)$ is called the Bayes risk. The Bayes risk is the minimum misclassification rate when distribution is known and is usually set as the benchmark when solving classification problems.

Let $f_k(\mathbf{x})$ be the conditional density of an observation $\mathbf{X}$ being in class k, and π_k be the prior probability of being in class k with $\sum_{i=1}^K \pi_i = 1$. Then by Bayes theorem it can be derived that the posterior probability of an observation $\mathbf{X}$ being in class k is

$$P(Y = k | \mathbf{X} = \mathbf{x}) = \frac{f_k(\mathbf{x})\pi_k}{\sum_{i=1}^K f_i(\mathbf{x})\pi_i}. \tag{2.5}$$

Using the above notation, it is easy to see that the Bayes classifier becomes

$$g^*(\mathbf{x}) = \arg\max_{k \in \mathcal{Y}} f_k(\mathbf{x})\pi_k. \tag{2.6}$$

For the following of this chapter, if not specified we shall consider the classification between two classes, that is, $K = 2$. The extension of various classification methods to the case where $K > 2$ will be discussed in the last section.

The *Fisher linear discriminant analysis* approaches the classification problem by assuming that both class densities are multivariate Gaussian $N(\boldsymbol{\mu}_1, \boldsymbol{\Sigma})$ and $N(\boldsymbol{\mu}_2, \boldsymbol{\Sigma})$, respectively, where $\boldsymbol{\mu}_k$, $k = 1, 2$ are the class mean vectors, and $\boldsymbol{\Sigma}$ is the common positive definite covariance matrix. If an observation $\mathbf{X}$ belongs to class k, then its density is

$$f_k(\mathbf{x}) = (2\pi)^{-p/2}(\det(\boldsymbol{\Sigma}))^{-1/2}\exp\left\{-\frac{1}{2}(\mathbf{x} - \boldsymbol{\mu}_k)^T\boldsymbol{\Sigma}^{-1}(\mathbf{x} - \boldsymbol{\mu}_k)\right\}, \tag{2.7}$$

where p is the dimension of the feature vectors $\mathbf{X}_i$. Under this assumption, the Bayes classifier assigns $\mathbf{X}$ to class 1 if

$$\pi_1 f_1(\mathbf{X}) \geqslant \pi_2 f_2(\mathbf{X}), \tag{2.8}$$

which is equivalent to

$$\log\frac{\pi_1}{\pi_2} + (\mathbf{X} - \boldsymbol{\mu})^T\boldsymbol{\Sigma}^{-1}(\boldsymbol{\mu}_1 - \boldsymbol{\mu}_2) \geqslant 0, \tag{2.9}$$

where $\boldsymbol{\mu} = \frac{1}{2}(\boldsymbol{\mu}_1 + \boldsymbol{\mu}_2)$. In view of (2.6), it is easy to see that the classification rule defined in (2.8) is the same as the Bayes classifier. The function

$$\delta_F(\mathbf{x}) = (\mathbf{x} - \boldsymbol{\mu})^T\boldsymbol{\Sigma}^{-1}(\boldsymbol{\mu}_1 - \boldsymbol{\mu}_2) \tag{2.10}$$

is called the *Fisher discriminant function*. It assigns $\mathbf{X}$ to class 1 if $\delta_F(\mathbf{X}) \geq \log \frac{\pi_2}{\pi_1}$; otherwise to class 2. It can be seen that the Fisher discriminant function is linear in $\mathbf{x}$. In general, a classifier is said to be linear if its discriminant function is a linear function of the feature vector. Knowing the discriminant function δ_F, the classification function of Fisher discriminant analysis can be written as $g_F(\mathbf{x}) = 2 - I(\delta_F(\mathbf{x}) \geq \log \frac{\pi_2}{\pi_1})$ with $I(\cdot)$ the indicator function. Thus the classification function is determined by the discriminant function. In the following, when we talk about a classification rule, it could be the classification function g or the corresponding discriminant function δ.

Denote by $\boldsymbol{\theta} = (\boldsymbol{\mu}_1, \boldsymbol{\mu}_2, \boldsymbol{\Sigma})$ the parameters of the two Gaussian distributions $N(\boldsymbol{\mu}_1, \boldsymbol{\Sigma})$ and $N(\boldsymbol{\mu}_2, \boldsymbol{\Sigma})$. Write $\overline{W}(\delta, \boldsymbol{\theta})$ as the misclassification rate of a classifier with discriminant function δ. Then the discriminant function δ_B of the Bayes classifier minimizes $\overline{W}(\delta, \boldsymbol{\theta})$. Let $\Phi(t)$ be the distribution function of a univariate standard normal distribution. If $\pi_1 = \pi_2 = \frac{1}{2}$, it can easily be calculated that the misclassification rate for Fisher discriminant function is

$$\overline{W}(\delta_F, \boldsymbol{\theta}) = \Phi\left(-\frac{d^2(\boldsymbol{\theta})}{2}\right), \tag{2.11}$$

where $d(\boldsymbol{\theta}) = \{(\boldsymbol{\mu}_1 - \boldsymbol{\mu}_2)^T \boldsymbol{\Sigma}^{-1} (\boldsymbol{\mu}_1 - \boldsymbol{\mu}_2)\}^{1/2}$ and is named as the *Mahalanobis distance* in the literature. It measures the distance between two classes and was introduced by Mahalanobis (1930). Since under the normality assumption the Fisher discriminant analysis is the Bayes classifier, the misclassification rate given in (2.11) is in fact the Bayes risk. It is easy to see from (2.11) that the Bayes risk is a decreasing function of the distance between two classes, which is consistent with our common sense.

Let Γ be some parameter space. With a slight abuse of the notation, we define the maximum misclassification rate of a discriminant function δ over Γ as

$$\overline{W}_\Gamma(\delta) = \sup_{\boldsymbol{\theta} \in \Gamma} \overline{W}(\delta, \boldsymbol{\theta}). \tag{2.12}$$

It measures the worst classification result of a classifier δ over the parameter space Γ. In some cases, we are also interested in the *minimax regret* of a classifier, which is the difference between the maximum misclassification rate and the minimax misclassification rate, that is,

$$R_\Gamma(\delta) = \overline{W}_\Gamma(\delta) - \sup_{\boldsymbol{\theta} \in \Gamma} \min_\delta \overline{W}(\delta, \boldsymbol{\theta}). \tag{2.13}$$

Since the Bayes classification rule δ_B minimizes the misclassification rate $\overline{W}(\delta, \boldsymbol{\theta})$, the minimax regret of δ can be rewritten as

$$R_\Gamma(\delta) = \overline{W}_\Gamma(\delta) - \sup_{\boldsymbol{\theta} \in \Gamma} \overline{W}(\delta_B, \boldsymbol{\theta}). \tag{2.14}$$

From (2.11) it is easy to see that for classification between two Gaussian distributions with common covariance matrix, the minimax regret of δ is

$$R_\Gamma(\delta) = \overline{W}_\Gamma(\delta) - \sup_{\boldsymbol{\theta} \in \Gamma} \Phi\left(-\frac{1}{2}d(\boldsymbol{\theta})\right). \tag{2.15}$$

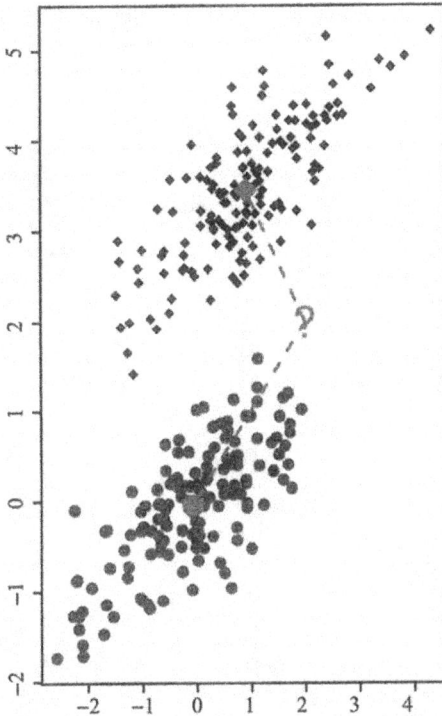

Figure 2.1   Illustration of distance-based classification. The centroid of each subsample in the training data is first computed by taking the sample mean or median. Then, for a future observation, indicated by query, it is classified according to its distances to the centroids.

The Fisher discriminant rule can be regarded as a specific method of distance-based classifiers, which have attracted much attention of researchers. Popularly used distance-based classifiers include support vector machine, naive Bayes classifier, and k-th-nearest-neighbor classifier. The distance-based classifier assigns a new observation $\mathbf{X}$ to class k if it is on average closer to the data in class k than to the data in any other classes. The "distance" and "average" are interpreted differently in different methods. Two widely used measures for distance are the Euclidean distance and the Mahalanobis distance. Assume that the center of class i distribution is $\boldsymbol{\mu}_i$ and the common convariance matrix is $\boldsymbol{\Sigma}$. Here "center" could be the mean or the median of a distribution. We use $\text{dist}(\mathbf{x}, \boldsymbol{\mu}_i)$ to denote the distance of a feature vector $\mathbf{x}$ to the centriod of class i. Then if the Euclidean distance is used,

$$\text{dist}_E(\mathbf{x}, \boldsymbol{\mu}_i) = \sqrt{(\mathbf{x} - \boldsymbol{\mu}_i)^T(\mathbf{x} - \boldsymbol{\mu}_i)}, \qquad (2.16)$$

and the Mahalanobis distance between a feature vector $\mathbf{x}$ and class i is

$$\text{dist}_M(\mathbf{x}, \boldsymbol{\mu}_i) = \sqrt{(\mathbf{x} - \boldsymbol{\mu}_i)^T \boldsymbol{\Sigma}^{-1}(\mathbf{x} - \boldsymbol{\mu}_i)}. \qquad (2.17)$$

Thus the distance-based classifier places a new observation $\mathbf{X}$ to class k if

$$\arg\min_{i\in\mathcal{Y}} \operatorname{dist}(\mathbf{X}, \boldsymbol{\mu}_i) = k. \tag{2.18}$$

Figure 2.1 illustrates the idea of distanced classifier classification.

When $\pi_1 = \pi_2 = 1/2$, the above defined Fisher discriminant analysis has the interpretation of distance-based classifier. To understand this, note that (2.9) is equivalent to

$$(\mathbf{X} - \boldsymbol{\mu}_1)^T \boldsymbol{\Sigma}^{-1}(\mathbf{X} - \boldsymbol{\mu}_1) \leqslant (\mathbf{X} - \boldsymbol{\mu}_2)^T \boldsymbol{\Sigma}^{-1}(\mathbf{X} - \boldsymbol{\mu}_2). \tag{2.19}$$

Thus δ_F assigns $\mathbf{X}$ to class 1 if its Mahalanobis distance to the center of class 1 is smaller than its Mahalanobis distance to the center of class 2. We will introduce in more details about distance-based classifiers in Section 4.

3 Impact of dimensionality on classification

A common feature of many contemporary classification problems is that the dimensionality p of the feature vector is much larger than the available training sample size n. Moreover, in most cases, only a fraction of these p features are important in classification. While the classical methods introduced in Section 2 are extremely useful, they no longer perform well or even break down in high dimensional setting. See Donoho (2000) and Fan and Li (2006) for challenges in high dimensional statistical inference. The impact of dimensionality is well understood for regression problems, but not as well understood for classification problems. In this section, we discuss the impact of high dimensionality on classification when the dimension p diverges with the sample size n. For illustration, we will consider discrimination between two Gaussian classes, and use the Fisher discriminant analysis and independence classification rule as examples. We assume in this section that $\pi_1 = \pi_2 = \frac{1}{2}$ and n_1 and n_2 are comparable.

3.1 Fisher discriminant analysis in high dimensions

Bickel and Levina (2004) theoretically studied the asymptotical performance of the sample version of Fisher discriminant analysis defined in (2.10), when both the dimensionality p and sample size n goes to infinity with p much larger than n. The parameter space considered in their paper is

$$\Gamma_1 = \{\boldsymbol{\theta} : d^2(\boldsymbol{\theta}) \geqslant c^2, c_1 \leqslant \lambda_{\min}(\boldsymbol{\Sigma}) \leqslant \lambda_{\max}(\boldsymbol{\Sigma}) \leqslant c_2, \boldsymbol{\mu}_k \in B, k = 1, 2\}, \tag{3.1}$$

where c, c_1 and c_2 are positive constants, $\lambda_{\min}(\boldsymbol{\Sigma})$ and $\lambda_{\max}(\boldsymbol{\Sigma})$ are the minimum and maximum eigenvalues of $\boldsymbol{\Sigma}$, respectively, and $B = B_{\mathbf{a},d} = \{\mathbf{u} : \sum_{j=1}^{\infty} a_j u_j^2 < d^2\}$ with d some constant, and $a_j \to \infty$ as $j \to \infty$. Here, the mean vectors $\boldsymbol{\mu}_k$, $k = 1, 2$ are viewed as points in l_2 by adding zeros at the end. The condition on eigenvalues ensures that $\frac{\lambda_{\max}(\boldsymbol{\Sigma})}{\lambda_{\min}(\boldsymbol{\Sigma})} \leqslant \frac{c_2}{c_1} < \infty$, and thus both $\boldsymbol{\Sigma}$ and $\boldsymbol{\Sigma}^{-1}$ are not ill-conditioned. The condition $d^2(\boldsymbol{\theta}) \geqslant c^2$ is to make sure that the Mahalanobis distance between two classes is at least c. Thus the smaller the value of c, the harder the classification problem is.

Given independent training data $(\mathbf{X}_i, Y_i)$, $i = 1, \ldots, n$, the common covariance matrix can be estimated by using the sample covariance matrix

$$\widehat{\boldsymbol{\Sigma}} = \sum_{k=1}^{K} \sum_{Y_i = k} (\mathbf{X}_i - \widehat{\boldsymbol{\mu}}_k)(\mathbf{X}_i - \widehat{\boldsymbol{\mu}}_k)^T / (n - K). \tag{3.2}$$

For the mean vectors, Bickel and Levina (2004) showed that there exist estimators $\widetilde{\boldsymbol{\mu}}_k$ of $\boldsymbol{\mu}_k$, $k = 1, 2$ such that

$$\max_{\Gamma_1} E_{\boldsymbol{\theta}} \|\widetilde{\boldsymbol{\mu}}_k - \boldsymbol{\mu}_k\|^2 = o(1). \tag{3.3}$$

Replacing the population parameters in the definition of δ_F by the above estimators $\widetilde{\boldsymbol{\mu}}_k$ and $\widehat{\boldsymbol{\Sigma}}$, we obtain the sample version of Fisher discriminant function $\hat{\delta}_F$.

It is well known that for fixed p, the worst case misclassification rate of $\hat{\delta}_F$ converges to the worst case Bayes risk over Γ_1, that is,

$$\overline{W}_{\Gamma_1}(\hat{\delta}_F) \to \overline{\Phi}(c/2), \quad \text{as } n \to \infty, \tag{3.4}$$

where $\overline{\Phi}(t) = 1 - \Phi(t)$ is the tail probability of the standard Gaussian distribution. Hence, $\hat{\delta}_F$ is asymptotically optimal for this low dimensional problem. However, in high dimensional setting, the result is very different.

Bickel and Levina (2004) studied the worst case misclassification rate of $\hat{\delta}_F$ when $n_1 = n_2$ in high dimensional setting. Specifically they showed that under some regularity conditions, if $p/n \to \infty$, then

$$\overline{W}_{\Gamma_1}(\hat{\delta}_F) \to \frac{1}{2}, \tag{3.5}$$

where the Moore-Penrose generalized inverse is used in the definition of $\hat{\delta}_F$. Note that $1/2$ is the misclassification rate of random guessing. Thus although Fisher discriminant analysis is asymptotically optimal and has Bayes risk when dimension p is fixed and sample size $n \to \infty$, it performs asymptotically no better than random guessing when the dimensionality p is much larger than the sample size n. This shows the difficulty of high dimensional classification. As have been demonstrated by Bickel and Levina (2004) and pointed out by Fan and Fan (2008), the bad performance of Fisher discriminant analysis is due to the diverging spectra (e.g., the condition number goes to infinity as dimensionality diverges) frequently encountered in the estimation of high-dimensional covariance matrices. In fact, even if the true covariance matrix is not ill conditioned, the singularity of the sample covariance matrix will make the Fisher discrimination rule inapplicable when the dimensionality is larger than the sample size.

3.2 Impact of dimensionality on independence rule

Fan and Fan (2008) studied the impact of high dimensionality on classification. They pointed out that the difficulty of high dimensional classification is intrinsically caused by the existence of many noise features that do not contribute to the reduction of classification error. For example, for the Fisher discriminant analysis discussed before, one needs to estimate the class mean vectors and covariance matrix. Although individually each parameter can be estimated accurately, aggregated estimation error over many features can be very large and this could significantly increase the misclassification rate. This is another important reason that causes the bad performance of Fisher discriminant analysis in high dimensional setting. Greenshtein and Ritov (2004) and Greenshtein (2006) introduced and studied the concept of persistence, which places more emphasis on misclassification rates or expected loss rather than the accuracy of estimated parameters. In high dimensional classification, since we care much more about the misclassification rate instead of the accuracy of the estimated parameters, estimating the full covariance matrix and the class mean vectors will result in very high accumulation error and thus low classification accuracy.

To formally demonstrate the impact of high dimensionality on classification, Fan and Fan (2008) theoretically studied the *independence rule*. The discriminant function of independence rule is

$$\delta_I(\mathbf{x}) = (\mathbf{x} - \boldsymbol{\mu})^T \boldsymbol{D}^{-1}(\boldsymbol{\mu}_1 - \boldsymbol{\mu}_2), \tag{3.6}$$

where $\boldsymbol{D} = \text{diag}\{\boldsymbol{\Sigma}\}$. It assigns a new observation $\mathbf{X}$ to class 1 if $\delta_I(\mathbf{X}) \geq 0$. Compared to the Fisher discriminant function, the independence rule pretends that features were independent and use the diagonal matrix $\boldsymbol{D}$ instead of the full covariance matrix $\boldsymbol{\Sigma}$ to scale the feature. Thus the aforementioned problems of diverging spectrum and singularity are avoided. Moreover, since there are far less parameters need to be estimated when implementing the independence rule, the error accumulation problem is much less serious when compared to the Fisher discriminant function.

Using the sample mean $\widehat{\boldsymbol{\mu}}_k = \frac{1}{n_k}\sum_{Y_i=k} \mathbf{X}_i$, $k = 1, 2$ and sample covariance matrix $\widehat{\boldsymbol{\Sigma}}$ as estimators and letting $\widehat{\boldsymbol{D}} = \text{diag}\{\widehat{\boldsymbol{\Sigma}}\}$, we obtain the sample version of independence rule

$$\hat{\delta}_I(\mathbf{x}) = (\mathbf{x} - \widehat{\boldsymbol{\mu}})^T \widehat{\boldsymbol{D}}^{-1}(\widehat{\boldsymbol{\mu}}_1 - \widehat{\boldsymbol{\mu}}_2). \tag{3.7}$$

Fan and Fan (2008) studied the theoretical performance of $\hat{\delta}_I(\mathbf{x})$ in high dimensional setting.

Let $\mathbf{R} = \boldsymbol{D}^{-1/2}\boldsymbol{\Sigma}\boldsymbol{D}^{-1/2}$ be the common correlation matrix and $\lambda_{\max}(\mathbf{R})$ be its largest eigenvalue, and write $\boldsymbol{\alpha} \equiv (\alpha_1, \ldots, \alpha_p)^T = \boldsymbol{\mu}_1 - \boldsymbol{\mu}_2$. Fan and Fan (2008) considered the parameter space

$$\Gamma_2 = \{(\boldsymbol{\alpha}, \boldsymbol{\Sigma}) : \boldsymbol{\alpha}'\boldsymbol{D}^{-1}\boldsymbol{\alpha} \geq C_p, \lambda_{\max}(\mathbf{R}) \leq b_0, \min_{1\leq j\leq p} \sigma_j^2 > 0\}, \tag{3.8}$$

where C_p is a deterministic positive sequence depending only on the dimensionality p, b_0 is a positive constant, and σ_j^2 is the j-th diagonal element of $\boldsymbol{\Sigma}$. The condition $\boldsymbol{\alpha}^T \boldsymbol{D} \boldsymbol{\alpha} \geq C_p$ is similar to the condition $d(\boldsymbol{\theta}) \geq c$ in Bickel and Levina (2004). In fact, $\boldsymbol{\alpha}'\boldsymbol{D}^{-1}\boldsymbol{\alpha}$ is the accumulated marginal signal strength of p individual features, and the condition $\boldsymbol{\alpha}'\boldsymbol{D}^{-1}\boldsymbol{\alpha} \geq C_p$ imposes a lower bound on it. Since there is no restriction on the smallest eigenvalue, the condition number of $\mathbf{R}$ can diverge with sample size. The last condition $\min_{1 \leqslant j \leqslant p} \sigma_j^2 > 0$ ensures that there are no deterministic features that make classification trivial and the diagonal matrix $\mathbf{D}$ is always invertible. It is easy to see that Γ_2 covers a large family of classification problems.

To access the impact of dimensionality, Fan and Fan (2008) studied the posterior misclassification rate and the worst case posterior misclassification rate of $\hat{\delta}_I$ over the parameter space Γ_2. Let $\mathbf{X}$ be a new observation from class 1. Define the posterior misclassification rate and the worst case posterior misclassification rate respectively as

$$W(\hat{\delta}_I, \boldsymbol{\theta}) = P(\hat{\delta}_I(\mathbf{X}) < 0 | (\mathbf{X}_i, Y_i), i = 1, \ldots, n), \tag{3.9}$$

$$W_{\Gamma_2}(\hat{\delta}_I) = \max_{\boldsymbol{\theta} \in \Gamma_2} W(\hat{\delta}_I, \boldsymbol{\theta}). \tag{3.10}$$

Fan and Fan (2008) showed that when $\log p = o(n)$, $n = o(p)$ and $nC_p \to \infty$, the following inequality holds

$$W(\hat{\delta}_I, \boldsymbol{\theta}) \leqslant \overline{\Phi}\left(\frac{\sqrt{\frac{n_1 n_2}{pn}} \boldsymbol{\alpha}' \mathbf{D}^{-1} \boldsymbol{\alpha}(1 + o_p(1)) + \sqrt{\frac{p}{n n_1 n_2}}(n_1 - n_2)}{2\sqrt{\lambda_{\max}(\mathbf{R})}\{1 + n_1 n_2/(pn)\boldsymbol{\alpha}' \mathbf{D}^{-1}\boldsymbol{\alpha}(1 + o_p(1))\}^{1/2}} \right). \tag{3.11}$$

This inequality gives an upper bound on the classification error. Since $\overline{\Phi}(\cdot)$ decreases with its argument, the right hand side decreases with the fraction inside $\overline{\Phi}$. The second term in the numerator of the fraction shows the influence of sample size on classification error. When there are more training data from class 1 than those from class 2, i.e., $n_1 > n_2$, the fraction tends to be larger and thus the upper bound is smaller. This is in line with our common sense, as if there are more training data from class 1, then it is less likely that we misclassify $\mathbf{X}$ to class 2.

Fan and Fan (2008) further showed that if $\sqrt{n_1 n_2/(np)} C_p \to C_0$ with C_0 some positive constant, then the worst case posterior classification error

$$W_{\Gamma_2}(\hat{\delta}_I) \xrightarrow{\text{P}} \overline{\Phi}\left(\frac{C_0}{2\sqrt{b_0}} \right). \tag{3.12}$$

We make some remarks on the above result (3.12). First of all, the impact of dimensionality is shown as $C_p/\sqrt{p}$ in the definition of C_0. As dimensionality p increases, so does the aggregated signal C_p, but a price of the factor $\sqrt{p}$ needs to be paid for using more features. Since n_1 and n_2 are assumed to be comparable, $n_1 n_2/(np) = O(n/p)$. Thus one can see that asymptotically $W_{\Gamma_2}(\hat{\delta}_I)$ increases with $\sqrt{n/p} C_p$. Note that $\sqrt{n/p} C_p$ measures the tradeoff between dimensionality p and the overall signal strength C_p. When the signal level is not strong enough to balance out the increase of dimensionality, i.e., $\sqrt{n/p} C_p \to 0$ as $n \to \infty$, then $W_{\Gamma_2}(\hat{\delta}_I) \xrightarrow{\text{P}} \frac{1}{2}$. This indicates that the independence rule $\hat{\delta}_I$ would be no better than the random guessing due to noise accumulation, and using less features can be beneficial.

The inequality (3.11) is very useful. Observe that if we only include the first m features $j = 1, \ldots, m$ in the independence rule, then (3.11) still holds with each term replaced by its truncated version and p replaced by m. The contribution of the j feature is governed by its marginal utility α_j^2/σ_j^2. Let us assume that the importance of the features is already ranked in the descending order of $\{\alpha_j^2/\sigma_j^2\}$. Then $m^{-1}\sum_{j=1}^{m}\alpha_j^2/\sigma_j^2$ will most possibly first increase and then decrease as we include more and more features, and thus the right hand side of (3.11) first decreases and then increases with m. Minimizing the upper bound in (3.11) can help us to find the optimal number of features m.

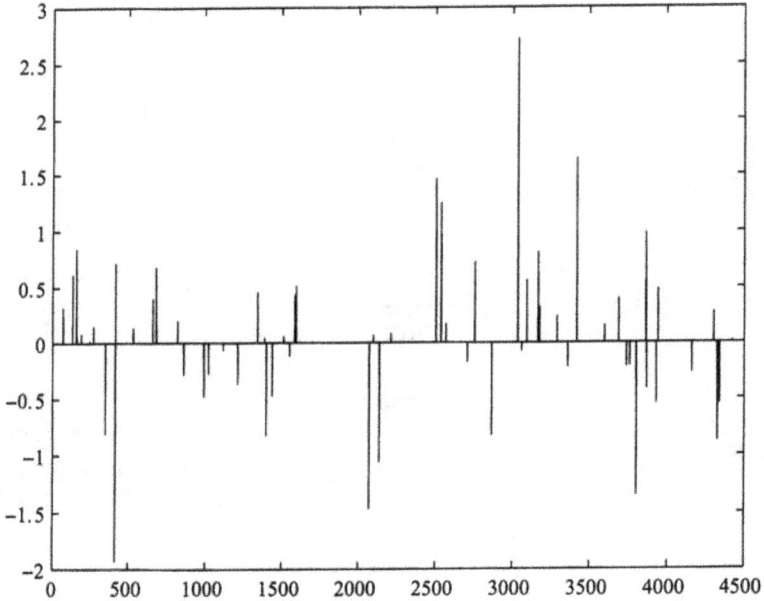

Figure 3.1   The centroid $\boldsymbol{\mu}_1$ of class 1. The heights indicate the values of non-vanishing elements.

To illustrate the impact of dimensionality, let us take $p = 4500$, $\boldsymbol{\Sigma}$ the identity matrix, and $\boldsymbol{\mu}_2 = 0$ whereas $\boldsymbol{\mu}_1$ has 98% of coordinates zero and 2% non-zero, generated from the double exponential distribution. Figure 3.1 illustrates the vector $\boldsymbol{\mu}_1$, in which the heights show the values of non-vanishing coordinates. Clearly, only about 2% of features have some discrimination power. The effective number of features that have reasonable discrimination power (excluding those with small values) is much smaller. If the best two features are used, it clearly has discrimination power, as shown in Figure 3.2(a), whereas when all 4500 features are used, they have little discrimination power (see Figure 3.2(d)) due to noise accumulation. When $m = 100$ (about 90 features are useful and 10 useless: the actual useful signals are less than 90 as many of them are weak) the signals are strong enough to overwhelm the noise accumulation, where as when $m = 500$ (at least 410 features are useless), the noise accumulation exceeds the strength of the signals so that there is no discrimination power.

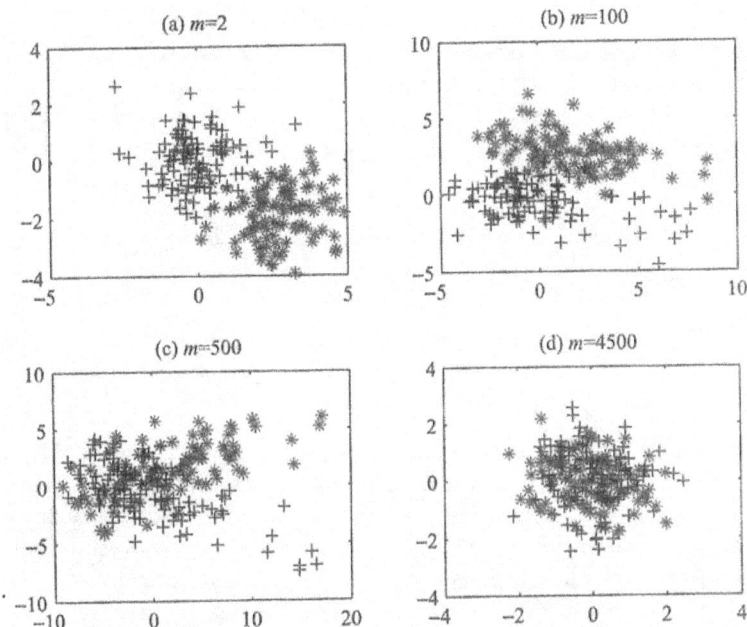

Figure 3.2   The plot of simulated data with "∗" indicated the first class and "+" the second class. The best m features are selected and the first two principal components are computed based on the sample covariance matrix. The data are then projected onto these two principal components and are shown in (a), (b) and (c). In (d), the data are projected on two randomly selected directions in the 4500-dimensional space.

3.3   Linear discriminants in high dimensions

From discussions in the previous two subsections, we see that in high dimensional setting, the performance of classifiers is very different from their performance when dimension is fixed. As we have mentioned earlier, the bad performance is largely caused by the error accumulation when estimating too many noise features with little marginal utility α_j^2/σ_j^2. Thus dimension reduction and feature selection are very important in high dimensional classification.

A popular class of dimension reduction methods is projection. See, for example, principal component analysis in Ghosh (2002), Zou et al. (2004), and Bair et al. (2006); partial least squares in Nguyen and Rocke (2002), Huang and Pan (2003), and Boulesteix (2004); and sliced inverse regression in Li (1991), Zhu et al. (2006), and Bura and Pfeiffer (2003). As pointed out by Fan and Fan (2008), these projection methods attempt to find directions that can result in small classification errors. In fact, the directions found by these methods put much more weight on features that have large classification power. In general, however, linear projection methods are likely to perform poorly unless the projection vector is sparse, namely, the effective number of selected features is small. This is due to the aforementioned noise accumulation prominently featured in high-dimensional problems.

To formally establish the result, let $\mathbf{a}$ be a p-dimensional unit random vector coming from a uniform distribution over a $(p-1)$-dimensional sphere. Suppose that we project all observations onto the vector $\mathbf{a}$ and apply the Fisher discriminant analysis to the projected data $\mathbf{a}^T\mathbf{X}_1, \ldots, \mathbf{a}^T\mathbf{X}_n$, that is, we use the discriminant function

$$\hat{\delta}_{\mathbf{a}}(\mathbf{x}) = (\mathbf{a}^T\mathbf{x} - \mathbf{a}^T\widehat{\mu})(\mathbf{a}^T\widehat{\mu}_1 - \mathbf{a}^T\widehat{\mu}_2). \tag{3.13}$$

Fan and Fan (2008) showed that under some regularity conditions, if $p^{-1}\sum_{i=1}^{p} \alpha_j^2/\sigma_j^2 \to 0$, then

$$P(\hat{\delta}_{\mathbf{a}}(\mathbf{X}) < 0 | (\mathbf{X}_i, Y_i), i = 1, \ldots, n) \xrightarrow{\text{P}} \frac{1}{2}, \tag{3.14}$$

where $\mathbf{X}$ is a new observation coming from class 1, and the probability is taken with respect to the random vector $\mathbf{a}$ and new observation $\mathbf{X}$ from class 1. The result demonstrates that almost all linear discriminants cannot perform any better than random guessing, due to the noise accumulation in the estimation of population mean vectors, unless the signals are very strong, namely the population mean vectors are very far apart. In fact, since the projection direction vector $\mathbf{a}$ is randomly chosen, it is nonsparse with probability one. When a nonsparse projection vector is used, one essentially uses all features to do classification, and thus the misclassification rate could be as high as random guessing due to the noise accumulation. This once again shows the importance of feature selection in high dimensionality classification. To illustrate the point, Figure 3.2(d) shows the projected data onto two randomly selected directions. Clearly, neither projections has discrimination power.

4 Distance-based classification rules

Many distance-based classifiers have been proposed in the literature to deal with classification problems with high dimensionality and small sample size. They intend to mitigate the "curse-of-dimensionality" in implementation. In this section, we will first discuss some specific distance-based classifiers, and then talk about the theoretical properties of general distance-based classifiers.

4.1 Naive Bayes classifier

As discussed in Section 2, the Bayes classifier predicts the class label of a new observation by comparing the posterior probabilities of the response. It follows from the Bayes theorem that

$$P(Y = k | \mathbf{X} = \mathbf{x}) = \frac{P(\mathbf{X} = \mathbf{x} | Y = k)\pi_k}{\sum_{i=1}^{K} P(\mathbf{X} = \mathbf{x} | Y = i)\pi_i}. \tag{4.1}$$

Since $P(\mathbf{X} = \mathbf{x} | Y = i)$ and π_i, $i = 1, \ldots, K$ are unknown in practice, to implement the Bayes classifier we need to estimate them from the training data. However, this method is impractical in high dimensional setting due to the curse of dimensionality and noise accumulation when estimating the distribution $P(\mathbf{X}|Y)$, as discussed in Section 3. The naive Bayes classifier, on the other hand, overcomes this difficulty by making a conditional independence assumption that dramatically reduces the number of parameters to be estimated when modeling $P(\mathbf{X}|Y)$. More specifically, the naive Bayes classifier uses the following calculation:

$$P(\mathbf{X} = \mathbf{x}|Y = k) = \prod_{j=1}^{p} P(X_j = x_j|Y = k), \tag{4.2}$$

where X_j and x_j are the j-th components of $\mathbf{X}$ and $\mathbf{x}$, respectively. Thus the conditional joint distribution of the p features depends only on the marginal distributions of them. So the naive Bayes rule utilizes the marginal information of features to do classification, which mitigates the "curse-of-dimensionality" in implementation. But, the dimensionality does have an impact on the performance of the classifier, as shown in the previous section. Combining (2.6), (4.1) and (4.2) we obtain that the predicted class label by naive Bayes classifier for a new observation is

$$g(\mathbf{x}) = \arg\max_{k \in \mathcal{Y}} \pi_k \prod_{j=1}^{p} P(X_j = x_j|Y = k). \tag{4.3}$$

In the case of classification between two normal distributions $N(\boldsymbol{\mu}_1, \boldsymbol{\Sigma})$ and $N(\boldsymbol{\mu}_2, \boldsymbol{\Sigma})$ with $\pi_1 = \pi_2 = \frac{1}{2}$, it can be derived that the naive Bayes classifier has the discriminant function

$$\delta_I(\mathbf{x}) = (\mathbf{x} - \boldsymbol{\mu})^T \boldsymbol{D}^{-1}(\boldsymbol{\mu}_1 - \boldsymbol{\mu}_2), \tag{4.4}$$

where $\boldsymbol{D} = \text{diag}(\boldsymbol{\Sigma})$, the same as the independence rule (3.7), which assigns a new observation $\mathbf{X}$ to class 1 if $\delta_I(\mathbf{X}) \geqslant 0$; otherwise to class 2. It is easy to see that $\delta_I(\mathbf{x})$ is a distance-based classifier with distance measure chosen to be the weighted L_2-distance: $\text{dist}_I(\mathbf{x}, \boldsymbol{\mu}_i) = (\mathbf{x} - \boldsymbol{\mu}_i)^T \boldsymbol{D}^{-1}(\mathbf{x} - \boldsymbol{\mu}_i)$.

Although in deriving the naive Bayes classifier, it is assumed that the features are conditionally independent, in practice it is widely used even when this assumption is violated. In other words, the naive Bayes classifier pretends that the features were conditionally independent with each other even if they are actually not. For this reason, the naive Bayes classifier is also called independence rule in the literature. In this chapter, we will interchangeably use the name "naive Bayes classifier" and "independence rule".

As pointed out by Bickel and Levina (2004), even when $\boldsymbol{\mu}$ and $\boldsymbol{\Sigma}$ are assumed known, the corresponding independence rule does not lose much in terms of classification power when compared to the Bayes rule defined in (2.10). To understand this, Bickel and Levina (2004) consider the errors of Bayes rule and independence rule, which can be derived to be

$$e_1 = P(\delta_B(\mathbf{X}) \leqslant 0) = \overline{\Phi}\left(\frac{1}{2}\{\boldsymbol{\alpha}^T \boldsymbol{\Sigma}^{-1} \boldsymbol{\alpha}\}^{1/2}\right) \text{ and }$$

$$e_2 = P(\delta_I(\mathbf{X}) \leqslant 0) = \overline{\Phi}\left(\frac{1}{2} \frac{\boldsymbol{\alpha}^T \boldsymbol{D}^{-1} \boldsymbol{\alpha}}{\{\boldsymbol{\alpha}^T \boldsymbol{D}^{-1} \boldsymbol{\Sigma} \boldsymbol{D}^{-1} \boldsymbol{\alpha}\}^{1/2}}\right),$$

respectively. Since the errors e_k, $k = 1, 2$ are both decreasing functions of the arguments of $\overline{\Phi}$, the efficiency of the independence rule relative to the Bayes rule is determined by the ratio r of the arguments of $\overline{\Phi}$. Bickel and Levina (2004) showed that the ratio r can be bounded as

$$r = \frac{\overline{\Phi}^{-1}(e_2)}{\overline{\Phi}^{-1}(e_1)} = \frac{\alpha^T D^{-1} \alpha}{\left\{(\alpha^T \Sigma^{-1} \alpha)(\alpha^T D^{-1} \Sigma D^{-1} \alpha)\right\}^{1/2}} \geqslant \frac{2\sqrt{K_0}}{1 + K_0}. \tag{4.5}$$

where $K_0 = \max_{\Gamma_1} \frac{\lambda_{\max}(\mathbf{R})}{\lambda_{\min}(\mathbf{R})}$ with $\mathbf{R}$ the common correlation matrix defined in Section 3.2. Thus the error e_2 of the independence rule can be bounded as

$$e_1 \leqslant e_2 \leqslant \overline{\Phi}\left(\frac{2\sqrt{K_0}}{1 + K_0} \overline{\Phi}^{-1}(e_1)\right). \tag{4.6}$$

It can be seen that for moderate K_0, the performance of independence rule is comparable to that of the Fisher discriminant analysis. Note that the bounds in (4.6) represents the worst case performance. The actual performance of independence rule could be better. In fact, in practice when α and Σ both need to be estimated, the performance of independence rule is much better than that of the Fisher discriminant analysis.

We use the same notation as that in Section 3, that is, we use $\hat{\delta}_F$ to denote the sample version of Fisher discriminant function, and use $\hat{\delta}_I$ to denote the sample version of the independence rule. Bickel and Levina (2004) theoretically compared the asymptotical performance of $\hat{\delta}_F$ and $\hat{\delta}_I$. The asymptotic performance of Fisher discriminant analysis is given in (3.5). As for the independence rule, under some regularity conditions, Bickel and Levina (2004) showed that if $\log p/n \to 0$, then

$$\lim_{n \to \infty} \sup \overline{W}_{\Gamma_1}(\hat{\delta}_I) = \overline{\Phi}\left(\frac{\sqrt{K_0}}{1 + \sqrt{K_0}} c\right), \tag{4.7}$$

where Γ_1 is the parameter set defined in Section 3.1. Recall that (3.5) shows that the Fisher discriminant analysis asymptotically performs no better than random guessing when the dimensionality p is much larger than the sample size n. While the above result (4.7) demonstrates that for the independence rule, the worst case classification error is better than that of the random guessing, as long as the dimensionality p does not grow exponentially faster than the sample size n and $K_0 < \infty$. This shows the advantage of independence rule in high dimensional classification. Note that the impact of dimensionality can not be seen in (4.7) whereas it can be seen from (3.12). This is due to the difference of Γ_2 from Γ_1.

On the practical side, Dudoit et al. (2002) compared the performance of various classification methods, including the Fisher discriminant analysis and the independence rule, for the classification of tumors based on gene expression data. Their results show that the independence rule outperforms the Fisher discriminant analysis.

Bickel and Levina (2004) also introduced a spectrum of classification rules which interpolate between $\hat{\delta}_F$ and $\hat{\delta}_I$ under the Gaussian coloured noise model assumption. They showed that the minimax regret of their classifier has asymptotic rate $O(n^{-\kappa} \log n)$ with κ some positive number defined in their paper. See Bickel and Levina (2004) for more details.

4.2   Centroid rule and k-nearest-neighbor rule

Hall et al. (2005) have given the geometric representation of high dimensional, low sample size data, and used it to analyze the performance of several distance-based classifiers, including the centroid rule and 1-nearest neighbor rule. In their analysis, the dimensionality $p \to \infty$ while the sample size n is fixed.

To appreciate their results, we first introduce some notations. Consider classification between two classes. Assume that within each class, observations are independent and identically distributed. Let $\mathbf{Z}_1 = (Z_{11}, Z_{12}, \ldots, Z_{1p})^T$ be an observation from class 1, and $\mathbf{Z}_2 = (Z_{21}, Z_{22}, \ldots, Z_{2p})^T$ be an observation from class 2. Assume the following results hold as $p \to \infty$

$$\frac{1}{p} \sum_{j=1}^{p} \mathrm{var}(Z_{1j}) \to \sigma^2, \quad \frac{1}{p} \sum_{j=1}^{p} \mathrm{var}(Z_{2j}) \to T^2,$$

$$\frac{1}{p} \sum_{j=1}^{p} [E(Z_{1j}^2) - E(Z_{2j}^2)] \to \kappa^2, \tag{4.8}$$

where σ, T and κ are some positive constants. Let C_k be the centroid of the training data from class k, where $k = 1, 2$. Here, the centroid C_k could be the mean or median of data in class k.

The "centroid rule" or "mean difference rule" classifies a new observation to class 1 or class 2 according to its distance to their centroids. This approach is popular in genomics. To study the theoretical property of this method, Hall et al. (2005) first assumed that $\sigma^2/n_1 \geqslant T^2/n_2$. They argued that if needed, the roles for class 1 and class 2 can be interchanged to achieve this. Then under some regularity conditions, they showed that if $\kappa^2 \geqslant \sigma^2/n_1 - T^2/n_2$, then the probability that a new datum from either class 1 or class 2 is correctly classified by the centroid rule converges to 1 as $p \to \infty$; If instead $\kappa^2 < \sigma^2/n_1 - T^2/n_2$, then with probability converging to 1, a new datum from either class will be classified by the centroid rule as belonging to class 2. This property is also enjoyed by the support vector machine method which to be discussed in a later section.

The nearest-neighbor rule uses those training data closest to $\mathbf{X}$ to predict the label of $\mathbf{X}$. Specifically, the k-nearest-neighbor rule predicts the class label of $\mathbf{X}$ as

$$\delta(\mathbf{X}) = \frac{1}{k} \sum_{\mathbf{X}_i \in N_k(\mathbf{X})} Y_i, \tag{4.9}$$

where $N_k(\mathbf{X})$ is the neighborhood of $\mathbf{X}$ defined by the k closest observations in the training sample. For two-class classification problems, $\mathbf{X}$ is assigned to class 1 if $\delta(\mathbf{X}) < 1.5$. This is equivalent to the majority vote rule in the "committee" $N_k(\mathbf{X})$. For more details on the nearest-neighbor rule, see Hastie et al. (2009).

Hall et al. (2005) also considered the 1-nearest-neighbor rule. They first assumed that $\sigma^2 \geqslant T^2$. The same as before, the roles of class 1 and class 2 can be interchanged to achieve this. They showed that if $\kappa^2 > \sigma^2 - T^2$, then the probability that a new datum from either class 1 or class 2 is correctly classified by the 1-nearest-neighbor rule converges to 1 as $p \to \infty$; if instead $\kappa^2 < \sigma^2 - T^2$, then with probability converging to 1, a new datum from either class will be classified by the 1-nearest-neighbor rule as belonging to class 2.

Hall et al. (2005) further discussed the contrasts between the centroid rule and the 1-nearest-neighbor rule. For simplicity, they assumed that $n_1 = n_2$. They pointed out that asymptotically, the centroid rule misclassifies data from at least one of the classes only when $\kappa^2 < |\sigma^2 - T^2|/n_1$, whereas the 1-nearest-neighbor rule leads to misclassification for data from at least one of the classes both in the range $\kappa^2 < |\sigma^2 - T^2|/n_1$ and when $|\sigma^2 - T^2|/n_1 \leqslant \kappa^2 < |\sigma^2 - T^2|$. This quantifies the inefficiency that might be expected from basing inference only on a single nearest neighbor. For the choice of k in the nearest neighbor, see Hall, Park and Samworth (2008).

For the properties of both classifiers discussed in this subsection, it can be seen that their performances are greatly determined by the value of κ^2. However, in view of (4.8), κ^2 could be very small or even 0 in high dimensional setting due the the existence of many noise features that have very little or no classification power (i.e. those with $EZ_{1j}^2 \approx EZ_{2j}^2$). This once again shows the difficulty of classification in high dimensional setting.

4.3 Theoretical properties of distance-based classifiers

Hall, Pittelkow and Ghosh (2008) suggested an approach to accessing the theoretical performance of general distance-based classifiers. This technique is related to the concept of "detection boundary" developed by Ingster and Donoho and Jin. See, for example, Donoho and Jin (2004); Hall and Jin (2008). Ingster (2002); and Jin (2006); Hall, Pittelkow and Ghosh (2008) studied the theoretical performance of a variety distance-based classifiers constructed from high dimensional data, and obtain the classification boundaries for them. We discuss their study in this subsection.

Let $g(\cdot) = g(\cdot|(\mathbf{X}_i, Y_i), i = 1, \ldots, n)$ be a distanced-based classifier which assigns a new observation $\mathbf{X}$ to either class 1 or class 2. Hall, Pittelkow and Ghosh (2008) argued that any plausible, distance-based classifier g should enjoy the following two properties:

(a) g assigns $\mathbf{X}$ to class 1 if it is closer to each of the $\mathbf{X}_i'$s in class 1 than it is to any of the $\mathbf{X}_j'$s in class 2;

(b) If g assigns $\mathbf{X}$ to class 1 then at least one of the $\mathbf{X}_i'$s in class 1 is closer to $\mathbf{X}$ than $\mathbf{X}$ is closer to the most distant $\mathbf{X}_j'$s in class 2.

These two properties together imply that

$$\pi_{k1} \leqslant P_k(g(\mathbf{X}) = k) \leqslant \pi_{k2}, \text{ for } k = 1, 2, \tag{4.10}$$

where P_k denotes the probability measure when assuming that $\mathbf{X}$ is from class k

with $k = 1, 2$, and π_{k1} and π_{k2} are defined as

$$\pi_{k1} = P_k \left(\max_{i \in \mathcal{G}_1} \|\mathbf{X}_i - \mathbf{X}\| \leqslant \min_{j \in \mathcal{G}_2} \|\mathbf{X}_j - \mathbf{X}\| \right) \text{ and} \tag{4.11}$$

$$\pi_{k2} = P_k \left(\min_{i \in \mathcal{G}_1} \|\mathbf{X}_i - \mathbf{X}\| \leqslant \max_{j \in \mathcal{G}_2} \|\mathbf{X}_j - \mathbf{X}\| \right) \tag{4.12}$$

with $\mathcal{G}_1 = \{i : 1 \leqslant i \leqslant n, Y_i = 1\}$ and $\mathcal{G}_2 = \{i : 1 \leqslant i \leqslant n, Y_i = 2\}$. Hall, Pittelkow and Ghosh (2008) considered a family of distance-based classifiers satisfying condition (4.10).

To study the theoretical property of these distance-based classifiers, Hall, Pittelkow and Ghosh (2008) considered the following model

$$X_{ij} = \mu_{kj} + \varepsilon_{ij}, \text{ for } i \in \mathcal{G}_k, \ k = 1, 2, \tag{4.13}$$

where X_{ij} denotes the j-th component of $\mathbf{X}_i$, μ_{kj} represents the j-th component of mean vector $\boldsymbol{\mu}_k$, and ε_{ij}'s are independent and identically distributed with mean 0 and finite fourth moment. Without loss of generality, they assumed that the class 1 population mean vector $\boldsymbol{\mu}_1 = \mathbf{0}$. Under this model assumption, they showed that if some mild conditions are satisfied, $\pi_{k1} \to 0$ and $\pi_{k2} \to 0$ if and only if $p = o(\|\boldsymbol{\mu}_2\|^4)$. Then using inequality (4.10), they obtained that the probability of the classifier g correctly classifying a new observation from class 1 or class 2 converges to 1 if and only if $p = o(\|\boldsymbol{\mu}_2\|^4)$ as $p \to \infty$. This result tells us just how fast the norm of the two class mean difference vector $\boldsymbol{\mu}_2$ must grow for it to be possible to distinguish perfectly the two classes using the distance-based classifier. Note that the above result is independent of the sample size n. The result is consistent with (a specific case of) (3.12) for independent rule in which $C_p = \|\boldsymbol{\mu}_2\|^2$ in the current setting and misclassification rate goes to zero when the signal is so strong that $C_p^2/p \to \infty$ (if n is fixed) or $\|\boldsymbol{\mu}_2\|^4/p \to \infty$. The impact of dimensionality is implied by the quantity $\|\boldsymbol{\mu}_2\|^2/\sqrt{p}$.

It is well known that the thresholding methods can improve the sensitivity of distance-based classifiers. The thresholding in this setting is a feature selection method, using only features with distant away from the other. Denote by $X_{ij}^{tr} = X_{ij} I(X_{ij} > t)$ the thresholded data, $i = 1, \ldots, n$, $j = 1, \ldots, p$, with t the thresholding level. Let $\mathbf{X}_i^{tr} = (X_{ij}^{tr})$ be the thresholded vector and g^{tr} be the version of the classifier g based on thresholded data. The case where the absolute values $|X_{ij}|$ are thresholded is very similar. Hall, Pittelkow and Ghosh (2008) studied the properties of the threshold-based classifier g^{tr}. For simplicity, they assumed that $\mu_{2j} = \nu$ for q distinct indices j, and $\mu_{2j} = 0$ for the remaining $p - q$ indices, where

(a) $\nu \geqslant t$,
(b) $t = t(p) \to \infty$ as p increases,
(c) $q = q(p)$ satisfies $q \to \infty$ and $1 \leqslant q \leqslant cp$ with $0 < c < 1$ fixed, and
(d) the errors ε_{ij} has a distribution that is unbounded to the right.

With the above assumptions and some regularity conditions, they proved that the general thresholded distance-based classifier g^{tr} has a property that is analogue to the standard distance-based classifier, that is, the probability that the classifier

g^{tr} correctly classifies a new observation from class 1 or class 2 tending to 1 if and only if $p = o(T)$ as $p \to \infty$, where $T = (q\nu^2)^2 / E[\varepsilon_{ij}^4 I(\varepsilon_{ij} > t)]$. Compared to the property of standard distance-based classifier, the thresholded classifier allows for higher dimensionality if $E[\varepsilon_{ij}^4 I(\varepsilon_{ij} > t)] \to 0$ as $p \to \infty$.

Hall, Pittelkow and Ghosh (2008) further compared the theoretical performance of standard distanced-based classifiers and thresholded distance-based classifiers by using the classification boundaries. To obtain the explicit form of classification boundaries, they assumed that for j-th feature, the class 1 distribution is $GN_\gamma(0, 1)$ and the class 2 distribution is $GN_\gamma(\mu_{2j}, 1)$, respectively. Here $GN_\gamma(\mu, \sigma^2)$ denotes the Subbotin, or generalized normal distribution with probability density

$$f(x|\gamma, \mu, \sigma) = C_\gamma \sigma^{-1} \exp\left(-\frac{|x - \mu|^\gamma}{\gamma \sigma^\gamma}\right), \qquad (4.14)$$

where $\gamma, \sigma > 0$ and C_γ is some normalization constant depending only on γ. It is easy to see that the standard normal distribution is just the standard Subbotin distribution with $\gamma = 2$. By assuming that $q = O(p^{1-\beta})$, $t = (\gamma r \log p)^{1/\gamma}$, and $\nu = (\gamma s \log p)^{1/\gamma}$ with $\frac{1}{2} < \beta < 1$ and $0 < r < s \leqslant 1$, they derived that the sufficient and necessary conditions for the classifiers g and g^{tr} to produce asymptotically correct classification results are

$$1 - 2\beta > 0 \text{ and} \qquad (4.15)$$
$$1 - 2\beta + s > 0, \qquad (4.16)$$

respectively. Thus the classification boundary of g^{tr} is lower than that of g, indicating that the distance-based classifier using truncated data are more sensible.

The classification boundaries for distance-based classifiers and for their thresholded versions are both independent of the training sample size. As pointed out by Hall, Pittelkow and Ghosh (2008), this conclusion is obtained from the fact that for fixed sample size n and for distance-based classifiers, the probability of correct classification converges to 1 if and only if the differences between distances among data have a certain extremal property, and that this property holds for one difference if and only if it holds for all of them. Hall, Pittelkow and Ghosh (2008) further compared the classification boundary of distance-based classifiers with that of the classifiers based on higher criticism. See their paper for more comparison results.

5 Feature selection by independence rule

As has been discussed in Section 3, classification methods using all features do not necessarily perform well due to the noise accumulation when estimating a large number of noise features. Thus, feature selection is very important in high dimensional classification. This has been advocated by Fan and Fan (2008) and many other researchers. In fact, the thresholding methods discussed in Hall, Pittelkow and Ghosh (2008) are also a type of feature selection.

5.1 Features annealed independence rule

Fan and Fan (2008) proposed the Features Annealed Independence Rule (FAIR) for feature selection and classification in high dimensional setting. We discuss their method in this subsection.

There is a huge literature on the feature selection in high dimensional setting. See, for example, Efron et al. (2004); Fan and Li (2001); Fan and Lv (2008, 2009); Fan et al. (2008); Lv and Fan (2009); Tibshirani (1996). Two sample t tests are frequently used to select important features in classification problems. Let $\bar{X}_{kj} = \sum_{Y_i=k} X_{ij}/n_k$ and $S^2_{kj} = \sum_{Y_i=k} (X_{kj} - \bar{X}_{kj})^2/(n_k - 1)$ be the sample mean and sample variance of j-th feature in class k, respectively, where $k = 1, 2$ and $j = 1, \ldots, p$. Then the two-sample t statistic for feature j is defined as

$$T_j = \frac{\bar{X}_{1j} - \bar{X}_{2j}}{\sqrt{S^2_{1j}/n_1 + S^2_{2j}/n_2}}, \quad j = 1, \ldots, p. \tag{5.1}$$

Fan and Fan (2008) studied the feature selection property of two-sample t statistic. They considered the model (3.14) and assumed that the error ε_{ij} satisfies the Cramér's condition and that the population mean difference vector $\boldsymbol{\mu} = \boldsymbol{\mu}_1 - \boldsymbol{\mu}_2 = (\mu_1, \ldots, \mu_p)^T$ is sparse with only the first s entries nonzero. Here, s is allowed to diverge to ∞ with the sample size n. They showed that if $\log(p - s) = o(n^\gamma)$, $\log s = o(n^{\frac{1}{2}-\beta_n})$, and $\min_{1 \leqslant j \leqslant s} \frac{|\mu_j|}{\sqrt{\sigma^2_{1j}+\sigma^2_{2j}}} = o(n^{-\gamma}\beta_n)$ with some $\beta_n \to \infty$ and $\gamma \in (0, \frac{1}{3})$, then with x chosen in the order of $O(n^{\gamma/2})$, the following result holds:

$$P(\min_{j \leqslant s} |T_j| > x, \max_{j>s} |T_j| < x) \to 1. \tag{5.2}$$

This result allows the lowest signal level $\min_{1 \leqslant j \leqslant s} \frac{|\mu_j|}{\sqrt{\sigma^2_{1j}+\sigma^2_{2j}}}$ to decay with sample size n. As long as the rate of decay is not too fast and the dimensionality p does not grow exponentially faster than n, the two-sample t-test can select all important features with probability tending to 1.

Although the theoretical result (5.2) shows that the t-test can successfully select features if the threshold is appropriately chosen, in practice it is usually very hard to choose a good threshold value. Moreover, even all revelent features are correctly selected by the two-sample t test, it may not necessarily be the best to use all of them, due to the possible existence of many faint features. Therefore, it is necessary to further single out the most important ones. To address this issue, Fan and Fan (2008) proposed the features annealed independence rule. Instead of constructing independence rule using all features, FAIR selects the most important ones and use them to construct independence rule. To appreciate the idea of FAIR, first note that the relative importance of features can be measured by the ranking of $\{|\alpha_j|/\sigma_j\}$. If such oracle ranking information is available, then one can construct the independence rule using m features with the largest $\{|\alpha_j|/\sigma_j\}$. The optimal value of m is to be determined. In this case, FAIR takes the following form:

$$\delta(\mathbf{x}) = \sum_{j=1}^{p} \alpha_j (x_j - \mu_j)/\sigma_j^2 1_{\{|\alpha_j|/\sigma_j > b\}}, \qquad (5.3)$$

where b is a positive constant chosen in a way such that there are m features with $|\alpha_j|/\sigma_j > b$. Thus choosing the optimal m is equivalent to selecting the optimal b. Since in practice such oracle information is unavailable, we need to learn it from the data. Observe that $|\alpha_j|/\sigma_j$ can be estimated by $|\hat{\alpha}_j|/\hat{\sigma}_j$. Thus the sample version of FAIR is

$$\hat{\delta}(\mathbf{x}) = \sum_{j=1}^{p} \hat{\alpha}_j (x_j - \hat{\mu}_j)/\hat{\sigma}_j^2 1_{\{|\hat{\alpha}_j|/\hat{\sigma}_j > b\}}. \qquad (5.4)$$

In the case where the two population covariance matrices are the same, we have

$$|\hat{\alpha}_j|/\hat{\sigma}_j = \sqrt{n/(n_1 n_2)}|T_j|.$$

Thus the sample version of the discriminant function of FAIR can be rewritten as

$$\hat{\delta}_{\text{FAIR}}(\mathbf{x}) = \sum_{j=1}^{p} \hat{\alpha}_j (x_j - \hat{\mu}_j)/\hat{\sigma}_j^2 1_{\{\sqrt{n/(n_1 n_2)}|T_j| > b\}}. \qquad (5.5)$$

It is clear from (5.5) that FAIR works the same way as that we first sort the features by the absolute values of their t-statistics in the descending order, and then take out the first m features to construct the classifier. The number of features m can be selected by minimizing the upper bound of the classification error given in (3.11). To understand this, note that the upper bound on the right hand side of (3.11) is a function of the number of features. If the features are sorted in the descending order of $|\alpha_j|/\sigma_j$, then this upper bound will first increase and then decrease as we include more and more features. The optimal m in the sense of minimizing the upper bound takes the form

$$m_{opt} = \arg \max_{1 \leqslant m \leqslant p} \frac{1}{\lambda_{\max}^m} \frac{[\sum_{j=1}^{m} \alpha_j^2/\sigma_j^2 + m(1/n_2 - 1/n_1)]^2}{nm/(n_1 n_2) + \sum_{j=1}^{m} \alpha_j^2/\sigma_j^2},$$

where $\lambda_{\max}^m$ is the largest eigenvalue of the correlation matrix $\mathbf{R}^m$ of the truncated observations. It can be estimated from the training data as

$$\begin{aligned}
\hat{m}_{opt} &= \arg \max_{1 \leqslant m \leqslant p} \frac{1}{\hat{\lambda}_{\max}^m} \frac{[\sum_{j=1}^{m} \hat{\alpha}_j^2/\hat{\sigma}_j^2 + m(1/n_2 - 1/n_1)]^2}{nm/(n_1 n_2) + \sum_{j=1}^{m} \hat{\alpha}_j^2/\hat{\sigma}_j^2} \\
&= \arg \max_{1 \leqslant m \leqslant p} \frac{1}{\hat{\lambda}_{\max}^m} \frac{n[\sum_{j=1}^{m} T_j^2 + m(n_1 - n_2)/n]^2}{mn_1 n_2 + n_1 n_2 \sum_{j=1}^{m} T_j^2}. \qquad (5.6)
\end{aligned}$$

Note that the above t-statistics are the sorted ones. Fan and Fan (2008) used simulation study and real data analysis to demonstrate the performance of FAIR. See their paper for the numerical results.

5.2   Nearest shrunken centroids method

In this section, we will discuss the nearest shrunken centroids (NSC) method proposed by Tibshirani et al. (2002). This method is used to identify a subset of features that best characterize each class and do classification. Compared to the centroid rule discussed in Section 3.2, it takes into account the feature selection. Moreover, it is general and can be applied to high-dimensional multi-class classification.

Define $\bar{X}_{kj} = \sum_{i \in \mathcal{G}_k} X_{ij}/n_k$ as the j-th component of the centroid for class k, and $\bar{X}_j = \sum_{i=1}^{n} X_{ij}/n$ as the j-th component of the overall centroid. The basic idea of NSC is to shrink the class centroids to the overall centroid. Tibshirani et al. (2002) first normalized the centroids by the within class standard deviation for each feature, i.e.,

$$d_{kj} = \frac{\bar{X}_{kj} - \bar{X}_j}{m_k(S_j + s_0)}, \tag{5.7}$$

where s_0 is a positive constant, and S_j is the pooled within class standard deviation for j-th feature with

$$S_j^2 = \sum_{k=1}^{K} \sum_{i \in \mathcal{G}_k} (X_{ij} - \bar{X}_{kj})^2/(n - K)$$

and $m_k = \sqrt{1/n_k - 1/n}$ the normalization constant. As pointed out by Tibshirani et al. (2002), d_{kj} defined in (5.7) is a t-statistic for feature j comparing the k-th class to the average. The constant s_0 is included to guard against the possibility of large d_{kj} simply caused by very small value of S_j. Then (5.7) can be rewritten as

$$\bar{X}_{kj} = \bar{X}_j + m_k(S_j + s_0)d_{kj}. \tag{5.8}$$

Tibshirani et al. (2002) proposed to shrink each d_{kj} toward zero by using soft thresholding. More specifically, they define

$$d'_{kj} = \text{sgn}(d_{ik})(|d_{kj}| - \Delta)_+, \tag{5.9}$$

where $\text{sgn}(\cdot)$ is the sign function, and $t_+ = t$ if $t > 0$ and $t_+ = 0$ otherwise. This yields the new shrunken centroids

$$\bar{X}'_{kj} = \bar{X}_j + m_k(S_j + s_0)d'_{kj}. \tag{5.10}$$

As argued in their paper, since many of $\bar{X}_{kj}$ are noisy and close to the overall mean $\bar{X}_j$, using soft thresholding produces more reliable estimates of the true means. If the shrinkage level Δ is large enough, many of the d_{kj} will be shrunken to zero and the corresponding shrunken centroid $\bar{X}'_{kj}$ for feature j will be equal to the overall centroid for feature j. Thus these features do not contribute to the nearest centroid computation.

To choose the amount of shrinkage Δ, Tibshirani et al. (2002) proposed to use the cross validation method. For example, if 10-fold cross validation is used, then the training data set is randomly split into 10 approximately equal-size subsamples. We first fit the model by using 90% of the training data, and then predict the class labels of the remaining 10% of the training data. This procedure is repeated 10 times for a fixed Δ, with each of the 10 sub-samples of the data used as the test sample to calculate the prediction error. The prediction errors on all 10 parts are then added together as the overall prediction error. The optimal Δ is then chosen to be the one that minimizes the overall prediction error.

After obtaining the shrunken centroids, Tibshirani et al. (2002) proposed to classify a new observation $\mathbf{X}$ to the class whose shrunken centroid is closest to this new observation. They define the discriminant score for class k as

$$\hat{\delta}_k(\mathbf{X}) = \sum_{j=1}^{p} \frac{(X_j - \bar{X}'_{kj})^2}{(S_j + s_0)^2} - 2\log\pi_k. \tag{5.11}$$

The first term is the standardized squared distance of $\mathbf{X}$ to the k-th shrunken centroid, and the second term is a correction based on the prior probability π_k. Then the classification rule is

$$g(\mathbf{X}) = \arg\min_k \hat{\delta}_k(\mathbf{X}). \tag{5.12}$$

It is clear that NSC is a type of distance-based classification method.

Compared to FAIR introduced in Section 5.1, NSC shares the same idea of using marginal information of features to do classification. Both methods conduct feature selection by t-statistic. But FAIR selects the number of features by using mathematical formula that is derived to minimize the upper bound of classification error, while NSC obtains the number of features by using cross validation. Practical implementation shows that FAIR is more stable in terms of the number of selected features and classification error. See Fan and Fan (2008).

6 Loss-based classification

Another popular class of classification methods is based on different (margin-based) loss functions. It includes many well known classification methods such as the support vector machine (SVM, Cristianini and Shawe-Taylor, 2000; Vapnik, 1998).

6.1 Support vector machine

As mentioned in Section 2, the zero-one loss is typically used to access the accuracy of a classification rule. Thus, based on the training data, one may ideally minimize $\sum_{i=1}^{n} I_{g(\mathbf{X}_i) \neq Y_i}$ with respect to $g(\cdot)$ over a function space to obtain an estimated classification rule $\hat{g}(\cdot)$. However the indicator function is neither convex nor smooth. The corresponding optimization is difficult, if not impossible, to solve. Alternatively, several convex surrogate loss functions have been proposed to replace the zero-one loss.

For binary classification, we may equivalently code the categorical response Y as either -1 or $+1$. The SVM replaces the zero-one loss by the hinge loss $H(u) = [1 - u]_+$, where $[u]_+ = \max\{0, u\}$ denotes the positive part of u. Note that the hinge loss is convex. Replacing the zero-one loss with the hinge loss, the SVM minimizes

$$\sum_{i=1}^{n} H(Y_i f(\mathbf{X}_i)) + \lambda J(f) \tag{6.1}$$

with respect to f, where the first term quantifies the data fitting, $J(f)$ is some roughness (complexity) penalty of f, and λ is a tuning parameter balancing the data fit measured by the hinge loss and the roughness of $f(\cdot)$ measured by $J(f)$. Denote the minimizer by $\hat{f}(\cdot)$. Then the SVM classification rule is given by $\hat{g}(\mathbf{x}) = \text{sign}(\hat{f}(\mathbf{x}))$. Note that the hinge loss is non-increasing. While minimizing the hinge loss, the SVM encourages positive $Yf(X)$ which corresponds to correct classification.

For linear SVM with $f(\mathbf{x}) = b + \mathbf{x}^T\boldsymbol{\beta}$, the standard SVM uses the 2-norm penalty $J(f) = \frac{1}{2}\sum_{j=1}^{p}\beta_j^2$. While in this exposition, we are formulating the SVM in the regularization framework. However it is worthwhile to point out that the SVM was originally introduced by V. Vapnik and his colleagues with the idea of searching for the optimal separating hyperplane. Interested readers may consult Boser, Guyon and Vapnik (1992) and Vapnik (1998) for more details. It was shown by Wahba (1998) that the SVM can be equivalently fit into the regularization framework by solving (6.1) as presented in the previous paragraph. Different from those methods that focus on conditional probabilities $P(Y|\mathbf{X} = \mathbf{x})$, the SVM targets at estimating the decision boundary $\{\mathbf{x} : P(Y = 1|\mathbf{X} = \mathbf{x}) = 1/2\}$ directly.

A general loss function $\ell(\cdot)$ is called Fisher consistent if the minimizer of $E[\ell(Yf(\mathbf{X})|\mathbf{X} = \mathbf{x})]$ has the same sign as $P(Y = +1|\mathbf{X} = \mathbf{x}) - 1/2$ (Lin, 2004). Fisher consistency is also known as classification-calibration (Bartlett, Jordan, and McAuliffe, 2006) and infinite-sample consistency (Zhang, 2004). It is a desirable property for a loss function.

Lin (2002) showed that the minimizer of $E[H(Yf(\mathbf{X})|\mathbf{X} = \mathbf{x})]$ is exactly $\text{sign}(P(Y = +1|\mathbf{X} = \mathbf{x}) - 1/2)$, the decision-theoretically optimal classification rule with the smallest risk, which is also known as the Bayes classification rule. Thus the hinge loss is Fisher consistent for binary classification.

When dealing with problems with many predictor variables, Zhu, Rosset, Hastie, and Tibshirani (2003) proposed the 1-norm SVM by using the L_1 penalty $J(f) = \sum_{j=1}^{p}|\beta_j|$ to achieve variable selection; Zhang, Ahn, Lin and Park (2006) proposed the SCAD SVM by using the SCAD penalty (Fan and Li, 2001); Liu and Wu (2007) proposed to regularize the SVM with a combination of the L_0 and L_1 penalties; and many others.

Either basis expansion or kernel mapping (Cristianini and Shawe-Taylor, 2000) may be used to accomplish nonlinear SVM. For the case of kernel learning, a bivariate kernel function $K(\cdot, \cdot)$, which maps from $\mathcal{X} \times \mathcal{X}$ to R, is employed. Then $f(\mathbf{x}) = b + \sum_{i=1}^{n} K(\mathbf{x}, \mathbf{X}_i)$ by the theory of reproducing kernel Hilbert spaces

(RKSH), see Wahba (1990). In this case, the 2-norm of $f(\mathbf{x}) - b$ in RKHS with $K(\cdot, \cdot)$ is typically used as $J(f)$. Using the representor theorem (Kimeldorf and Wahba, 1971), $J(f)$ can be represented as $J(f) = \frac{1}{2} \sum_{i=1}^{n} \sum_{j=1}^{n} c_i K(\mathbf{X}_i, \mathbf{X}_j) c_j$.

6.2 ψ-learning

The hinge loss $H(u)$ is unbounded and shoots to infinity when t goes to negative infinity. This characteristic makes the SVM tend to be sensitive to noisy training data. When there exist points far away from their own classes (namely, "outliers" in the training data), the SVM classifier tends to be strongly affected by such points due to the unboundedness of the hinge loss. In order to improve over the SVM, Shen, Tseng, Zhang, and Wong (2003) proposed to replace the convex hinge loss by a nonconvex ψ-loss function. The ψ-loss function $\psi(u)$ satisfies

$$U \geqslant \psi(u) > 0 \text{ if } u \in [0, T];$$
$$\psi(u) = 1 - \text{sign}(u) \text{ otherwise,}$$

where $0 < U \leqslant 2$ and $T > 0$ are constants. The positive values of $\psi(u)$ for $u \in [0, T]$ eliminate the scaling issue of the sign function and avoid too many points piling around the decision boundary. Their method was named the ψ-learning. They showed that the ψ-learning can achieve more accurate class prediction.

Similarly motivated, Wu and Liu (2007a) proposed to truncate the hinge loss function by defining $H_s(u) = \min(H(s), H(u))$ for $s \leqslant 0$ and worked on the more general multi-category classification. According to their Proposition 1, the truncated hinge loss is also Fisher consistent for binary classification.

6.3   AdaBoost

Boosting is another very successful algorithm for solving binary classification. The basic idea of boosting is to combine weaker learners to improve performance (Freund, 1995; Schapire, 1990). The AdaBoost algorithm, a special boosting algorithm, was first introduced by Freund and Schapire (1996). It constructs a "strong" classifier as a linear combination

$$f(\mathbf{x}) = \sum_{t=1}^{T} \alpha_t h_t(\mathbf{x})$$

of "simple", "weak" classifiers $h_t(\mathbf{x})$. The "weak" classifiers $h_t(\mathbf{x})$'s can be thought of as features and $H(\mathbf{x}) = \text{sign}(f(\mathbf{x}))$ is called "strong" or final classifier. It works by sequentially reweighing the training data, applying a classification algorithm (weaker learner) to the reweighed training data, and then taking a weighted majority vote of the thus-obtained classifier sequence. This simple reweighing strategy improves performance of many weaker learners. Freund and Schapire (1996) and Breiman (1997) tried to provide a theoretic understanding based on game theory. Another attempt to investigate its behavior was made by Breiman (1998) using bias and variance tradeoff. Later Friedman, Hastie, and Tibshirani (2000) provided a new statistical perspective, namely using additively modeling and maximum likelihood, to understand why this seemingly mysterious AdaBoost algorithm works so well. They showed that AdaBoost is equivalent to using the exponential loss $\ell(u) = e^{-u}$.

6.4   Other loss functions

There are many other loss functions in this regularization framework. Examples include the squared loss $\ell(u) = (1-u)^2$ used in the proximal SVM (Fung and Mangasarian, 2001) and the least square SVM (Suykens and Vandewalle, 1999), the logistic loss $\ell(u) = \log(1+e^{-u})$ of the logistic regression, and the modified least squared loss $\ell(u) = ([1-u]_+)^2$ proposed by Zhang and Oles (2001). In particular, the logistic loss is motivated by assuming that the probability of $Y = +1$ given $\mathbf{X} = \mathbf{x}$ is given by $e^{f(\mathbf{x})}/(1+e^{f(\mathbf{x})})$. Consequently the logistic regression is capable of estimating the conditional probability.

7   Feature selection in loss-based classification

As mentioned above, variable selection-capable penalty functions such as the L_1 and SCAD can be applied to the regularization framework to achieve variable selection when dealing with data with many predictor variables. Examples include the L_1 SVM (Zhu, Rosset, Hastie, and Tibshirani, 2003), SCAD SVM (Zhang, Ahn, Lin and Park, 2006), SCAD logistic regression (Fan and Peng, 2004). These methods work fine for the case with a fair number of predictor variables. However the remarkable recent development of computing power and other technology has allowed scientists to collect data of unprecedented size and complexity. Examples include data from microarrays, proteomics, functional MRI, SNPs and others. When dealing with such high or ultra-high dimensional data, the usefulness of these methods becomes limited.

In order to handle linear regression with ultra-high dimensional data, Fan and Lv (2008) proposed the sure independence screening (SIS) to reduce the dimensionality from ultra-high p to a fairly high d. It works by ranking predictor variables according to the absolute value of the marginal correlation between the response variable and each individual predictor variable and selecting the top ranked d predictor variables. This screening step is followed by applying a refined method such as the SCAD to these d predictor variables that have been selected. In a fairly general asymptotic framework, this simple but effective correlation learning is shown to have the sure screening property even for the case of exponentially growing dimensionality, that is, the screening retains the true important predictor variables with probability tending to one exponentially fast.

The SIS methodology may break down if a predictor variable is marginally unrelated, but jointly related with the response, or if a predictor variable is jointly uncorrelated with the response but has higher marginal correlation with the response than some important predictors. In the former case, the important feature has already been screened out at the first stage, whereas in the latter case, the unimportant feature is ranked too high by the independent screening technique.

Iterative SIS (ISIS) was proposed to overcome these difficulties by using more fully the joint covariate information while retaining computational expedience and stability as in SIS. Basically, ISIS works by iteratively applying SIS to recruit a small number of predictors, computing residuals based on the model fitted using these recruited variables, and then using the working residuals as the response variable to continue recruiting new predictors. Numerical examples in Fan and Lv (2008) have demonstrated the improvement of ISIS. The crucial step is to compute the working residuals, which is easy for the least-squares regression problem but not obvious for other problems. By sidestepping the computation of working residuals, Fan et al. (2008) has extended (I)SIS to a general pseudo-likelihood framework, which includes generalized linear models as a special case. Roughly they use the additional contribution of each predictor variable given the variables that have been recruited to rank and recruit new predictors.

In this section, we will elaborate (I)SIS in the context of binary classification using loss functions presented in the previous section. While presenting the (I)SIS methodology, we use a general loss function $\ell(\cdot)$. The R-code is publicly available at cran.r-project.org.

7.1 Feature ranking by marginal utilities

By assuming a linear model $f(\mathbf{x}) = b + \mathbf{x}^T \boldsymbol{\beta}$, the corresponding model fitting amounts to minimizing

$$Q(b, \boldsymbol{\beta}) = \frac{1}{n} \sum_{i=1}^{n} \ell(Y_i(b + \mathbf{X}_i^T \boldsymbol{\beta})) + \lambda J(f),$$

where $J(f)$ can be the 2-norm or some other penalties that are capable of variable selection. The marginal utility of the j-th feature is

$$\ell_j = \min_{b, \beta_j, \beta_{M_1}} \sum_{i=1}^{n} \ell(Y_i(b + X_{ij}^T \beta_j)).$$

For some loss functions such as the hinge loss, another term $\frac{1}{2}\beta_j^2 + \frac{1}{2} \sum_{m \in M_1} \beta_{m^2}$ may be required to avoid possible identifiability issue. In that case

$$\ell_j = \min_{b, \beta_j} \left\{ \sum_{i=1}^{n} \ell(Y_i(b + X_{ij}^T \beta_j)) + \frac{1}{2}\beta_j^2 \right\}. \tag{7.1}$$

The idea of SIS is to compute the vector of marginal utilities $\boldsymbol{\ell} = (\ell_1, \ell_2, \ldots, \ell_p)^T$ and rank predictor variables according to their corresponding marginal utilities. The smaller the marginal utility is the more important the corresponding predictor variable is. We select d variables corresponding to the d smallest components of $\boldsymbol{\ell}$. Namely, variable j is selected if ℓ_j is one of the d smallest components of $\boldsymbol{\ell}$. A typical choice of d is $\lfloor n/\log n \rfloor$. Fan and Song (2009) provided an extensive account on the sure screening property of the independence learning and on the capacity of the model size reduction.

7.2   Penalization

With the d variables crudely selected by SIS, parameter estimation and variable selection can be further carried out simultaneously using a more refined penalization method. This step takes joint information into consideration. By reordering the variables if necessary, we may assume without loss of generality that X_1, X_2, ..., X_d are the variables that have been recruited by SIS. In the regularization framework, we use a penalty that is capable of variable selection and minimize

$$\frac{1}{n}\sum_{i=1}^{n}\ell(Y_i(b+\sum_{j=1}^{d}X_{ij}\beta_j))+\sum_{j=1}^{d}p_\lambda(|\beta_j|), \qquad (7.2)$$

where $p_\lambda(\cdot)$ denotes a general penalty function and $\lambda > 0$ is a regularization parameter. For example, $p_\lambda(\cdot)$ can be chosen to be the L_1 (Tibshirani, 1996), SCAD (Fan and Li, 2001), adaptive L_1 (Zhang and Lu, 2007; Zou, 2006), or some other penalty.

7.3   Iterative feature selection

As mentioned before, the SIS methodology may break down if a predictor is marginally unrelated, but jointly related with the response, or if a predictor is jointly uncorrelated with the response but has higher marginal correlation with the response than some important predictors. To handle such difficult scenario, iterative SIS may be required. ISIS seeks to overcome these difficulties by using more fully the joint covariate information.

The first step is to apply SIS to select a set $\mathcal{A}_1$ of indices of size d, and then employ (7.2) with the L_1 or SCAD penalty to select a subset $\mathcal{M}_1$ of these indices. This is our initial estimate of the set of indices of important variables.

Next, we compute the conditional marginal utility

$$\ell_j^{(2)} = \min_{b,\beta_j}\sum_{i=1}^{n}\ell(Y_i(b+\mathbf{X}_{i,\mathcal{M}_1}^T\boldsymbol{\beta}_{\mathcal{M}_1}+X_{ij}^T\beta_j)) \qquad (7.3)$$

for any $j \in \mathcal{M}_1^c = \{1,2,\ldots,p\}\backslash\mathcal{M}_1$, where $\mathbf{X}_{i,\mathcal{M}_1}$ is the sub-vector of $\mathbf{X}_i$ consisting of those elements in $\mathcal{M}_1$. If necessary, the term of $\frac{1}{2}\beta_j^2$ may be added in (7.3) to avoid identifiability issue just as the case of defining the marginal utilities in (7.1). The conditional marginal utility $\ell_j^{(2)}$ measures the additional contribution of variable X_j given that the variables in $\mathcal{M}_1$ have been included. We then rank variables in $\mathcal{M}_1^c$ according to their corresponding conditional marginal utilities and form the set $\mathcal{A}_2$ consisting of the indices corresponding to the smallest $d - |\mathcal{M}_1|$ elements.

The above prescreening step using the conditional utility is followed by solving

$$\min_{b,\,\boldsymbol{\beta}_{\mathcal{M}_1},\,\boldsymbol{\beta}_{\mathcal{A}_2}}\frac{1}{n}\sum_{i=1}^{n}\ell(Y_i(b+\mathbf{X}_{i,\mathcal{M}_1}^T\boldsymbol{\beta}_{\mathcal{M}_1}+\mathbf{X}_{i,\mathcal{A}_2}^T\boldsymbol{\beta}_{\mathcal{M}_2})+\sum_{j\in\mathcal{M}_1\cup\mathcal{A}_2}p_\lambda(|\beta_j|). \qquad (7.4)$$

The penalty $p_\lambda(\cdot)$ leads to a sparse solution. The indices in $\mathcal{M}_1 \cup \mathcal{A}_2$ that have non-zero β_j yield a new estimate $\mathcal{M}_2$ of the active indices.

This process of iteratively recruiting and deleting variables may be repeated until we obtain a set of indices $\mathcal{M}_k$ which either reaches the prescribed size d or satisfies convergence criterion $\mathcal{M}_k = \mathcal{M}_{k-1}$.

7.4 Reducing false discovery rate

Sure independence screening is a simple but effective method to screen out irrelevant variables. They are usually conservative and include many unimportant variables. Next we present two possible variants of (I)SIS that have some attractive theoretical properties in terms of reducing the false discovery rate (FDR).

Denote $\mathcal{A}$ to be the set of active indices, namely the set containing those indices j for which $\beta_j \neq 0$ in the true model. Denote $\mathbf{X}_{\mathcal{A}} = \{X_j, j \in \mathcal{A}\}$ and $\mathbf{X}_{\mathcal{A}^c} = \{X_j, j \in \mathcal{A}^c\}$ to be the corresponding sets of active and inactive variables respectively.

Assume for simplicity that n is even. We randomly split the sample into two halves. Apply SIS separately to each half with $d = \lfloor n/\log n \rfloor$ or larger, yielding two estimates $\hat{\mathcal{A}}^{(1)}$ and $\hat{\mathcal{A}}^{(2)}$ of the set of active indices $\mathcal{A}$. Both $\hat{\mathcal{A}}^{(1)}$ and $\hat{\mathcal{A}}^{(2)}$ may have large FDRs because they are constructed by SIS, a crude screening method. Assume that both $\hat{\mathcal{A}}^{(1)}$ and $\hat{\mathcal{A}}^{(2)}$ have the sure screening property, $P(\mathcal{A} \subset \hat{\mathcal{A}}^{(j)}) \rightarrow 1$, for $j = 1$ and 2. Then

$$P(\mathcal{A} \subset \hat{\mathcal{A}}^{(1)} \cap \hat{\mathcal{A}}^{(2)}) \rightarrow 1.$$

Thus motivated, we define our first variant of SIS by estimating $\mathcal{A}$ with $\hat{\mathcal{A}} = \hat{\mathcal{A}}^{(1)} \cap \hat{\mathcal{A}}^{(2)}$.

To provide some theoretical support, we make the following assumption:
Exchangeability Condition: Let $r \in \mathbb{N}$, the set of natural numbers. The model satisfies the exchangeability condition at level r if the set of random vectors

$$\{(Y, \mathbf{X}_{\mathcal{A}}, X_{j_1}, \ldots, X_{j_r}) : j_1, \ldots, j_r \text{ are distinct elements of } \mathcal{A}^c\}$$

is exchangeable.

The Exchangeability Condition ensures that each inactive variable has the same chance to be recruited by SIS. Then we have the following nonasymptotic probabilistic bound.

Let $r \in \mathbb{N}$, and assume that the model satisfies the Exchangeability Condition at level r. For $\hat{\mathcal{A}} = \hat{\mathcal{A}}^{(1)} \cap \hat{\mathcal{A}}^{(2)}$ defined above, we have

$$P(|\hat{\mathcal{A}} \cup \mathcal{A}^c| \geqslant r) \leqslant \frac{\binom{d}{r}^2}{\binom{p - |\mathcal{A}|}{r}} \leqslant \frac{1}{r!}\left(\frac{d^2}{p - |\mathcal{A}|}\right)^r,$$

where there second inequality requires $d^2 \leqslant p - |\mathcal{A}|$.

When $r = 1$, the above probabilistic bound implies that, when the number of selected variables $d \leqslant n$, we have with high probability $\hat{\mathcal{A}}$ reports no 'false positives' if the exchangeability condition is satisfied at level 1 and if p is large by comparison with n^2. It means that it is very likely that any index in the estimated active set also belongs to the active set in the true model, which, together with sure screening assumption, implies the model selection consistency. The nature of this result is somehow unusual in that it suggests that a 'blessing of dimensionality' the probability bound one false positives decreases with p. However, this is only part of the full store, because the probability of missing elements of the true active set is expected to increase with p.

The iterative version of the first variant of SIS can be defined analogously. We apply SIS to each partition separately to get two estimates of the active index set $\hat{A}_1^{(1)}$ and $\hat{A}_1^{(2)}$, each having d elements. After forming the intersection $\hat{A}_1 = \hat{A}_1^{(1)} \cap \hat{A}_1^{(2)}$, we carry out penalized estimation with all data to obtain a first approximation $\hat{\mathcal{M}}_1$ to the true active index set. We then perform a second stage of the ISIS procedure to each partition separately to obtain sets of indices $\hat{\mathcal{M}}_1 \cup \hat{A}_2^{(1)}$ and $\hat{\mathcal{M}}_1 \cup \hat{A}_2^{(2)}$. Take their intersection and re-estimate parameters using penalized estimation to get a second approximation $\hat{\mathcal{M}}_2$ to the true active set. This process can be continued until convergence criterion is met as in the definition of ISIS.

8   Multi-category classification

Sections 6 and 7 focus on binary classifications. In this section, we will discuss how to handle classification problems with more than two classes.

When dealing with classification problems with a multi-category response, one typically label the response as $Y \in \{1, 2, \ldots, K\}$, where K is the number of classes. Define conditional probabilities $p_j(\mathbf{x}) = P(Y = j | \mathbf{X} = \mathbf{x})$ for $j = 1, 2, \ldots, K$. The corresponding Bayes rule classifies a test sample with predictor vector $\mathbf{x}$ to the class with the largest $p_j(\mathbf{x})$. Namely the Bayes rule is given by $\underset{j}{\operatorname{argmax}}\, p_j(\mathbf{x})$.

Existing methods for handling multi-category problems can be generally divided into two groups. One is to solve the multi-category classification by solving a series of binary classifications while the other considers all the classes simultaneously. Among the first group, both methods of constructing either pairwise classifiers (Krefsel, 1998; Schmidt and Gish, 1996) or one-versus-all classifiers (Hsu and Lin, 2002; Rifkin and Klautau, 2004) are popularly used. In the one-versus-all approach, one is required to train K distinct binary classifiers to separate one class from all others and each binary classifier uses all training samples. For the pairwise approach, there are $K(K-1)/2$ binary classifier to be trained with one for each pair of classes. Comparing to the one-versus-all approach, the number of classifiers is much larger for the pairwise approach but each one involves only a subsample of the training data and thus is easier to train. Next we will focus on the second group of methods.

Weston and Watkins (1999) proposed the k-class support vector machine. It solves

$$\min \frac{1}{n} \sum_{i=1}^{n} \sum_{j \neq Y_i} (2 - [f_{Y_i}(\mathbf{X}_i) - f_j(\mathbf{X}_i)])_+ + \lambda \sum_{j=1}^{K} \| f_j \| . \qquad (8.1)$$

The linear classifier takes the form $f_j(\mathbf{x}) = b_j + \boldsymbol{\beta}_j^T \mathbf{x}$, whereas the penalty in (8.1) can be taken as the L_2-norm $\|f_j\| = w_j \|\boldsymbol{\beta}_j\|^2$ for some weight w_j. Let $\hat{f}_j(\mathbf{x})$ be the solution to (8.1). Then the classifier assigns a new observation $\mathbf{x}$ to class $\hat{k} = $ argmax$_x \hat{f}_j(\mathbf{x})$. Zhang (2004) generalized this loss to $\sum_{k \neq Y} \phi(f_Y(\mathbf{X}) - f_k(\mathbf{X}))$ and called it pairwise comparison method. Here $\phi(\cdot)$ can be any decreasing function so that a large value $f_Y(\mathbf{X}) - f_k(\mathbf{X})$ for $k \neq Y$ is favored while optimizing. In particular Weston and Watkins (1999) essentially used the hinge loss up to a scale of factor 2. By assuming the differentiability of $\phi(\cdot)$, Zhang (2004) showed that the desirable property of order preserving. See Theorem 5 of Zhang (2004). However the differentiability condition on $\phi(\cdot)$ rules out the important case of hinge loss function.

Lee, Lin, and Wahba (2004) proposed a nonparametric multi-category SVM by minimizing

$$\frac{1}{n} \sum_{i=1}^{n} \sum_{j \neq Y_i} \left(f_j(\mathbf{X}_i) + \frac{1}{k-1} \right)_+ + \lambda \sum_{j=1}^{K} \| f_j \| \qquad (8.2)$$

subject to the sum-to-zero constraint in the reproducing kernel Hilbert space. Their loss function works with the sum-to-zero constraint to encourage $f_Y(\mathbf{X}) = 1$ and $f_k(\mathbf{X}) = -1/(k-1)$ for $k \neq Y$. For their loss function, they obtained Fisher consistency by proving that the minimizer of $E \sum_{j \neq Y} (f_j(\mathbf{X}) - 1/(k-1))_+$ under the sum-to-zero constraint at $\mathbf{X} = \mathbf{x}$ is given by $\hat{f}_j(\mathbf{x}) = 1$ if $j = $ argmax$_m p_m(\mathbf{x})$ and $-1/(k-1)$ otherwise. This formulation motivated the constrained comparison method in Zhang (2004). The constrained comparison method use the loss function $\sum_{k \neq Y} \phi(-f_k(\mathbf{X}))$. Zhang (2004) showed that this loss function in combination with the sum-to-zero constraint has the order preserving property as well (Theorem 7, Zhang 2004).

Liu and Shen (2006) proposed one formulation to extend the ψ-learning from binary to multicategory. Their loss performs multiple comparisons of class Y versus other classes in a more natural way by solving

$$\min \frac{1}{n} \sum_{i=1}^{n} \psi(\min_{j \neq Y_i} (f_{Y_i}(\mathbf{X}_i) - f_j(\mathbf{X}_i))) + \lambda \sum_{j=1}^{K} \| f_j \| \qquad (8.3)$$

subject to the sum-to-zero constraint. Note that the ψ loss function is non-increasing. The minimization in (8.3) encourages $f_{Y_i}(\mathbf{X}_i)$ to be larger than $f_j(\mathbf{X}_i)$ for all $j \neq Y_i$ thus leading to correct classification. They provided some statistical learning theory for the multicategory ψ-learning methodology and obtained fast convergence rates for both linear and nonlinear learning examples.

Similarly motivated as Liu and Shen (2006), Wu and Liu (2007a) proposed the robust truncated hinge loss support vector machines. They define the truncated hinge loss function to be $H_s(u) = \min\{H(u), H(s)\}$ for some $s \leqslant 0$. The robust truncated hinge loss support vector machine solves

$$\min \frac{1}{n} \sum_{i=1}^{n} H_s(\min_{j \neq Y_i}(f_{Y_i}(\mathbf{X}_i) - f_j(\mathbf{X}_i))) + \lambda \sum_{j=1}^{K} \| f_j \| . \qquad (8.4)$$

Wu and Liu (2007a) used the idea of support vectors to show that the robust truncated hinge loss support vector machine is less sensitive to outliers than the SVM. Note that $H_s(u) = H(u) - [s - u]_+$. This decomposition makes it possible to use the difference convex algorithm (An and Tao, 1997) to solve (8.4). In this way, they showed that the robust truncated hinge loss support vector machine removes some support vectors form the SVM and consequently its corresponding support vectors are a subset of the support vectors of the SVM. Fisher consistency is also established for the robust truncated hinge loss support vector machine when $s \in [-1/(K-1), 0]$. Recall that K is the number of classes. This tells us that more truncation is needed to guarantee consistency for larger K.

The truncation idea is in fact very general. It can be applied to other loss functions such as the logistic loss in logistic regression and the exponential loss in AdaBoost. Corresponding Fisher consistency is also available. Wu and Liu (2007a) only used the hinge loss to demonstrate how the truncation works. In another work, Wu and Liu (2007b) studied the truncated hinge loss function using the formulation of Lee, Lin, and Wahba (2004).

Other formulations of multicategory classification includes those of Vapnik (1998), Bredensteiner and Bennett (1999), Crammer and Singer (2001) among many others. Due to limited space, we cannot list all of them here. Interested readers may read those papers and references therein for more formulations.

In the aforementioned different formulations of multicategory classification with linear assumption that $f_k(\mathbf{x}) = b_k + \boldsymbol{\beta}_k^T \mathbf{x}$ for $k = 1, 2, \ldots, K$, variable selection-capable penalty function can be used in place of $\| f_k \|$ to achieve variable selection. For example Wang and Shen (2007) studied the L_1 norm multiclass support vector machine by using penalty $\sum_{k=1}^{K} \sum_{j=1}^{p} |\beta_{jk}|$. Note that the L_1 norm treats all the coefficients equally. It ignores the fact that the group of $\beta_{j1}, \beta_{j2}, \ldots, \beta_{jK}$ corresponds to the same predictor variable X_j. As a result the L_1 norm SVM is not efficient in achieving variable selection. By including this group information into consideration, Zhang, Liu, Wu, and Zhu (2008) proposed the adaptive super norm penalty for multi-category SVM. They use the penalty $\sum_{j=1}^{p} w_j \max_{k=1,2,\ldots,K} |\beta_{jk}|$, where the adaptive weight w_j is based on a consistent estimate in the same way as the adaptive L_1 penalty (Zhang and Lu, 2007; Zou, 2006) does. Note that the super norm penalty encourages the entire group $\beta_{j1}, \beta_{j2}, \ldots, \beta_{jK}$ to be exactly zero for any noise variable X_j and thus achieves more efficient variable selection.

Variable selection-capable penalty works effectively when the dimensionality is fairly high. However when it comes to ultrahigh dimensionality, things may get complicated. For example, the computational complexity grows with the dimensionality. In this case, the (I)SIS method may be extended to aforementioned multi-category classifications as they are all given in loss function based formulations. Fan et al. (2008) considered (I)SIS for the formulation by Lee, Lin, and Wahba (2004). They used a couple of microarray datasets to demonstrated its practical utilities.

References

[1] An L. T. H. and Tao P. D. (1997). Solving a class of linearly constrained indefinite quadratic problems by D.C. algorithms. *Journal of Global Optimization* **11**, 253-285.

[2] Bair E., Hastie T., Paul D. and Tibshirani R. (2006). Prediction by supervised principal components. *J. Amer. Statist. Assoc.* **101**, 119-137.

[3] Bartlett P., Jordan M., and McAuliffe J. (2006). Convexity, classification, and risk bounds. *Journal of the American Statistical Association* **101**, 138-156.

[4] Bickel P. J. and Levina E. (2004). Some theory for Fisher's linear discriminant function, "naive Bayes", and some alternatives when there are many more variables than observations. *Bernoulli* **10**, 989-1010.

[5] Boser B., Guyon I. and Vapnik V. N. (1992). A training algorithm for optimal margin classifiers. In *Proceedings of the Fifth Annual Conference on Computational Learning Theory*, 144-152. ACM Press, Pittsburgh, PA.

[6] Boulesteix A. L. (2004). PLS dimension reduction for classification with microarray data. *Stat. Appl. Genet. Mol. Biol.* **3**, 1-33.

[7] Breiman L. (1997). Prediction games and arcing algorithms. *Technical Report* **504**, Dept. Statistics, Univ. California, Berkeley.

[8] Breiman L. (1998). Arcing classifiers (with discussion). *Ann. Statist.* **26**, 801-849.

[9] Bura E. and Pfeiffer R. M. (2003). Graphical methods for class prediction using dimension reduction techniques on DNA microarray data. *Bioinformatics* **19**, 1252-1258.

[10] Cristianini N. and Shawe-Taylor, J. (2000). *An Introduction to Support Vector Machines*. Cambridge University Press, Cambridge.

[11] Donoho D. L. (2000). High-dimensional data analysis: The curses and blessings of dimensionality. *Aide-Memoire of a Lecture at AMS Conference on Math Challenges of the 21st Century*.

[12] Donoho D. L. and Jin J. (2004). Feature Selection by Higher Criticism Thresholding: Optimal Phase Diagram. *Manuscript*.

[13] Dudoit S., Fridlyand J. and Speed T. P. (2002). Comparison of discrimination methods for the classification of tumors using gene expression data. *J. Amer. Statist. Assoc.* **97**, 77-87.

[14] Efron B., Hastie T., Johnstone I. and Tibshirani R. (2004). Least angle regression (with discussion). *Ann. Statist.* **32**, 407-499.

[15] Fan J. and Fan Y. (2008). High-dimensional classification using features annealed independence rules. *Ann. Statist.* **36**, 2605-2637.

[16] Fan J. and Li R. (2001). Variable selection via nonconcave penalized likelihood and its oracle properties. *J. Amer. Statist. Assoc.* **96**, 1348-1360.

[17] Fan J. and Li R. (2006). Statistical challenges with high dimensionality: Feature selection in knowledge discovery. *Proceedings of the International Congress of Mathematicians* (M. Sanz-Sole, J. Soria, J.L. Varona, J. Verdera, eds.), Vol. III, 595-622. European Mathematical Society, Zurich.

[18] Fan J. and Lv J. (2008). Sure independence screening for ultrahigh dimensional feature space (with discussion). *J. Roy. Statist. Soc. Ser. B* **70**, 849-911.

[19] Fan J. and Lv J. (2009). Properties of Non-concave Penalized Likelihood with NP-dimensionality. *Manuscript.*

[20] Fan J. and Peng H. (2004). Nonconcave penalized likelihood with diverging number of parameters. *Ann. Statist.* **32**, 928-961.

[21] Fan J., Samworth R. and Wu Y. (2008). Ultrahigh dimensional variable selection: Beyond the linear model. *Journal of Machine Learning Research.* To appear.

[22] Fan J. and Song R. (2009). Sure Independence Screening in Generalized Linear Models with NP-dimensionality. *Manuscript.*

[23] Freund Y. (1995). Boosting a weak learning algorithm by majority. *Inform. and Comput.* **121**, 256-285.

[24] Freund Y. and Schapire R. E. (1996). Experiments with a new boosting algorithm. In *Machine Learning: Proceedings of the Thirteenth International Conference*, 148-156. Morgan Kaufman, San Francisco.

[25] Friedman J., Hastie T. and Tibshirani R. (2000). Additive logistic regression: a statistical view of boosting. *The Annals of Statistics* **28**, 337-407.

[26] Fung G. and Mangasarian O. L. (2001). Proximal support vector machine classifiers. In *Proceedings KDD-2001: Knowledge Discovery and Data Mining* (Provost F. and Srikant F., eds), 77-86. Asscociation for Computing Machinery.

[27] Ghosh D. (2002). Singular value decomposition regression modeling for classification of tumors from microarray experiments. *Proceedings of the Pacific Symposium on Biocomputing.* 11462-11467.

[28] Greenshtein E. (2006). Best subset selection, persistence in high-dimensional statistical learning and optimization under l1 constraint. *Ann. Statist.* **34**, 2367-2386.

[29] Greenshtein E. and Ritov Y. (2004). Persistence in high-dimensional linear predictor selection and the virtue of overparametrization. *Bernoulli* **10**, 971-988.

[30] Hall P. and Jin J. (2008). Properties of higher criticism under strong dependence. *Ann. Statist.* **1**, 381-402.

[31] Hall, P., Marron, J. S. and Neeman, A. (2005). Geometric representation of high dimension, low sample size data. *J. R. Statist. Soc. B* **67**, 427-444.

[32] Hall P., Park B. and Samworth R. (2008). Choice of neighbor order in nearest-neighbor classification. *Ann. Statist.* **5**, 2135-2152.

[33] Hall P., Pittelkow Y. and Ghosh M. (2008). Theoretical measures of relative performance of classifiers for high dimensional data with small sample sizes. *J. R. Statist. Soc. B* **70**, 159-173.

[34] Hastie T., Tibshirani R. and Friedman J. (2009). *The Elements of Statistical Learning: Data Mining, Inference, and Prediction (2nd edition)*. Springer-Verlag, New York.

[35] Hsu C. and Lin C. (2002). A comparison of methods for multi-class support vector machines. *IEEE Trans. Neural Netw.* **13**, 415-425.

[36] Huang X. and Pan W. (2003). Linear regression and two-class classification with gene expression data. *Bioinformatics* **19**, 2072-2978.

[37] Ingster Yu. I. (2002). Adaptive detection of a signal of growing dimension: II. *Math. Meth. Statist.* **11**, 37-68.

[38] Jin J. (2006). Higher criticism statistic: theory and applications in non-Gaussian detection. In *Proc. PHYSTAT 2005: Statistical Problems in Particle Physics, Astrophysics and Cosmology* (L. Lyons and M. K. ünel, eds). World Scientific Publishing, Singapore.

[39] Kimeldorf G. and Wahba G. (1971). Some results on Tchebycheffian spline functions, *J. Math. Anal. Applic.* **33**, 82-95.

[40] Krefsel U. (1998). Pairwise classification and support vector machines. In *Advances in Kernel Methods - Support Vector Learning* (B. Schölkopf, C. Burges, and A. Smola, eds.) MIT Press, Cambridge, MA.

[41] Lee Y., Lin Y. and Wahba, G. (2004). Multicategory support vector machines, theory, and application to the classification of microarray data and satellite radiance data. *Journal of the American Statistical Association* **99**, 67-81.

[42] Li K.-C. (1991). Sliced inverse regression for dimension reduction. *Journal of the American Statistical Association* **414**, 316-327.

[43] Lin Y. (2002). Support vector machines and the Bayes rule in classification. *Data Mining and Knowledge Discovery* **6**, 259-275

[44] Lin Y. (2004). A note on margin-based loss functions in classification. *Statistics and Probability Letters* **68**, 73-82

[45] Liu Y. and Shen X. (2006). Multicategory ψ-learning. *Journal of the American Statistical Association* **101**, 500-509.

[46] Liu Y. and Wu Y. (2007). Variable selection via a combination of the L0 and L1 penalties. *Journal of Computational and Graphical Statistics* **16** (4), 782-798.

[47] Lv J. and Fan Y. (2009). A unified approach to model selection and sparse recovery using regularized least squares. *Ann. Statist.* **37**, 3498-3528.

[48] Mahalanobis P. C. (1930). On tests and measures of group divergence. *Journal of the Asiatic Society of Bengal* **26**, 541-588.

[49] Nguyen D. V. and Rocke D. M. (2002). Tumor classification by partial least squares using microarray gene expression data. *Bioinformatics* **18**, 39-50.

[50] Rifkin R. and Klautau A. (2004). In defence of one-versus-all classificaiton. *Journal of Machine Learning Research* **5**, 101-141.

[51] Schapire R. E. (1990). The strength of weak learnability. *Machine Learning* **5**, 197-227.

[52] Schmidt M. S. and Gish H. (1996). Speaker identification via support vector classifiers. In *Proceedings of the 21st IEEE International Conference Conference on Acoustics, Speech, and Signal Processing (ICASSP-96)* 105-108, Atlanta, GA.

[53] Shen X., Tseng G. C., Zhang X. and Wong W. H. (2003). On ψ-Learning. *Journal of the American Statistical Association* **98**, 724-734.

[54] Suykens J. A. K. and Vandewalle J. (1999). Least squares support vector machine classifiers. *Neural Processing Letters* **9**, 293-300.

[55] Tibshirani R. (1996). Regression shrinkage and selection via the LASSO. *J. Roy. Statist. Soc. Ser. B* **58**, 267-288.

[56] Tibshirani R., Hastie T., Narasimhan B. and Chu G. (2002). Diagnosis of multiple cancer types by shrunken centroids of gene expression. *Proc. Natl. Acad. Sci.* **99**, 6567-6572.

[57] Vapnik V. (1998). *Statistical Learning Theory.* Wiley, New York.

[58] Wahba G. (1990). Spline models for observational data. *CBMS-NSF Regional Conference Series in Applied Mathematics* **59**. SIAM, Philadelphia.

[59] Wahba G. (1998). Support Vector Machines, Reproducing Kernel Hilbert Spaces, and Randomized GACV. In *Advances in Kernel Methods: Support Vector Learning* (Schökopf, Burges, and Smola, eds.), 125-143. MIT Press, Cambridge, MA.

[60] Wang L. and Shen X. (2007). On L1-norm multi-class support vector machines: methodology and theory. *Journal of the American Statistical Association* **102**, 595-602.

[61] Weston J., and Watkins C. (1999). Support vector machines for multi-class pattern recognition. In *Proceedings of the 7th European Symposium on Artificial Neural Networks (ESANN-99)*, 219-224.

[62] Wu Y. and Liu Y. (2007a). Robust truncated-hinge-loss support vector machines. *Journal of the American Statistical Association* **102** (479) 974-983.

[63] Wu Y. and Liu Y. (2007b). On multicategory truncated-hinge-loss support vector machines. *Contemporary Mathematics* **443**, 49-58.

[64] Zhang T. (2004), Statistical analysis of some multi-category large margin classification methods. *Journal of Machine Learning Research* **5**, 1225-1251.

[65] Zhang H. H., Ahn J., Lin X. and Park C. (2006). Gene selection using support vector machines with nonconvex penalty. *Bioinformatics* **22**, 88-95.

[66] Zhang H. H., Liu Y., Wu. and Zhu J. (2008). Variable selection for the multicategory SVM via sup-norm regularization. *Electronic Journal of Statistics* **2**, 149-167.

[67] Zhang H. H. and Lu W. (2007). Adaptive-LASSO for Cox's proportional hazard model. *Biometrika* **94**, 691-703.

[68] Zhang T. and Oles F. J. (2001). Text categorization based on regularized linear classification methods. *Information Retrieval* **4**, 5-31.

[69] Zhu L., Miao B. and Peng H. (2006). On sliced inverse regression with high-dimensional covariates. *Journal of the American Statistical Association.* **101**, 630-643.

[70] Zhu J., Rosset S., Hastie T. and Tibshirani R. (2003). 1-norm support vector machines. *Neural Information Processing Systems* **16**.

[71] Zou H. (2006). The adaptive Lasso and its oracle properties. *Journal of the American Statistical Association* **101**, 1418-1429.

[72] Zou H., Hastie T. and Tibshirani. R. (2004). Sparse principal component analysis. *Technical Report.*

Chapter 2

Flexible Large Margin Classifiers*

Yufeng Liu[†] and Yichao Wu[‡]

Abstract

Classification is an important tool for statistical analysis. Among numerous classification methods, margin-based techniques have attracted a lot of attention due to its competitive performance and ability in handling complex and high dimensional data. In this chapter, we review some recent advances of large margin classifiers. We start with the Support Vector Machine in terms of margin and maximum separation. Then we view the SVM in the regularization framework and compare it with several other existing classifiers. Recent extensions of the SVM such as ψ-learning, robust SVM (RSVM), bounded constraint machine (BCM), and balancing SVM (BSVM) are discussed. Issues on multicategory classification and various extensions of binary classifiers to the multicategory case are explored. Finally, issues on hard classifiers and the corresponding class probability estimation problem are briefly mentioned.

Keywords: Classification, Fisher consistency, L_1-norm penalty, large margin, multicategory, probability estimation, sup-norm, SVM.

1 Background on classification

The statistical community has witnessed a boom in statistical learning. Statistical learning techniques have been widely applied in various disciplines such as biology, computer science, and finance. Rapid development of the theory and methods of statistical learning, especially in supervised learning, has been made in recent years.

The goal of supervised learning is to predict an output variable based on one or multiple associated covariates. Classification, as an important example of supervised learning, is a procedure that builds a model based on a training dataset to predict the class memberships for new examples with only covariates available.

---

*The authors are partly supported by US NSF grants DMS-0606577, 0747575, and 0905561 and NIH/NCI grant IROICA149569.

†Department of Statistics and Operations Research, Carolina Center for Genome Sciences, University of North Carolina, Chapel Hill, NC 27599, USA, E-mail: yfliu@email.unc.edu

‡Department of Statistics, North Carolina State University, Raleigh, NC 27695, USA, E-mail: wu@stat.ncsu.edu

It can be understood as a special form of regression with the response variable being categorical. When the response variable is binary, the learning problem is known as a binary classification. If there are more than two classes, we have the multicategory classification.

There are a large number of methods for classification in the literature. Examples include Fisher Linear Discrimination Analysis (LDA), logistic regression, k-nearest neighbor (Dasarathy 1991), decision trees (Breiman et al., 1984), neural networks (Bishop, 1995), boosting (Freund and Schapire, 1996), and many more. See Duda et al. (2001) and Hastie et al. (2001) for more comprehensive reviews of various classification methods.

Among numerous classification techniques, margin-based classifiers have attracted tremendous attentions in recent years due to their competitive performance and ability in handling high dimensional data. One of the most well-known large margin classifiers is the Support Vector Machine (SVM). Since its introduction, the SVM has gained much popularity in both machine learning and statistics. The seminal work by Vapnik (1995, 1998) has laid the foundation for the general statistical learning theory for the SVM, which furthermore inspired various extensions on the SVM. For other references on the binary SVM, see Christianini and Shawe-Taylor (2000), Schölkopf and Smola (2002), and references therein. Besides the SVM, there are a number of other large margin classifiers introduced in the literature. Examples include the Penalized Logistic Regression (PLR) (Wahba, 1999; Lin et al., 2000), ψ-learning (Shen et al., 2003, Liu and Shen, 2006), the robust SVM (Wu and Liu, 2007a), Distance-Weighted Discrimination (DWD) (Marron et al., 2007; Liu, Zhang, and Wu, 2009), and so on.

In this chapter, we review some recent advances of large margin classifiers. We start with the SVM using the interpretation of margin and maximum separation in Section 2. Then in Section 3, we view the SVM in the regularization framework and compare it with several other existing classifiers. In Section 4, we discuss a few recent extensions of the SVM in terms of the optimization criteria. In particular, the bounded constraint machine (BCM) and the balancing SVM (BSVM) are discussed. When the number of classes is more than two, the corresponding classification can be complicated. In Section 5, we discuss issues on multicategory classification and various extensions of binary classifiers to the multicategory case. In Section 6, we explain some issues on hard classifiers and the corresponding class probability estimation problem is briefly mentioned. Some conclusions and discussions are contained in Section 7.

We would like to point out that our review is based on some selected view of large margin classifiers. We do not attempt to include all existing work in this direction. More references can be found in the cited work of this review.

2   The support vector machine: the margin formulation and the SV interpretation

The SVM is a powerful classification tool and has enjoyed great success in many applications (Cristianini and Shawe-Taylor, 2000; Vapnik, 1998). It was first proposed using the idea of seeking the optimal separating hyperplane with elegant margin interpretation.

To learn a classification rule, we are typically given a training sample $\{(\mathbf{x}_i, y_i) : i = 1, 2, \cdots, n\}$ which is generated according to some unknown probability distribution function $P(\mathbf{x}, y)$. Here $\mathbf{x}_i \in \mathcal{S} \subset \mathbb{R}^p$ and y_i denote the input feature vector and categorical output label, respectively. The sample size is denoted by n and the dimensionality of the input feature space is given by p.

When the output is binary, it is typically coded as -1 or $+1$. The goal of binary classification is to estimate a function $f(\cdot)$, a mapping from $\mathcal{S}$ to $\mathbb{R}$, whose sign will be used as the classification rule.

2.1   Linearly-separable SVM

For simplicity, we start with the linearly separable SVM. In this case, the two classes are linearly separable in the feature space. Then there are many hyperplanes that are capable of separating the two classes. Among all possible separating hyperplanes, the SVM tries to seek the one that leads to a maximal separation between these two classes. Assume $f(\mathbf{x}) = \mathbf{x}^T \mathbf{w} + b$ with $\mathbf{w} = (w_1, w_2, \ldots, w_p)^T$. The SVM can be formulated as the following optimization problem

$$\min \frac{1}{2} \parallel \mathbf{w} \parallel^2 \tag{2.1}$$

subject to

$$y_i f(\mathbf{x}_i) \geqslant 1,$$

where $\parallel \mathbf{w} \parallel = \sqrt{\sum_{j=1}^p w_j^2}$ denotes the two-norm of $\mathbf{w}$. Here the constraint in (2.1) ensures that each observation is correctly classified, which is achievable due to the linearly separable assumption. Pictorially this means that it enforces all observations staying outside the region between two hyperplanes $\{\mathbf{x} : \mathbf{x}^T \mathbf{w} + b = 1\}$ and $\{\mathbf{x} : \mathbf{x}^T \mathbf{w} + b = -1\}$. The distance between these two hyperplanes is $2/\parallel \mathbf{w} \parallel$, which is inversely proportional to the objective function of (2.1). This explains that the SVM seeks a maximal separation between these two classes. It is pictorially displayed in Figure 2.1.

The SVM (2.1) involves quadratic optimization, which is typically solved in terms of its dual problem. The Lagrangian primal function of (2.1) is given by

$$L_P(\mathbf{w}, b, \boldsymbol{\alpha}) = \frac{1}{2} \parallel \mathbf{w} \parallel^2 - \sum_{i=1}^n \alpha_i [y_i(\mathbf{x}_i^T \mathbf{w} + b) - 1], \tag{2.2}$$

where Lagrangian multipliers are non-negative, i.e., $\alpha_i \geqslant 0$ for $i = 1, 2, \ldots, n$. Maximizing the Lagrangian primal (2.2) with respect to primal variables $\mathbf{w}$ and b, we get the solution representation $\mathbf{w} = \sum_{i=1}^n \alpha_i y_i \mathbf{x}_i$ and the Karush-Kuhn-Tucker (KKT) condition $\sum_{i=1}^n \alpha_i y_i = 0$. Plugging the solution representation into the Lagrangian primal function (2.2), we obtain the Lagrangian dual function $L_D(\boldsymbol{\alpha}) = \sum_{i=1}^n \alpha_i - \frac{1}{2} \sum_{1 \leqslant i,j \leqslant n} \alpha_i \alpha_j y_i y_j \mathbf{x}_i^T \mathbf{x}_j$. Then in the dual space, the SVM solves

$$\max L_D(\boldsymbol{\alpha}) = \sum_{i=1}^n \alpha_i - \frac{1}{2} \sum_{1 \leqslant i,j \leqslant n} \alpha_i \alpha_j y_i y_j \mathbf{x}_i^T \mathbf{x}_j \tag{2.3}$$

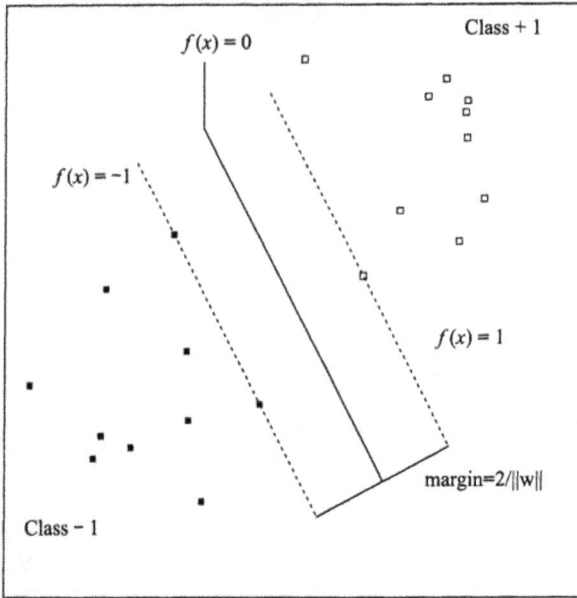

Figure 2.1   Plot of the SVM decision boundary defined by $\mathbf{x}^T\mathbf{w}+b=0$, the geometric margin, and three SVs on hyperplanes $\mathbf{x}^T\mathbf{w}+b=\pm 1$.

subject to

$$\sum_{i=1}^{n}\alpha_i y_i = 0, \quad \alpha_i \geqslant 0; i = 1, 2, \ldots, n.$$

Denote the optimal solution to (2.3) by $\hat{\alpha}_i$, $i = 1, 2, \ldots, n$. Then the optimal solution to the SVM is given by $\hat{\mathbf{w}} = \sum_{i=1}^{n}\hat{\alpha}_i y_i \mathbf{x}_i$. Most of $\hat{\alpha}_i$, $i = 1, 2, \ldots, n$ are exactly equal to zero, except a few of them. Based on the solution representation $\hat{\mathbf{w}} = \sum_{i=1}^{n}\hat{\alpha}_i y_i \mathbf{x}_i$, the sparsity of $\hat{\alpha}_i$s means that the optimal solution to the SVM depends only on a few $\mathbf{x}_i$ and y_i corresponding to $\hat{\alpha}_i > 0$. Those pairs of observations $(\mathbf{x}_i, y_i)$ with $\hat{\alpha}_i > 0$ are called the support vectors (SVs) and hence comes the name of the support vector machine.

The optimization problem for the SVM is completely solved once the intercept b is recovered. By properties of Lagrangian multiplier, $\hat{\alpha}_i > 0$ implies that the corresponding constraint in (2.1) holds with "$\geqslant$" being replaced by "$=$", namely, $y_i(\mathbf{x}_i^T\mathbf{w}+b) = 1$. By this we can recover the intercept b, whose estimate is denoted by $\hat{b}$. Then the SVM boundary is given by $\{\mathbf{x} : \mathbf{x}^T\hat{\mathbf{w}}+\hat{b} = 0\}$ and the classification rule is $\operatorname{sign}(\mathbf{x}^T\hat{\mathbf{w}} + \hat{b})$.

In this formulation, the term $yf(\mathbf{x})$ is called the margin by which the input feature $\mathbf{x}$ is correctly classified. Positive margin means correct classification while negative margin corresponds to misclassification. The larger the margin is, the more confident we are in making class prediction.

From Figure 2.1, we can see that the separating hyperplane is completely determined by a subset of the sample, namely these SV points on hyperplanes

$\{\mathbf{x} : \mathbf{x}^T\mathbf{w} + b = 1\}$ and $\{\mathbf{x} : \mathbf{x}^T\mathbf{w} + b = -1\}$.

As a remark, we note that the optimization problem (2.3) is of size n. Consequently, it can be very efficient for high dimensional low sample size problems with $p \gg n$ such as microarray gene expression data.

2.2   Linearly nonseparable SVM

When two classes are not linearly separable, it is not possible for us to find a linear hyperplane to perfectly separate these two classes from each other. In this case, one approach is to allow constraints in (2.1) to be potentially violated by some observations (Bennett and Mangasarian, 1992; Cortes and Vapnik, 1995; Vapnik, 1995, 1998). This potential constraint violation can be handled by introducing slack variables $\xi_i \geqslant 0$, $i = 1, 2, \cdots, n$, and relaxing to constraints $\mathbf{x}_i^T\mathbf{w} + b \geqslant 1 - \xi_i$ for $y_i = +1$ and $\mathbf{x}_i^T\mathbf{w} + b \leqslant -1 + \xi_i$ for $y_i = -1$. In this case, a misclassification occurs whenever $\xi_i > 1$. Consequently the number of misclassified points is given by $\sum_{i=1}^{n} I(\xi_i > 1)$. Recall that a very important goal of classification is to minimize the misclassification rate. However the indication function $I(\cdot)$ is neither smooth nor convex. Thus it is challenging, if not impossible, to include $\sum_{i=1}^{n} I(\xi_i > 1)$ in the objective function. The SVM replaces $I(\xi_i > 1)$ by ξ_i, which is a convex upper bound of $I(\xi_i > 1)$ and serves as a surrogate of the 0-1 loss function.

For the linearly nonseparable case, the SVM solves the following optimization

$$\min \ C \sum_{i=1}^{n} \xi_i + \frac{1}{2} \| \mathbf{w} \|^2 \tag{2.4}$$

$$\text{subject to} \quad \xi_i \geqslant 0; \ \ i = 1, 2, \ldots, n,$$
$$y_i(\mathbf{x}_i^T\mathbf{w} + b) \geqslant 1 - \xi_i; \ \ i = 1, 2, \ldots, n$$

with some $C > 0$. Here (2.4) controls how much misclassification is allowed while maximizing the separation margin by changing the regularization parameter C.

A similar primal-dual derivation as in the linearly separable case shows that the solution to (2.4) can also be represented as $\mathbf{w} = \sum_{i=1}^{n} \alpha_i y_i \mathbf{x}_i$. The corresponding KKT conditions are

$$\sum_{i=1}^{n} \alpha_i y_i = 0,$$

$$\alpha_i[y_i(\mathbf{x}_i^T\mathbf{w} + b) - 1 + \xi_i] = 0, i = 1, 2, \ldots, n, \tag{2.5}$$

$$(C - \alpha_i)\xi_i = 0, i = 1, 2, \ldots, n. \tag{2.6}$$

The corresponding dual problem is the same as in the linearly separable case except for the additional constraints $\alpha_i \leqslant C$ for $i = 1, 2, \ldots, n$.

Recall that SVs are those observation pairs associated with nonzero Lagrangian multipliers. In the linearly nonseparable case, SVs can be divided into two types. One type of SVs are observation pairs satisfying $0 < \alpha_i < C$ while the other type corresponds to $\alpha_i = C$. The first type of SVs sit either on the hyperplane $\mathbf{x}^T\mathbf{w} + b = 1$ when $y_i = 1$ or $\mathbf{x}^T\mathbf{w} + b = -1$ when $y_i = -1$ because

$0 < \alpha_i < C$ together with (2.6) implies that $\xi_i = 0$. This property enables us to recover the intercept b using SVs with $0 < \alpha_i < C$ when necessary. The other type of SVs satisfying $\alpha_i = C$ includes two subtypes: either misclassification ($\xi_i > 1$) or correct classification with margin less than one ($0 < \xi_i \leqslant 1$). Here correct classification with margin less than one implies $y_i(\mathbf{x}_i^T\mathbf{w} + b) \geqslant 0$ (correct classification) and $y_i(\mathbf{x}_i^T\mathbf{w} + b) < 1$ (the margin is less than one).

## 2.3	Nonlinear SVM

So far we have focused on linear learning by assuming $f(\mathbf{x}) = \mathbf{x}^T\mathbf{w} + b$ whose usage can be limited in real applications. Depending on the particular problem one is working with, sometime nonlinear learning may be more appropriate. Basis expansion is a simple method to achieve nonlinear learning. For example for each component X_j, one may also include X_j^2, X_j^3, or even high order polynomials and interactions. In this case, linear learning in this expanded feature space will lead to a nonlinear classifier in the original feature space spanned by X_j's. Spline basis is another example.

Here we introduce another method for nonlinear learning using the kernel trick. For a Mercer kernel function $K(\cdot, \cdot)$, function $f(\mathbf{x})$ can be represented as $\sum_{i=1}^{n} c_i K(\mathbf{x}_i, \mathbf{x}) + b$ due to the representor theorem (Kimeldorf and Wahba, 1971; Wahba, 1990). Using the norm $\frac{1}{2}\sum_{1 \leqslant i,j \leqslant n} c_i K(\mathbf{x}_i, \mathbf{x}_j)c_j$, the corresponding kernel SVM solves

$$\min C \sum_{i=1}^{n} \xi_i + \frac{1}{2} \sum_{1 \leqslant i,j \leqslant n} c_i K(\mathbf{x}_i, \mathbf{x}_j)c_j \tag{2.7}$$

subject to $\xi_i \geqslant 0;\ i = 1, 2, \ldots, n,$

$$y_i\left[\sum_{j=1}^{n} c_j K(\mathbf{x}_i, \mathbf{x}_j) + b\right] \geqslant 1 - \xi_i;\ i = 1, 2, \ldots, n.$$

As in the linear case, by using Lagrangian multipliers, the dual optimization of (2.7) is given by

$$\max \sum_{i=1}^{n} \alpha_i - \frac{1}{2} \sum_{1 \leqslant i,j \leqslant n} \alpha_i \alpha_j y_i y_j K(\mathbf{x}_i, \mathbf{x}_j) \tag{2.8}$$

subject to	$0 \leqslant \alpha_i \leqslant C, i = 1, 2, \ldots, n,$

$$\sum_{i=1}^{n} \alpha_i y_i = 0.$$

The only difference from the linear case is to replace the inner product $\mathbf{x}_i^T\mathbf{x}_j$ by $K(\mathbf{x}_i, \mathbf{x}_j)$. Once the solution to (2.7) is obtained, either a KKT condition or linear programming can be used to recover the intercept b. Denote estimates by $\hat{\alpha}_i$ and $\hat{b}$. Then the classifier is given by $\text{sign}(\sum_{i=1}^{n} \hat{\alpha}_i y_i K(\mathbf{x}_i, \mathbf{x}) + \hat{b})$.

3   Regularization framework

It is now well known that the SVM can be fit into the regularization framework (Wahba, 1998). Define the hinge loss function by $H_1(u) = [1 - u]_+$, where $[u]_+ = \max\{0, u\}$. Then (2.4) is equivalent to

$$\min C \sum_{i=1}^{n} H_1(y_i(\mathbf{x}_i^T \mathbf{w} + b)) + \frac{1}{2} \parallel \mathbf{w} \parallel^2 . \qquad (3.1)$$

Define $0 \times \infty = 0$. Then (3.1) includes the linearly separable SVM (2.1) as a special case with $C = \infty$. Here $C = \infty$ corresponds to the case with no constraint violation allowed. It is worthwhile to point out that (3.1) can also be extended accordingly to nonlinear SVM using the kernel trick.

3.1   Choice of the loss function

Within the regularization framework, there are many other loss functions available for classification. Since the functional margin $y_i f(\mathbf{x}_i)$ indicates correctness of classification, one often uses a loss function $V(\cdot)$ with $y_i f(\mathbf{x}_i)$ as its argument.

Note that the hinge loss function is an upper bound of the 0-1 loss function. It is unbounded and shoots to infinity when the margin $yf(\mathbf{x})$ goes to negative infinity. This property implies that the optimal solution can be heavily affected by outliers. By outliers, we refer to observations that are far away from their own classes. This potential heavy dependency on outliers is an undesirable property of a classification method in that it may give bad prediction accuracy.

ψ-learning

In order to alleviate the heavy dependency on outliers, Shen, Tseng, Zhang, and Wong (2003) proposed ψ-learning. Their method replaces the hinge loss with the ψ loss function which satisfies

$$0 < \psi(u) \leqslant U \text{ if } u \in [0, T];$$
$$\psi(u) = I(u < 0) \text{ otherwise}$$

with two parameters $0 < U \leqslant 1$ and $T > 0$. The ψ loss function is defined to have positive value for $u \in [0, T]$. This helps to solve the scaling issue of the 0-1 loss function. However a general ψ loss function is not convex and this makes it difficult to solve the corresponding optimization problem. For some special cases, both difference convex (d.c.) algorithms and mixed integer programming techniques are adopted to solve the corresponding optimization problem (Liu, Shen, and Doss, 2006, Liu and Wu, 2006). Numerical improvements over the SVM have been observed for ψ-learning (Shen et al., 2003, Liu and Wu, 2006).

Robust truncated hinge-loss SVM

Similarly motivated, Wu and Liu (2007a) proposed to truncate the hinge loss function and defined the truncated hinge loss $T_s(u) = H_1(u) - H_s(u)$, where

$H_s(u) = [s - u]_+$ for $s \leqslant 0$. The corresponding method is named as the robust truncated-hinge-loss SVM (RSVM). Similar as the ψ loss, the truncated hinge loss function is non-convex. The d.c. algorithm is adopted to solve the corresponding optimization. Interestingly, the continuous ψ loss considered in Liu et al. (2005) can be viewed as a special truncated hinge loss at $s = 0$. More discussion is provided in Section 5.

AdaBoost

Boosting (Freund, 1995; Schapire, 1990) is another machine learning algorithm that can be used for classification. It targets to address the question raised by Kearns (1988): "Can a set of weak learners create a single strong learner?" A weak learner can be any classifier that does slightly better than random guessing. Boosting tries to aggregate weak learners to form a strong learner, which leads to good classification performance. Many different boosting algorithms have been studied. They are based on different methods of weighting training data points and different weak learners. As a special case, AdaBoost (Freund and Schapire, 1996) is very popular among many different boosting algorithms. It has attracted lots of attentions on theoretical investigations to understand why AdaBoost performs very well. Interested readers may consult Breiman (1997, 1998); Friedman, Hastie, and Tibshirani (2000); Freund and Schapire (1996). In particular, Friedman, Hastie, and Tibshirani (2000) pointed out that AdaBoost can be interpreted as a large margin classifier with the exponential loss function, namely replacing the hinge loss of the SVM by the exponential loss e^{-u}.

Logistic regression

These aforementioned different loss functions were introduced to achieve good classification performance. In contrast, logistic regression is a traditional classification method and it was originally motivated from a likelihood point view (McCullagh and Nelder, 1989). It uses the logistic loss $\log(1 + e^{-u})$. In terms of likelihood, (linear) logistic model assumes that the conditional class probability $P(Y = +1|\mathbf{X} = \mathbf{x})$ is equal to

$$e^{\mathbf{x}^T\mathbf{w}+b}/(1 + e^{\mathbf{x}^T\mathbf{w}+b}). \tag{3.2}$$

Then the 2-norm regularized logistic regression uses $\| \mathbf{w} \|^2$ to regularize the corresponding log-likelihood and solves

$$\min C \sum_{i=1}^{n} \log(1 + e^{-y_i(\mathbf{x}_i^T\mathbf{w}+b)}) + \| \mathbf{w} \|^2 .$$

We slightly abuse our notation and denote the optimal solution by $\hat{\mathbf{w}}$ and $\hat{b}$. In addition to the logistic classification rule given by $\mathrm{sign}(\mathbf{x}^T\hat{\mathbf{w}} + \hat{b})$, the logistic regression can be used to estimate the conditional class probability by plugging $\hat{\mathbf{w}}$ and $\hat{b}$ into (3.2).

Here we have listed several loss functions in the literature. There are many other existing loss functions, including the squared loss $V(u) = (1 - u)^2$ in the proximal SVM (Fung and Mangasarian, 2001) and the least square SVM (Suykens and Vandewalle, 1999), the modified least squared loss $V(u) = [1 - u]_+^2$ (Zhang and Oles, 2001), and many others. Figure 3.1 shows some of these aforementioned loss functions.

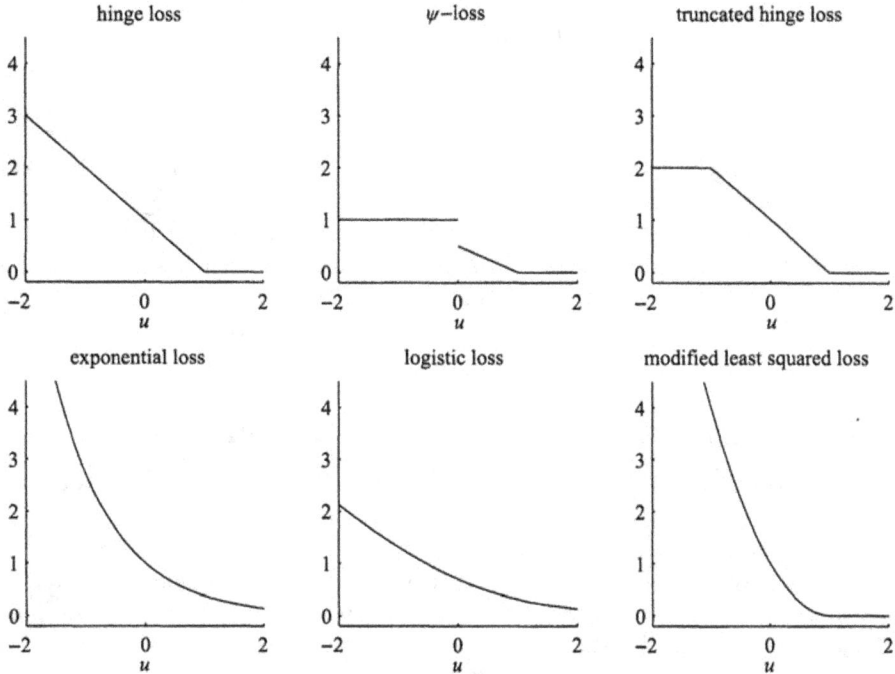

Figure 3.1   Plots of different loss functions.

3.2   Choice of penalty

For the linear function $\mathbf{x}^T\mathbf{w} + b$, the standard SVM uses the 2-norm penalty $\| \mathbf{w} \|^2 = \sum_{j=1}^{p} w_j^2$ as we discussed earlier. When the 2-norm penalty is used, it leads to properties similar as those of the ridge regression in addition to the nice margin interpretation. The 2-norm penalty shrinks the coefficients and helps to make the solution more stable. However it does not produce sparse $\mathbf{w}$. Sparsity can be very desirable when dealing with high dimensional problems.

High dimensional data are very common these days with the recent advance of technology. Examples include microarray and proteomics data which may contain thousands or tens of thousands of predictor variables. While dealing with such kind of data, we are not only interested in looking for classifiers with good performance but also like to select important predictor variables.

In order to achieve variable selection, Bradley and Mangasarian (1998) demonstrated usefulness of the L_1 penalty $\sum_{j=1}^{p} |w_j|$. The L_1 penalty can effectively shrink small or redundant coefficients to zero and is also known as the LASSO penalty in the regression setting (Tibshirani, 1996). Zhu, Rosset, Hastie, and Tibshirani (2003) proposed a solution path algorithm for the 1-norm SVM. The SCAD penalty is another penalty function that is capable of variable selection due to Fan and Li (2001). It is defined as

$$
p_{\lambda,a}(\theta) = \begin{cases} \lambda|\theta|, & |\theta| \leqslant \lambda, \\ -\frac{\theta^2 - 2a\lambda\theta + \lambda^2}{2(a-1)}, & \lambda < |\theta| \leqslant a\lambda, \\ (a+1)\lambda^2/2, & |\theta| > a\lambda, \end{cases}
$$

where $a > 0$ and $\lambda > 0$ are two parameters. From a Bayesian point view, Fan and Li (2001) recommended to choose $a = 3.7$. By using the penalty $\sum_{j=1}^{p} p_{\lambda,a}(w_j)$, Zhang, Ahn, Lin, and Park (2006) studied the SCAD SVM. In another paper, Liu and Wu (2007) studied the combination of the L_0 and L_1 penalties and used this combined penalty to regularize the SVM. Other penalties include the F_∞-norm (Zou and Yuan, 2008) and many others (Zhao et al., 2006; Zou, 2006).

4   Some extensions of the SVM: Bounded constraint machine and the balancing SVM

Due to the design of the SVM, its solution only depends on the set of SVs. This helps to simplify the solution. However, in some situations, it may be better to use more data information to determine the classification boundary. For example, if the training dataset is noisy with outliers, the solution can be deteriorated if it is determined by a SV set containing outliers. Liu (2007) and Park and Liu (2009) proposed the bounded constraint machine (BCM) and the balancing support vector machine (BSVM) as alternatives of the SVM.

4.1   The bounded constraint machine

Liu (2007) and Park and Liu (2009) proposed a different optimization criterion. In particular, they proposed to minimize the sum of signed distances to the boundary and solve the following problem

$$
\min_{f} J(f) - C \sum_{i=1}^{n} y_i f(\mathbf{x}_i)
$$

subject to

$$
-1 \leqslant f(\mathbf{x}_i) \leqslant 1, \forall i = 1, \ldots, n.
$$

That is, they proposed to maximize $\sum_{i=1}^{n} y_i f(\mathbf{x}_i)$, while forcing all the training data to stay between the hyperplanes $f(\mathbf{x}) = \pm 1$. One can view that the BCM uses the hinge loss of the SVM with $y_i f(\mathbf{x}_i) \in [-1, 1]$. In contrast of the SVM, the BCM makes use of all training points to obtain the resulting classifier.

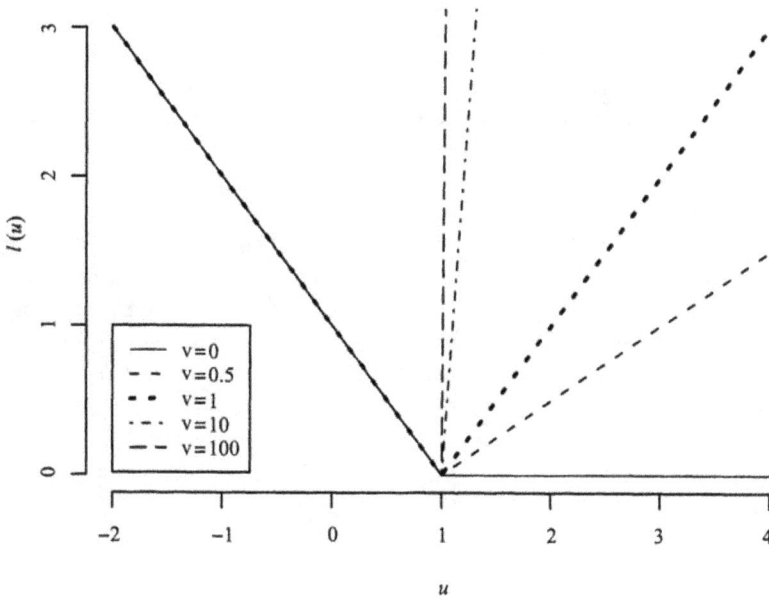

Figure 4.1   Plot of loss function $V(u) = g(u)$ for the BSVM with different values of v.

4.2   The balancing support vector machine

The SVM only uses the SV set to calculate its solution, while the BCM utilizes all training points. To connect these two, Park and Liu (2009) studied the BSVM using the following loss function

$$g(u) = \begin{cases} 1 - u & \text{if } u \leqslant 1, \\ v(u - 1) & \text{otherwise,} \end{cases} \tag{4.1}$$

where v is the slope of the loss function when $u \in (1, \infty)$, as shown in Figure 4.1. Note that v determines how much the solution will rely on the data points with $y_i f(\mathbf{x}_i) \geqslant 1$, and the problem becomes equivalent to the SVM when $v = 0$. The BSVM can be viewed as a bridge to connect the SVM with the proposed BCM.

Note that when $v = \infty$, the BSVM becomes equivalent to solving

$$\min_{(b, \mathbf{w})} J(f) - C \sum_{i=1}^{n} y_i f(\mathbf{x}_i)$$

subject to

$$f(\mathbf{x}_i) \leqslant 1, \forall i = 1, \dots, n. \tag{4.2}$$

Comparing to the BCM in (4.1), the only difference is that the BCM has the constraint $f(\mathbf{x}_i) \geqslant -1$ but the BSVM with $v = \infty$ does not. Typically this difference does not matter since the solution of (4.2) usually satisfies $f(\mathbf{x}_i) \geqslant -1$. The only case that the BCM actually works differently from the BSVM is when a data point moves far away from its own class, even further than the other class. This rarely happens in practice. Thus, the BSVM with $v = \infty$ can be viewed as a good approximation of the BCM. Overall, the BSVM builds a continuum from the standard SVM ($v = 0$) to the BCM ($v = \infty$).

4.3   Interpretation of the BSVM

Since the loss $g(u)$ for the BSVM is not a decreasing function and it imposes big loss values even on the correctly classified data points as well as misclassified observations, it might seem counterintuitive. However, the increasing part with $y_i f(\mathbf{x}_i) > 1$ may help to bring the decision boundary towards the correctly classified points, which can be desirable in some situations. To understand the behavior of the BSVM further, Park and Liu (2009) gave its primal problem as follows

$$\min_{(b,\mathbf{w})} \frac{1}{2}\|\mathbf{w}\|^2 + C\sum_{i=1}^{n} \xi_i$$

subject to

$$\xi_i \geqslant 1 - y_i f(\mathbf{x}_i); \quad \xi_i \geqslant v(y_i f(\mathbf{x}_i) - 1), \forall i = 1, \ldots, n.$$

The corresponding Lagrange primal can be written as

$$L(\mathbf{w}, b, \boldsymbol{\alpha}) = \frac{1}{2}\|\mathbf{w}\|^2 + C\sum_{i=1}^{n}\xi_i + \sum_{i=1}^{n}\gamma_i[1 - y_i f(\mathbf{x}_i) - \xi_i] + \sum_{i=1}^{n}\delta_i[vy_i f(\mathbf{x}_i) - v - \xi_i].$$
(4.3)

Setting derivatives to zero gives the KKT conditions

$$\gamma_i(1 - y_i f(\mathbf{x}_i) - \xi_i) = 0, \tag{4.4}$$
$$\delta_i(vy_i f(\mathbf{x}_i) - v - \xi_i) = 0. \tag{4.5}$$

Then the corresponding dual problem becomes

$$\min_{\boldsymbol{\alpha}} \frac{1}{2}\sum_{i,j=1}^{n} y_i y_j \alpha_i \alpha_j \langle \mathbf{x}_i, \mathbf{x}_j \rangle - \sum_{i=1}^{n} \alpha_i$$

subject to

$$\sum_{i=1}^{n} y_i \alpha_i = 0; \quad -Cv \leqslant \alpha_i \leqslant C, \forall i = 1, \ldots, n. \tag{4.6}$$

Once the solution of (4.6) is obtained, $\mathbf{w}$ can be calculated as $\sum_{i=1}^{n} \alpha_i y_i \mathbf{x}_i$ and b can be determined by KKT conditions. This problem is almost identical to the SVM problem (2.3). The difference is on the constraint. In particular, we have $0 \leqslant \alpha_i \leqslant C$ for the SVM, but $-Cv \leqslant \alpha_i \leqslant C$ for the BSVM. This helps to explain the difference in behaviors between the SVM and the BSVM. In contrast to the SVM, the BSVM with $v > 0$ makes use of all data points to determine

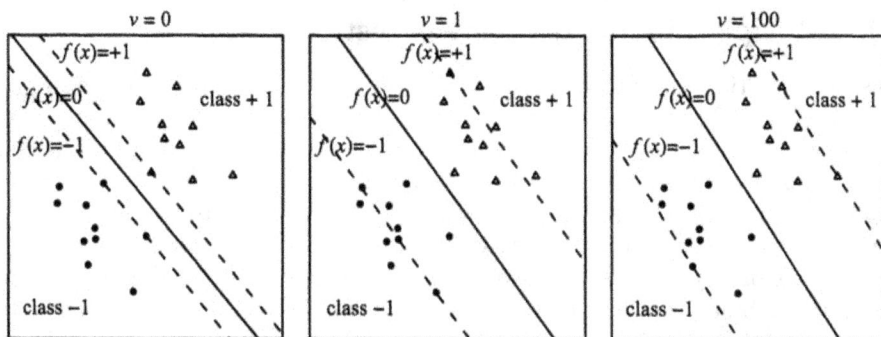

Figure 4.2   Plots of the effect of different values of v on the BSVM.

the solution. Points with $y_i f_i \leqslant 1$ may help to reduce the effect of outliers and consequently the BSVM classifier can be more robust against outliers.

In the separable case, the standard SVM, i.e. the BSVM with $v = 0$, finds the decision boundary which maximizes the distance from the decision boundary to the nearest data point, i.e., the distance between $f(\mathbf{x}) = \pm 1$ is maximized. Here, the soft margins $f(\mathbf{x}) = \pm 1$ are the hyperplanes that bound the data points of each class, so that the observations are forced to lie outside of the soft margins. The BSVM with $v > 0$ maximizes the distance between $f(\mathbf{x}) = \pm 1$ as well, but the observations are clustered around the hyperplanes $f(\mathbf{x}) = \pm 1$ without being forced to be outside of the margin lines. When $v = 1$, the BSVM minimizes $\sum_i |1 - y_i f(\mathbf{x}_i)|$, resulting data points laid inside and outside of $f(\mathbf{x}) = \pm 1$ evenly as shown in the middle panel of the Figure 4.2. As the value of v increases, the value of $v[y_i f(\mathbf{x}_i) - 1]_+$, the distance between the hyperplanes $f(\mathbf{x}) = \pm 1$ and the observations outside of them, increases as well. Thus the hyperplanes $f(\mathbf{x}) = \pm 1$ move towards outside to reduce it. As v goes to infinity, the BSVM reduces to the BCM and the hyperplanes $f(\mathbf{x}) = \pm 1$ go far enough to bound all data points. The right panel of the Figure 4.2 illustrates the behavior of the BCM with large v. More discussions can be found in Park and Liu (2009).

5   Multicategory classifiers

So far we have only focused on binary classification problems. However, it is very often for one to encounter multicategory problems in practice. To solve a multicategory problem using large margin classifiers, one have two common approaches. The first approach is to solve the multicategory problem via a sequence of binary problems, e.g., one-versus-rest and one-versus-one. The second approach is to generalize the binary classifier into a simultaneous multicategory formulation.

5.1   Mutiple binary: one-versus-rest, one-versus-one

To solve a multicategory problem, a natural and direct way is to implement multiple binary classifiers. For example, one can use the one-versus-rest or one-versus-one approach. The one-versus-rest approach relabels the training data in the class j as the positive class and data which are not in the class j as the negative class, for each $j = 1, \ldots, k$. Then, one can employ a sequence of k binary classifiers for the membership of each data point, which can possibly give contradictory results among the k binary classifiers. The one-versus-one approach applies a given binary classifier to a binary problem of the class j_1 and the class j_2 for each of all possible pairs $j_1, j_2 \in \{f_1, \ldots, f_k\}$. Overall, $\binom{k}{2}$ binary classifications are performed. For each binary problem, the dataset can be very small.

When there is no dominating class, in the SVM context, the one-versus-rest approach can be self-contradicted (Lee et al., 2004) and Fisher consistency is not guaranteed (Liu, 2007). Thus, it is necessary to generalize binary classification methods to multicategory versions which consider all classes simultaneously and retain good properties of the original methods.

5.2   Simultaneous loss formulation

Consider a k-class classification problem with $k \geqslant 2$. When $k = 2$, the methodology to be discussed here reduces to the binary counterpart in Section 3. Let $\boldsymbol{f} = (f_1, f_2, \ldots, f_k)$ be the decision function vector, where each component represents one class and maps from S to $\mathbb{R}$. To remove redundant solutions, a sum-to-zero constraint $\sum_{j=1}^{k} f_j = 0$ is employed. For any new input vector $\mathbf{x}$, its label is estimated via a decision rule $\hat{y} = \text{argmax}_{j=1,2,\ldots,k} f_j(\mathbf{x})$. Clearly, the argmax rule is equivalent to the sign function used in the binary case in Section 3.

For simplicity, we only focus our discussion on standard learning where all types of misclassification are treated equally. These techniques, however, can be extended to more general settings with unequal costs.

Multicategory SVM

The extension of the SVM from the binary to multicategory case is nontrivial. One can work on the separation idea as in the binary case. For example, one can extend the concept of maximum separation as shown in Figure 5.1 (Liu and Shen, 2006). However, the extension to nonseparable case is not unique and becomes more involved. It can be relatively easier to work with the loss formulation directly.

Before we discuss the detailed formulation of multicategory hinge loss, we first discuss Fisher consistency for multicategory problems. Consider $y \in \{1, \ldots, k\}$ and let $P_j(\mathbf{x}) = P(Y = j|\mathbf{x})$. Suppose $V(\boldsymbol{f}(\mathbf{x}), y)$ is a multicategory loss function. Then in this context, Fisher consistency requires that $\text{argmax}_j f_j^* = \text{argmax}_j P_j$, where $\boldsymbol{f}^*(\mathbf{x}) = (f_1^*(\mathbf{x}), \ldots, f_k^*(\mathbf{x}))$ denotes the minimizer of $E[V(\boldsymbol{f}(\mathbf{X}), Y)|\mathbf{X} = \mathbf{x}]$. Fisher consistency is a desirable condition of a loss function, although a consistent loss may not always translate into better classification accuracy (Hsu and Lin, 2002, Rifkin and Klautau, 2004).

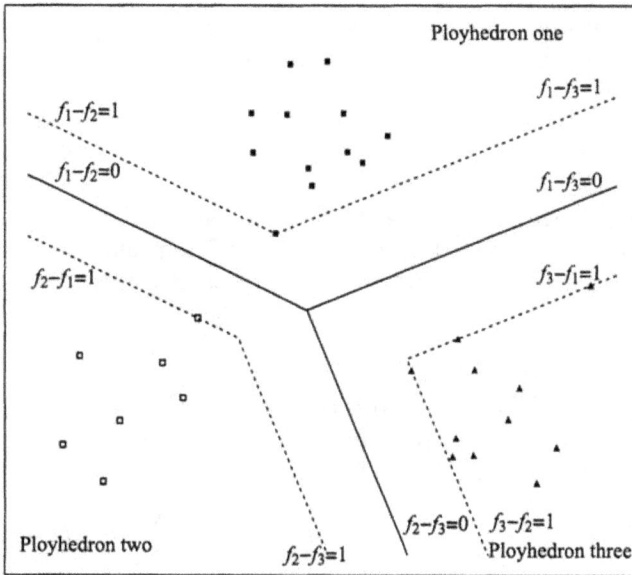

Figure 5.1   Illustration of the concept of margins and support vectors in a 3-class separable example: The instances for classes 1-3 fall respectively into the polyhedrons D_j; $j = 1, 2, 3$, where D_1 is $\{\mathbf{x} : f_1(\mathbf{x}) - f_2(\mathbf{x}) \geqslant 1, f_1(\mathbf{x}) - f_3(\mathbf{x}) \geqslant 1\}$, D_2 is $\{\mathbf{x} : f_2(\mathbf{x}) - f_1(\mathbf{x}) \geqslant 1, f_2(\mathbf{x}) - f_3(\mathbf{x}) \geqslant 1\}$, and D_3 is $\{\mathbf{x} : f_3(\mathbf{x}) - f_1(\mathbf{x}) \geqslant 1, f_3(\mathbf{x}) - f_2(\mathbf{x}) \geqslant 1\}$. The generalized geometric margin γ defined as $\min\{\gamma_{12}, \gamma_{13}, \gamma_{23}\}$ is maximized to obtain the decision boundary. There are five support vectors on the boundaries of the three polyhedrons. Among the five support vectors, one is from class 1, one is from class 2, and the other three are from class 3.

Note that a point $(\mathbf{x}, y)$ is misclassified by $\boldsymbol{f}$ if $y \neq \mathrm{argmax}_j f_j(\mathbf{x})$. Thus a sensible loss V should try to force f_y to be the maximum among k functions. Once V is chosen, multicategory SVM solves the following problem

$$\min_{\boldsymbol{f}} \sum_{j=1}^{k} J(f_j) + C \sum_{i=1}^{n} V(\boldsymbol{f}(\mathbf{x}_i), y_i) \qquad (5.1)$$

subject to

$$\sum_{j=1}^{k} f_j(\mathbf{x}) = 0.$$

The key of extending the SVM from binary to multicategory is the choice of loss V. There are a number of extensions of the binary hinge loss to the multicategory case proposed in the literature. We consider the following four commonly used extensions:

(a) (Naive hinge loss) $[1 - f_y(\mathbf{x})]_+$;
(b) (Lee et al., 2004) $\sum_{j \neq y} [1 + f_j(\mathbf{x})]_+$;
(c) (Vapnik, 1998; Weston and Watkins, 1999; Bredensteiner and Bennett, 1999; Guermeur, 2002) $\sum_{j \neq y} [1 - (f_y(\mathbf{x}) - f_j(\mathbf{x}))]_+$;
(d) (Crammer and Singer, 2001; Liu and Shen, 2006) $[1 - \min_j (f_y(\mathbf{x}) - f_j(\mathbf{x}))]_+$.

Note that the constant 1 in these losses can be changed to a general positive value. However, the resulting losses will be equivalent to the current ones by re-scaling $\boldsymbol{f}$.

As a remark, we note that the sum-to-zero constraint is essential for Losses (a) and (b), not so for Losses (c) and (d). Losses (c) and (d) only involve differences of functions and thus they are location invariant. It is easy to see that all these losses try to encourage f_y to be the maximum among k functions, either explicitly or implicitly. Interestingly, although these four different multicategory extensions are all reasonable, only (b) is always Fisher consistent (Liu, 2007).

Reinforced SVM

Although both losses (a) and (b) aim to encourage $f_y(\mathbf{x})$ to be the maximum, loss (a) is inconsistent while loss (b) is consistent. Based on (a) and (b), Liu and Yuan (2009) proposed the reinforced multicategory hinge loss function as a convex combination of losses (a) and (b) as follows:

$$V(\boldsymbol{f}(\mathbf{x}), y) = \gamma[(k-1) - f_y(\mathbf{x})]_+ + (1 - \gamma) \sum_{j \neq y} [1 + f_j(\mathbf{x})]_+ \qquad (5.2)$$

subject to $\sum_{j=1}^{k} f_j(\mathbf{x}) = 0$, where $\gamma \in [0, 1]$ and $a > 0$. They call the loss function (5.2) the reinforced hinge loss function since there are two terms in the loss and both terms try to enforce f_y to be the maximum, either explicitly or implicitly.

Note that if $f_j = -1$ for $\forall j \neq y$, $f_y = k - 1$ using the sum-to-zero constraint. Liu and Yuan (2009) showed that the reinforced hinge loss is Fisher consistent for any $0 \leqslant \gamma \leqslant 1/2$. Thus, the Fisher consistent loss by Lee, Lin and Wahba (2004) is only a special case with $\gamma = 0$ among this group of novel Fisher consistent loss functions. Furthermore, Liu and Yuan (2009) demonstrated that different values of γ affect the performance of the reinforced SVM. They recommended to use $\gamma = 0.5$, which appears to work the best.

Bounded Constraint Machine

Although the loss (a) is inconsistent while the loss (b) is consistent, their asymptotic forms are very similar as shown by Liu (2007). In fact, the only difference is that the loss (a) has the constraint $f_l(\mathbf{x}) \leqslant 1$ for $\forall l$, and in contrast, the loss (b) has the constraint $f_l(\mathbf{x}) \geqslant -1$ for $\forall l$. Therefore, the key is the corresponding constraints for the minimizer $\boldsymbol{f}^*$. Liu (2007) pointed out that one can add additional constraints on $\boldsymbol{f}$ to force the loss (a) to be consistent. In fact, if additional constraints $f_j \geqslant -1/(k-1)$ for $\forall j$ are imposed on (a), then the minimizer becomes $f_j^*(\mathbf{x}) = 1$ if $j = \text{argmax}_j P_j(\mathbf{x})$ and $-1/(k-1)$ otherwise. Consequently, the corresponding loss is Fisher consistent. Thus, adding new constraints to the loss (a) or (b), after proper rescaling, yield the following new loss:

- $-f_y(\mathbf{x})$ subject to $\sum_{j}^{k} f_j(\mathbf{x}) = 0$ and $-1 \leqslant f_l(\mathbf{x}) \leqslant k - 1$ for $\forall l$.

This loss is the multicategory extension of the binary BCM discussed in Section 4.1.

Liu (2007) pointed out that the constraint $-1 \leqslant f_l(\mathbf{x}) \leqslant k - 1$ for $\forall l$ can be difficult to implement for all $\mathbf{x} \in \mathcal{S}$. For simplicity of learning, to solve the problem, he suggested to relax such constraints on all training points only, that is $-1 \leqslant f_l(\mathbf{x}_i) \leqslant k - 1$ for $i = 1, \ldots, n$ and $l = 1, \ldots, k$. Then the multicategory BCM solves

$$\min_{\mathbf{f}} \sum_{j=1}^{k} J(f_j) - C \sum_{i=1}^{n} f_{y_i}(\mathbf{x}_i)$$

subject to

$$\sum_{j}^{k} f_j(\mathbf{x}_i) = 0; f_l(\mathbf{x}_i) \geqslant -1; l = 1, \ldots, k, i = 1, \ldots, n. \tag{5.3}$$

ψ-learning and robust support vector machine

From the argmax classification rule, we know that a point $(\mathbf{x}, y)$ is misclassified by $\mathbf{f}$ if $y \neq \operatorname{argmax}_j f_j(\mathbf{x})$, that is if $\min g(\mathbf{f}(\mathbf{x}), y) \leqslant 0$, where $g(\mathbf{f}(\mathbf{x}), y) = \{f_y(\mathbf{x}) - f_j(\mathbf{x}), j \neq y\}$. The quantity $\min g(\mathbf{f}(\mathbf{x}), y)$ is the generalized functional margin and it reduces to $yf(\mathbf{x})$ in the binary case with $y \in \{\pm 1\}$ (Liu and Shen, 2006).

If we use the hinge loss $H_1(u)$ with $\min g(\mathbf{f}(\mathbf{x}), y)$ as its argument, then we get one version of the multicategory SVM, that is the loss (d). Similar to the binary case, a point with large $1 - \min g(\mathbf{f}(\mathbf{x}), y)$ results in large H_1 and, as a consequence, greatly influences the final solution, as discussed in the binary case. Such points are typically far away from their own classes and tend to deteriorate the SVM performance. Similar observations can be made for some other large-margin classifiers with unbounded loss functions.

To gain robustness and achieve better generalization, Liu and Shen (2006) proposed multicategory ψ-learning. In particular, they defined multivariate ψ-functions on $k - 1$ arguments as follows:

$$U \geqslant \psi(\mathbf{u}) > 0 \text{ if } \mathbf{u}_{\min} \in (0, T);$$
$$\psi(\mathbf{u}) = 1 - \operatorname{sign}(\mathbf{u}_{\min}) \text{ otherwise,} \tag{5.4}$$

where $0 < T \leqslant 1$ and $0 < U \leqslant 2$ are some constants, and $\psi(\mathbf{u})$ is non-increasing in $\mathbf{u}_{\min}$. This multivariate version preserves the desired properties of its univariate counterpart. Particularly, the multivariate ψ assigns a positive penalty to any instance with $\min(\mathbf{g}(\mathbf{f}(\mathbf{x}_i), y_i)) \in (0, T)$ to eliminate the scaling problem. To utilize the d.c. algorithm based on a difference convex (d.c.) decomposition, Liu and Shen (2006) suggested a specific ψ in implementation:

$$\psi(\mathbf{u}) = 0 \text{ if } \mathbf{u}_{\min} \geqslant 1; \quad 2 \text{ if } \mathbf{u}_{\min} < 0; \quad 2(1 - \mathbf{u}_{\min}) \text{ if } 0 \leqslant \mathbf{u}_{\min} < 1. \tag{5.5}$$

A plot of this ψ function for $k = 3$ is displayed in Figure 5.2.

Wu and Liu (2007a) proposed to reduce the influence of outliers via truncating the unbounded large-margin loss functions. For a given non-increasing large margin loss $V(\cdot)$, denote $V_{T_s}(\cdot) = \min(V(\cdot), V(s))$ as the corresponding truncated loss of $V(\cdot)$. The value of $s \leqslant 0$ specifies the location of truncation. Note that the truncated hinge loss function $T_s(u)$ is a special example of V_{T_s}.

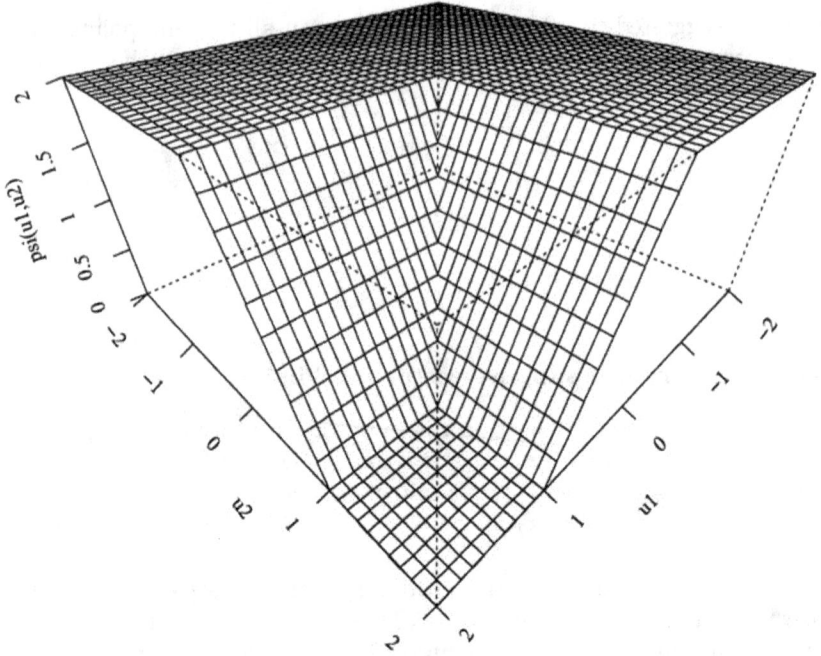

Figure 5.2   Perspective plot of the 3-class ψ function defined in (5.5)

For multicategory problems with $k > 2$, the issue of Fisher consistency becomes more complex. Zhang (2004) and Tewari and Bartlett (2005) pointed out Fisher inconsistency of $H_1(\min g(f(x), y))$. Wu and Liu (2007a) showed that under certain conditions of V, $V(\min g(f(x), y))$ is Fisher consistent only when $\max_j P_j > 1/2$, i.e., when there is a dominating class. For a problem with $k > 2$, existence of a dominating class may not be guaranteed (Lee, Lin and Wahab, 2004). Interestingly, truncating $V(\min g(f(x), y))$ can make it Fisher consistent even in the situation of no dominating class as shown in Wu and Liu (2007a). Interestingly, the truncation value s depends on the class number k. Numerical results suggest that the larger k is, the more truncation is needed to ensure Fisher consistency.

Truncating a convex loss produces a nonconvex loss and, as a result, the optimization problem in (5.1) with $V = V_{T_s}$ involves nonconvex minimization. Compared to convex minimization, nonconvex minimization problems are more challenging. In the literature, Fan and Li (2001) proposed local quadratic approximation (LQA) to handle some non-convex penalized likelihood problem. Hunter and Li (2005) showed that the LQA is a special instance of the minorize-maximize or majorize-minimize (MM) algorithm.

Notice that for a given convex loss function $V(u)$, its truncated version $V_{T_s}(u)$ can be decomposed as the difference of two convex functions, $V(u)$ and $[V(u) - V(s), 0]_+$. Using this property, Wu and Liu (2007a) proposed to apply the d.c. algorithm (An and Tao, 1997; Liu, Shen, and Doss, 2005) to solve the nonconvex optimization problem involving a truncated loss function. The d.c. algorithm solves the nonconvex minimization problem via minimizing a sequence of convex subproblems. Interestingly, the d.c. algorithm is also an instance of the generic MM algorithm.

Here we briefly discuss the d.c. algorithm for robust SVM (RSVM) using the linear learning. One interesting point is that the algorithm retains the SV interpretation. The readers are referred to Wu and Liu (2007a) for more detailed derivation and discussion.

Let $f_j(\mathbf{x}) = \mathbf{w}_j^T \mathbf{x} + b_j$; $\mathbf{w}_j \in \Re^d$, $b_j \in \Re$, and $\boldsymbol{b} = (b_1, b_2, \ldots, b_k)^T \in \Re^k$, where $\mathbf{w}_j = (w_{1j}, w_{2j}, \ldots, w_{dj})^T$, and $\mathbf{W} = (\mathbf{w}_1, \mathbf{w}_2, \ldots, \mathbf{w}_k)$. With $V = T_s$, the robust SVM solves

$$\min_{\mathbf{W}, \boldsymbol{b}} \frac{1}{2} \sum_{j=1}^{k} \|\mathbf{w}_j\|_2^2 + C \sum_{i=1}^{n} T_s(\min \boldsymbol{g}(\boldsymbol{f}(\mathbf{x}_i), y_i)) \quad (5.6)$$

$$\text{subject to } \sum_{j=1}^{k} w_{mj} = 0; m = 1, 2, \ldots, d; \sum_{j=1}^{k} b_j = 0,$$

where the constraints are adopted to avoid non-identifiability issue of the solution.

Denote Θ as $(\mathbf{W}, \boldsymbol{b})$. Applying the fact that $T_s = H_1 - H_s$, the objective function in (5.6) can be decomposed as

$$Q^s(\Theta) = \frac{1}{2} \sum_{j=1}^{k} \|\mathbf{w}_j\|_2^2 + C \sum_{i=1}^{n} H_1(\min \boldsymbol{g}(\boldsymbol{f}(\mathbf{x}_i), y_i)) - C \sum_{i=1}^{n} H_s(\min \boldsymbol{g}(\boldsymbol{f}(\mathbf{x}_i), y_i))$$

$$= Q_{vex}^s(\Theta) + Q_{cav}^s(\Theta),$$

where $Q_{vex}^s(\Theta) = \frac{1}{2} \sum_{j=1}^{k} \|\mathbf{w}_j\|_2^2 + C \sum_{i=1}^{n} H_1(\min \boldsymbol{g}(\boldsymbol{f}(\mathbf{x}_i), y_i))$ and $Q_{cav}^s(\Theta) = Q^s(\Theta) - Q_{vex}^s(\Theta)$ denote the convex and concave parts respectively.

It can be shown that the convex dual problem at the $(t+1)$-th iteration, given the solution $\boldsymbol{f}^t$ at the t-th iteration, is as follows

$$\min_{\alpha} \frac{1}{2} \sum_{j=1}^{k} \| \sum_{i:\, y_i = j} \sum_{j' \neq y_i} (\alpha_{ij'} - \beta_{ij'}) \mathbf{x}_i^T - \sum_{i:\, y_i \neq j} (\alpha_{ij} - \beta_{ij}) \mathbf{x}_i^T \|_2^2 - \sum_{i=1}^{n} \sum_{j' \neq y_i} \alpha_{ij'}$$

$$\text{subject to } \sum_{i:\, y_i = j} \sum_{j' \neq y_i} (\alpha_{ij'} - \beta_{ij'}) - \sum_{i:\, y_i \neq j} (\alpha_{ij} - \beta_{ij}) = 0, \ j = 1, 2, \ldots, k \quad (5.7)$$

$$0 \leqslant \sum_{j \neq y_i} \alpha_{ij} \leqslant C, \ i = 1, 2, \ldots, n \quad (5.8)$$

$$\alpha_{ij} \geqslant 0, \ i = 1, 2, \ldots, n; \ j \neq y_i, \quad (5.9)$$

where $\beta_{ij} = C$ if $f_{y_i}^t(\mathbf{x}_i) - f_j^t(\mathbf{x}_i) < s$ with $j = \operatorname{argmax}(f_{j'}^t(\mathbf{x}_i) : j' \neq y_i)$, and 0 otherwise.

This dual problem is a quadratic programming (QP) problem similar to that of the standard SVM and can be solved by many optimization softwares. Once its solution is obtained, the coefficients $\mathbf{w}_j$'s can be recovered as follows,

$$\mathbf{w}_j = \sum_{i:\, y_i=j}\sum_{j'\neq y_i}(\alpha_{ij'} - \beta_{ij'})\mathbf{x}_i - \sum_{i:\, y_i\neq j}(\alpha_{ij} - \beta_{ij})\mathbf{x}_i. \tag{5.10}$$

It is interesting to note that the representation of $\mathbf{w}_j$'s given in (5.10) automatically satisfies $\sum_{j=1}^{k} w_{mj} = 0$ for each $1 \leqslant m \leqslant d$. Moreover, we can see that coefficients $\mathbf{w}_j$'s are determined only by those data points whose corresponding $\alpha_{ij} - \beta_{ij}$ is not zero for some $1 \leqslant j \leqslant k$ and these data points are the SVs of the RSVM. The set of SVs of the RSVM using the d.c. algorithm is only a subset of the set of SVs of the original SVM. The RSVM tries to remove points satisfying $f_{y_i}^t(\mathbf{x}_i) - f_j^t(\mathbf{x}_i) < s$ with $j = \operatorname{argmax}(f_{j'}^t(\mathbf{x}_i) : j' \neq y_i)$ from the original set of SVs and consequently eliminate the effects of outliers. This provides an intuitive algorithmic explanation of the robustness of the RSVM to outliers.

After the solution of $\mathbf{W}$ is derived, b can be obtained via solving either a sequence of KKT conditions as used in the standard SVM or a linear programming (LP) problem.

As a remark, we note that our discussion of RSVM focuses on the loss (d). One can perform the truncation operation on other multicategory versions as well. For example, Wu and Liu (2007b) studied the robust version of loss (b) via truncation.

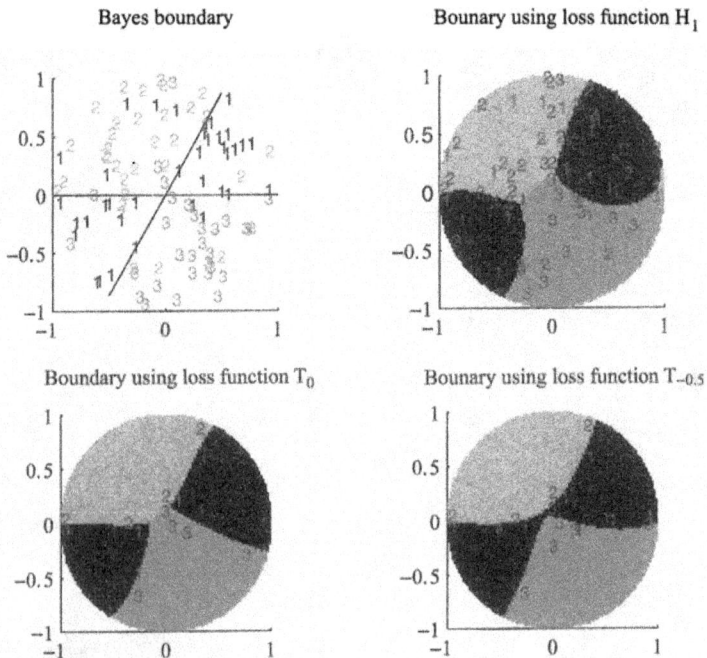

Figure 5.3   An illustrating plot of robust SVM: the top-left panel plots all observations with the black straight lines shown as the Bayes boundary. The remaining three panels display the classification boundaries obtained by using different loss functions with SVs shown in red.

Wu and Liu (2007a) demonstrated that RSVMs give smaller testing errors while using fewer SVs than the standard SVM. Figure 5.3 illustrates decision boundaries and SVs of the original SVM and RSVMs. The top-left panel shows the observations as well as the Bayes boundary. In the remaining three panels, boundaries using nonlinear learning with different loss functions H_1, T_0, and $T_{-0.5}$ are plotted and their corresponding SVs are labelled in red. From the plots, we can see that the RSVMs use much fewer SVs and at the same time yield more accurate classification boundaries than the standard SVM.

Finally, we want to point out that both the BSVM (Park and Liu, 2009) discussed in Section 4.2 and the RSVM (Wu and Liu, 2007a) discussed in this section can help to deliver robust classifiers. The RSVM achieves robustness via removing potential outliers from the set of SVs for the standard SVM. Consequently, the RSVM gains robustness by using a smaller but more robust set of observations. In contrast, the BSVM tries to reduce the impact of outliers by making use of more data points. Both methods are reasonable, however, they use different philosophies in using the training data to obtain robustness. More discussion and comparisons can be found in Park and Liu (2009).

5.3   Choice of penalty: L_1, L_1-max

As discussed earlier, selecting relevant variables is an important goal for classification, especially in high dimensional problems. One can achieve variable selection by modifying the penalty function. Our discussion emphasizes the SVM. However, the methods discussed here are not limited to the SVM.

For the binary SVM, Bradley and Mangasarian (1998) demonstrated the utility of the L_1 penalty, which can effectively select variables by shrinking small or redundant coefficients to zero. Zhu et al. (2003a) provides an efficient algorithm to compute the entire solution path for the L_1-norm SVM. Other forms of penalty have been also studied in the context of binary SVMs, such as the L_0 penalty (Weston et al., 2003), the SCAD penalty (Zhang et al., 2006), the L_q penalty (Liu et al., 2007), the combination of L_0 and L_1 penalty (Liu and Wu, 2007), the combination of L_1 and L_2 penalty (Wang et al. 2006), the F_∞ norm (Zou and Yuan, 2008), and others (Zhao et al., 2006; Zou, 2006).

For multiclass problems, variable selection becomes more complex than the binary case, since the MSVM requires estimation of multiple discriminating functions, among which each function has its own subset of important predictors. One natural idea is to extend the L_1 SVM to L_1 MSVM, as done in Lee et al. (2006) and Wang and Shen (2007). However, the L_1 penalty does not distinguish the source of coefficients. It treats all the coefficients equally, no matter whether they correspond to the same variable or different variables, or they are more likely to be relevant or irrelevant. Zhang, Liu, Wu, and Zhu (2008) proposed a new regularized MSVM for more effective variable selection. In contrast to the L_1 MSVM, which imposes a penalty on the sum of absolute values of all coefficients, the sup-norm SVM penalizes the sup-norm of the coefficients associated with each variable. Zhang et al. (2008) showed that their proposed method is able to achieve a higher degree of model parsimony than the L_1 MSVM without compromising classification accuracy.

The L_1 MSVM treats all w_{lj}'s equally without distinction. As opposed to this, the sup-norm SVM takes into account the fact that some of the coefficients are associated with the same covariate, therefore it is more natural to treat them as a group rather than separately.

Define the coefficient matrix W of size $k \times d$ such that its (l,j) entry is w_{lj}. The structure of W is shown in Table 5.1. Define $\mathbf{w}_j = (w_{j1}, \ldots, w_{jd})^T$ to represent the jth row vector of W, and $\mathbf{w}_{(j)} = (w_{1j}, \ldots, w_{kj})^T$ for the jth column vector of W. According to Crammer and Singer (2001), the value $b_j + \mathbf{w}_j^T \mathbf{x}$ defines the similarity score of the class j, and the predicted label is the index of the row attaining the highest similarity score with $\mathbf{x}$. Zhang et al. (2008) defined the sup-norm for the coefficient vector $\mathbf{w}_{(j)}$ as

$$\|\mathbf{w}_{(j)}\|_\infty = \max_{l=1,\ldots,k} |w_{lj}|. \tag{5.11}$$

Table 5.1   Linear multicategory SVM coefficient structure

	x_1	$\cdots$	x_j	$\cdots$	x_d
Class 1	w_{11}	$\cdots$	w_{1j}	$\cdots$	w_{1d}
	$\vdots$		$\vdots$		$\vdots$
Class l	w_{l1}	$\cdots$	w_{lj}	$\cdots$	w_{ld}
	$\vdots$		$\vdots$		$\vdots$
Class k	w_{k1}	$\cdots$	w_{kj}	$\cdots$	w_{kd}

In this way, the importance of each covariate x_j is directly controlled by its largest absolute coefficient. The sup-norm regularization for MSVM solves:

$$\min_{v,\mathbf{w}} \quad \frac{1}{n} \sum_{i=1}^n \sum_{l=1}^k I(y_i \neq l)[b_l + \mathbf{w}_l^T \mathbf{x}_i + 1]_+ + \lambda \sum_{j=1}^d \|\mathbf{w}_{(j)}\|_\infty,$$

$$\text{subject to} \quad \mathbf{1}^T v = 0, \quad \mathbf{1}^T \mathbf{w}_{(j)} = 0, \qquad \text{for } j = 1, \ldots, d, \tag{5.12}$$

where $v = (b_1, \ldots, b_k)^T$.

The sup-norm MSVM encourages more sparse solutions than the L_1 MSVM, and identifies important variables more precisely. To further understand the main motivation of the sup-norm MSVM, we note that with a sup-norm penalty, a noise variable is removed if and only if all corresponding k estimated coefficients are 0. On the other hand, if a variable is important with a positive sup-norm, the sup-norm penalty, unlike the L_1 penalty, does not put any additional penalties on the other $k-1$ coefficients. This is desirable since a variable will be kept in the model as long as the sup-norm of the k coefficients is positive. No further shrinkage is needed for the remaining coefficients in terms of variable selection.

In the standard L_1 penalty and the sup-L_1 penalty, the same weights are used for different variables in the penalty terms, which may be too restrictive, since a smaller penalty may be more desired for those variables which are so important that we want to retain them in the model. To cope with the problem, Zhang et al. (2008) suggested that different variables should be penalized differently according to their relative importance. Ideally, large penalties should be imposed on redundant variables in order to eliminate them from models more easily; and small penalties should be used on important variables in order to retain them in the final classifier. Motivated by this, Zhang et al. (2008) proposed the following adaptive L_1 MSVM:

$$\min_{v,w} \frac{1}{n} \sum_{i=1}^{n} \sum_{l=1}^{k} I(y_i \neq l)[b_l + \mathbf{w}_l^T \mathbf{x}_i + 1]_+ + \lambda \sum_{l=1}^{k} \sum_{j=1}^{d} T_{lj} |w_{lj}|,$$

subject to

$$\mathbf{1}^T \boldsymbol{v} = 0, \quad \mathbf{1}^T \mathbf{w}_{(j)} = 0, \qquad \text{for } j = 1, \ldots, d, \qquad (5.13)$$

where $T_{lj} > 0$ represents the weight for coefficient w_{lj}.

Adaptive shrinkage for each variable has been proposed and studied in various contexts of regression problems, including the adaptive LASSO for linear regression (Zou 2006), proportional hazard models (Zhang and Lu 2007), and quantile regression (Wang et al. 2007, Wu and Liu, 2009a). In particular, Zou (2006) has established the oracle property of the adaptive LASSO and justified the use of different amounts of shrinkage for different variables. For simplicity of discussion, we only focus on loss (d) here. However, these penalties can be used for other loss functions as well. Due to the special form of the sup-norm SVM, the following two ways to employ the adaptive penalties were considered by Zhang et al. (2008):

[I]

$$\min_{v,w} \frac{1}{n} \sum_{i=1}^{n} \sum_{l=1}^{k} I(y_i \neq l)[b_l + \mathbf{w}_l^T \mathbf{x}_i + 1]_+ + \lambda \sum_{j=1}^{d} T_j \|\mathbf{w}_{(j)}\|_\infty,$$

subject to

$$\mathbf{1}^T \boldsymbol{v} = 0, \quad \mathbf{1}^T \mathbf{w}_{(j)} = 0, \qquad \text{for } j = 1, \ldots, d, \qquad (5.14)$$

[II]

$$\min_{v,w} \frac{1}{n} \sum_{i=1}^{n} \sum_{l=1}^{k} I(y_i \neq l)[b_l + \mathbf{w}_l^T \mathbf{x}_i + 1]_+ + \lambda \sum_{j=1}^{d} \|(\boldsymbol{T}\mathbf{w})_{(j)}\|_\infty,$$

subject to

$$\mathbf{1}^T \boldsymbol{v} = 0, \quad \mathbf{1}^T \mathbf{w}_{(j)} = 0, \qquad \text{for } j = 1, \ldots, d, \qquad (5.15)$$

where the vector $(\boldsymbol{T}\mathbf{w})_{(j)} = (T_{1j} w_{1j}, \ldots, T_{kj} w_{kj})^T$ for $j = 1, \ldots, d$.

In (5.13), (5.14), and (5.15), the weights can be regarded as leverage factors, which are adaptively chosen such that large penalties are imposed on coefficients of unimportant covariates and small penalties on coefficients of important ones. Let $\tilde{w}$ be the solution to standard MSVM using loss (d) with the L_2 penalty. Empirical experience suggests that

$$T_{lj} = \frac{1}{|\tilde{w}_{lj}|}$$

is a good choice for (5.13) and (5.15), and

$$T_j = \frac{1}{\|\tilde{\mathbf{w}}_{(j)}\|_\infty}$$

is a good choice for (5.14). If $\tilde{w}_{lj} = 0$, which implies the infinite penalty on w_{lj}, we set the corresponding coefficient solution $\hat{w}_{lj}$ to be zero.

In terms of computational issues, all three problems (5.13), (5.14), and (5.15) can be solved as LP problems.

6   Probability estimation

Besides good classification accuracy, one may be also interested in estimating conditional probabilities $P(\mathbf{x}) = P(Y = 1|\mathbf{X} = \mathbf{x})$ for the binary case and $P_j(\mathbf{x}) = P(Y = j|\mathbf{X} = \mathbf{x})$, $j = 1, 2, \ldots, k$ for the multi-category case.

Classifiers can be generally divided in two groups depending on whether they can estimate the conditional class probabilities. A classifier is called a hard classifier if it targets directly at the classification boundary without estimating the conditional class probabilities. For example, the SVM is a typical example of the hard classifier in that it can only tell whether the conditional probability $P(Y = 1|\mathbf{X} = \mathbf{x})$ is larger than a half in the binary case (Lin, 2002). On the other hand, a soft classifier provides an estimate of the conditional probabilities and the estimated probabilities can be used to make class prediction using the argmax rule. Logistic regression is a well known soft classifier.

Steinwart (2003) and Tewari and Bartlett (2005) pointed out that the solution of the SVM is sparse due to the fact that the hinge loss is piecewise linear. This property disables the SVM to estimate the conditional probability. In general it was shown by Steinwart (2003) and Tewari and Bartlett (2005) that conditional class probability estimation is achievable if the hinge loss is replaced by some differentiable loss function. For example the logistic loss function is differentiable and it is capable of estimating the conditional class probability.

With the question "Can hard classifiers be used to estimate conditional probabilities?" in mind, we next discuss how to extract conditional class probability information from hard classifiers.

6.1   Platt's method

Platt (1999) provided a method to estimate the conditional class probability based on the optimal solution of the SVM or the more general large-margin classifier. Platt (1999) makes the following sigmoid model assumption

$$P(Y = 1|\mathbf{X} = \mathbf{x}) = \frac{1}{1 + e^{Af(x)+B}}, \tag{6.1}$$

where $f(\cdot)$ is any large-margin classifier. He proposes to minimize the cross-entropy to estimate these two constants A and B in (6.1).

Once parameters A and B are estimated, (6.1) can be used to provide an estimate of the conditional class probability. Although Platt's method of probability is easy to use and empirically useful, it lacks any theoretical justification. Furthermore, Lin (2002) showed that the SVM only estimates the Bayes boundary by telling whether the conditional class probability is larger than $1/2$. Thus the sigmoid model assumption used in the Platt's metohd is hard to justify.

6.2   Probability estimation for binary hard classifiers

Lin (2002) showed that the (binary) SVM classifier is consistent for estimating the Bayes classification rule and targets at the Bayes classification boundary $\{\mathbf{x} : P(\mathbf{x}) = \frac{1}{2}\}$. It means that the SVM addresses the question whether the conditional class probability is larger than $\frac{1}{2}$.

By using different weights for observations in different classes, Wang, Shen, and Liu (2008) proposed to solve the weighted SVM

$$C\left[(1-\pi)\sum_{i:y_i=1} H_1(f(\mathbf{x}_i)) + \pi \sum_{i:y_i=-1} H_1(-f(\mathbf{x}_i))\right] + J(f), \qquad (6.2)$$

where $0 < \pi < 1$ and $f(\cdot)$ can be either linear or nonlinear with an appropriately chosen penalty $J(f)$. For the weighted SVM (6.2), its classification boundary is shown to consistently estimate the boundary $\{\mathbf{x} : p(\mathbf{x}) = \pi\}$. Basically the weighted classifier with weight $(1-\pi)$ for the positive class and π for the negative class delivers information regarding wether the conditional class probability is larger than π. This means that the weighted SVM contains some information regarding the conditional probabilities. But the information is limited in the sense that it is only capable of telling whether the conditional probability is larger than π. Wang, Shen, and Liu (2008) proposed to solve weighted SVMs for different weights $\pi \in (0,1)$. Then an interval estimate for the conditional class probability is obtained.

Suppose we solve (6.2) over a grid of weights $0 = \pi_1 < \pi_2 < \cdots < \pi_M = 1$. The optimal solution corresponding to π_m is denoted by $\hat{f}_{\pi_m}(\cdot)$ for $m = 1, 2, \ldots, M$. For any $\mathbf{x}$, define $\overline{m}(\mathbf{x}) = \max\{m : \hat{f}_{\pi_m}(\mathbf{x}) \geq 0\}$ and $\underline{m}(\mathbf{x}) = \min\{m : \hat{f}_{\pi_m}(\mathbf{x}) \leq 0\}$. Then a point estimate of the conditional class probability $P(\mathbf{x})$ is given by $\frac{1}{2}(\pi_{\overline{m}(\mathbf{x})} + \pi_{\underline{m}(\mathbf{x})})$. The corresponding algorithm for estimating the conditional class probability can be summarized as follows.

Wang et al. (2008)'s Algorithm:

1. Initialize $\pi_m = (m-1)/(M-1)$ for $m = 1, 2, \ldots, M$.
2. Train a weighted large-margin classifier for π_m by solving (6.2) and denote the solution as $\hat{f}_{\pi_m}(\mathbf{x})$ for $m = 1, 2, \ldots, M$.
3. For each $\mathbf{x}$, compute $\overline{m} = \overline{m}(\mathbf{x})$ and $\underline{m} = \underline{m}(\mathbf{x})$. The estimated conditional class probability is given by $\hat{p}(\mathbf{x}) = \frac{1}{2}(\pi_{\overline{m}} + \pi_{\underline{m}})$.

6.3 Noncrossing probability estimation

Theoretically Bayes boundaries $\{\mathbf{x} : p(\mathbf{x}) - p = 0\}$ for different p should not cross each other. To bracket the conditional class probability, we estimate Bayes boundaries using classification boundaries of the weighted margin-based classifiers based on a finite training data set. Thus the individually estimated boundaries for different π's may unavoidably cross each other, especially when the training sample size is small. This crossing phenomenon was also pointed out by Wang et al. (2008). Their probability bracketing method cannot avoid crossing. Wu and Liu (2009b) targeted at this crossing issue and provided a technique to enforce non-crossing of the estimated classification boundaries. With this technique, one can ensure non-crossing of the estimated boundaries, at least over the convex hull of the training data points.

For simplicity, we discuss the estimation scheme for the linear case. Details on the nonlinear case cay be found in Wu and Liu (2009b). Denote m_0 to be the index such that π_{m_0} is the closest to $1/2$ among all different π_m's, namely, $|\pi_{m_0} - 1/2| = \min\limits_{1 \leqslant m \leqslant M} |\pi_m - 1/2|$. By making the linearity assumption on the classification boundaries, we assume that $f_m(\mathbf{x}) = \mathbf{x}^T \mathbf{w}_m + b_m$ corresponds to π_m for $m = 1, 2, \ldots, M$.

The non-crossing algorithm by Wu and Liu (2009b) begins with estimating the classification boundary corresponding to π_{m_0} by solving the following optimization problem

$$\min \left\{ (1 - \pi_{m_0}) \sum_{i:y_i=1} [1 - (\mathbf{x}_i^T \mathbf{w} + b)]_+ + \pi_{m_0} \sum_{i:y_i=-1} [1 + (\mathbf{x}_i^T \mathbf{w} + b)]_+ + \lambda \sum_{j=1}^{p} w_j^2 \right\},$$
(6.3)

where the regularization parameter can be tuned either using a separate independent tuning data set or cross-validation. Denote the corresponding estimator by $\hat{\mathbf{w}}_{m_0}$ and $\hat{b}_{m_0}$.

For any $m > m_0$, we require that the classification boundary corresponding to π_m does not cross that of π_{m-1} within the convex hull of the data points. To achieve this goal, one can solve the following constrained optimization problem

$$\min \left\{ (1 - \pi_m) \sum_{i:y_i=1} [1 - (\mathbf{x}_i^T \mathbf{w} + b)]_+ \right.$$
$$\left. + \pi_m \sum_{i:y_i=-1} [1 + (\mathbf{x}_i^T \mathbf{w} + b)]_+ + \lambda \sum_{j=1}^{p} w_j^2 \right\},$$

subject to

$$\mathbf{x}^T \mathbf{w} + b \leqslant 0,$$
$$\forall \, \mathbf{x} \in \{\mathbf{x} : \mathbf{x}^T \hat{\mathbf{w}}_{m-1} + \hat{b}_{m-1} = 0\} \cap H_{convex}(\mathbf{x}_1, \mathbf{x}_2, \ldots, \mathbf{x}_n).$$
(6.4)

Note that the restriction is satisfied as long as $\mathbf{x}^T \mathbf{w} + b \leqslant 0$ holds for any $\mathbf{x}$ in the intersection of the surface $\mathbf{x} \in \{\mathbf{x} : \mathbf{x}^T \hat{\mathbf{w}}_{m-1} + \hat{b}_{m-1} = 0\}$ and the edges of the convex hull $H_{convex}(\mathbf{x}_1, \mathbf{x}_2, \ldots, \mathbf{x}_n)$. Points in the intersection are enumerable. Thus the above constrained weighted SVM can be implemented using quadratic programming once the interaction points are available. We denote the optimizer to be $\hat{\mathbf{w}}_m$ and $\hat{b}_m$.

Similarly one can estimate the classification boundary corresponding to π_m for $m < m_0$ by solving

$$
\min \left\{ (1 - \pi_m) \sum_{i:y_i=1} [1 - (\mathbf{x}_i^T \mathbf{w} + b)]_+ \right.
$$
$$
\left. + \pi_m \sum_{i:y_i=-1} [1 + (\mathbf{x}_i^T \mathbf{w} + b)]_+ + \lambda \sum_{j=1}^p w_j^2 \right\},
$$

subject to

$$
\mathbf{x}^T \mathbf{w} + b \geqslant 0, \forall \{\mathbf{x} : \mathbf{x}^T \hat{\mathbf{w}}_{m+1} + \hat{b}_{m+1} = 0\} \cap H_{convex}(\mathbf{x}_1, \mathbf{x}_2, \ldots, \mathbf{x}_n). \quad (6.5)
$$

The corresponding optimizers are denoted by $\hat{\mathbf{w}}_m$ and $\hat{b}_m$.

Note that we begin with the center π_{k_0} and move to the two ends sequentially to estimate the classification boundaries. By using constraints (6.4) and (6.5), it is guaranteed that the estimated classification boundaries do not cross each other within the convex hull of our data. Then Step 3 of Wang et al. (2008)'s algorithm can be applied to estimate the conditional class probability. The main difference of the new algorithm is to enforce non-crossing constraints on the corresponding classification boundaries.

Wu and Liu (2009b) also proposed to make use of the estimated class probabilities to precondition the training data. This preconditioning operation may help to remove potential outliers and consequently yield more robust classifiers.

For a training data set with a low noise level, one would expect that most data points fall on their corresponding class sets. A training point is likely to be an outlier if it is far from its own class. Specifically, for a non-outlier data point $\mathbf{x}_i$ with $y_i = 1$ ($y_i = -1$), its estimated conditional class probability $\hat{p}(\mathbf{x}_i)$ should be likely to be larger (smaller) than 0.5. This helps to identify outliers. If there are some data points $\mathbf{x}_i$ with $y_i = 1$ and $\hat{p}(\mathbf{x}_i)$ close to zero or some data points $\mathbf{x}_i$ with $y_i = -1$ and $\hat{p}(\mathbf{x}_i)$ close to one, then these points $(\mathbf{x}_i, y_i)$'s are likely to be outliers.

Once the potential outliers are identified, one can correct them by flipping their binary response from y_i to $-y_i$ and train the standard SVM with the modified training data set. In this way, one may be able to remove the effect of these outliers from deteriorating the performance of the SVM. Wu and Liu (2009b) named this technique preconditioning. In the literature, preconditioning has been applied in the regression problems. In particular, Paul, Bair, Hastie and Tibshirani (2008) proposed to first estimate the regression function to yield a preconditioned response variable. Then they applied standard variable selection techniques to the preconditioned response variable. They argued that the preconditioning step helps to get better variable selection performance. The preconditioning method by Wu and Liu (2009b) is similar to that of Paul et al. (2008) in the sense that they proposed to first generate a more "clean" binary response variable first and then apply the standard large-margin classifiers to achieve robustness.

6.4   Issues on multi-category classifiers

Conditional class probability estimation becomes more complicated for multi-category problems in that we need to estimate multiple functions $P(Y = j|\mathbf{X} = \mathbf{x})$ for $j = 1, 2, \ldots, K$. Existing approaches includes multiple logistic regression, linear/quadratic discriminant analysis, and others. See Agresti and Coull (1998) for more details. Wu, Lin, and Weng (2004) studied pairwise coupling for estimating probabilities in the multicategory setting. Essentially they first estimate the pairwise conditional probabilities $P(Y = j|\mathbf{X} = \mathbf{x}, Y \in \{i, j\})$ for any pair of $i, j \in \{1, 2, \ldots, K\}$. Once these pairwise conditional class probability estimates are obtained, different coupling methods are used to combine them to estimate the conditional class probability $P(Y = j|\mathbf{X} = \mathbf{x})$ for any $j \in \{1, 2, \ldots, K\}$.

In the multicategory setting, weighted classifiers with weight π_j for observations in class j using the truncated hinge loss are shown to be consistent for the weighted Bayes rule $\text{argmax}_j \pi_j P(Y = j|\mathbf{X} = \mathbf{x})$ (Wu, Zhang, and Liu, 2009). Based on this consistency property of weighted truncated-hinge-loss SVM, Wu, Zhang, and Liu (2009) extend the probability estimation scheme of Section 6.2 to the multi-category case. See Wu, Zhang, and Liu (2009) for more details.

7   Conclusions and discussions

In this book chapter, we provide some selected review on large margin classifiers. We have focused on the SVM and its extensions in various aspects. We have discussed the choice of loss functions and some of the computational issues involved. Both binary and multicategory classification problems are considered. Furthermore, we discuss the issue of probability estimation for hard margin classifiers. Variable selection is discussed as well. In general, the question that "Which one is better between hard and soft classification?" is an interesting unsolved classification problem. A recent paper by Liu, Zhang, and Wu (2009) attempted to bridge hard and soft classification via a new family, namely the large margin unified machine (LUM). The LUM family helps to shed some light on the problem of classification.

One important issue in large margin classifiers is the corresponding statistical learning theory. A number of papers on the convergence of large-margin classifiers have appeared in the literature. To list a few, Lin (2000) investigated rates of convergence of binary SVMs with a spline kernel. Shen et al. (2003); Liu and Shen (2006) provided learning theories for binary and multicategory ψ-learning respectively. Tarigan and van de Geer (2004); Wang and Shen (2007) derived rates of convergence for the L_1 SVMs. Steinwart and Scovel (2006) studied the convergence rate of the SVM using Gaussian kernels. Recently, Shen and Wang (2007) studied rates of convergence of the generalization error of a class of multicategory margin classifiers. We have not included much discussion on these theoretical developments of large margin classifiers. Readers are referred to these related papers on this aspect.

Our focus here is only on standard classification in supervised learning. There exists other types of classification problems. For example, in semi-supervised learning, one may have an unlabeled dataset for training besides the labeled training set. Then the question is how to make use of both labeled and unlabeled data to maximize the classification accuracy. More details can be found in a survey paper by Zhu (2008). Wang, Shen and Pan (2009a) proposed a new technique for semi-supervised learning. Another interesting classification problem is hierarchical classification where the class label has a tree structure (Wang et al., 2009b).

References

[1] An L. T. H. and Tao P. D. (1997). Solving a class of linearly constrained indefinite quadratic problems by D.C. algorithms. *Journal of Global Optimization* **11**, 253-285.

[2] Agresti A. and Coull B. (1998). Approximate is better than "exact" for interval estimation of binomial proportions. *The American Statistician* **52**, 119-126.

[3] Bennett K. P. and Mangasarian O. L. (1992). Robust linear programming discrimination of two linearly inseparable sets. *Optimization Methods and Software* **1**, 23-34.

[4] Bishop C. M. (1995). *Neural Networks for Pattern Recognition*. Oxford University Press, Cambridge.

[5] Bradley P. and Mangasarian O. L.(1998). Feature selection via concave minimization and support vector machines. In *ICML98*(J. Shavlik ed.). Morgan Kaufmann, San Francisco.

[6] Bredensteiner E. and Bennett K. (1999). Multicategory classification by support vector machines. *Computational Optimizations and Applications* **12**, 53-79.

[7] Breiman L. (1997). Prediction games and arcing algorithms. *Technical Report* **504**. Dept. Statistics, Univ. California, Berkeley.

[8] Breiman L. (1998). Arcing classifiers (with discussion). *Annals of Statistics* **26**, 801-849.

[9] Breiman L., Friedman J., Stone C. J. and Olshen, R.A. (1984). *Classification and Regression Trees*. Wadsworth, New York.

[10] Cortes C. and Vapnik V. (1995). Support vector networks. *Machine Learning* **20**, 273-297.

[11] Crammer K. and Singer Y. (2001). On the algorithmic implementation of multiclass Kernel-based vector machines. *Journal of Machine Learning Research* **2**, 265-292.

[12] Cristianini N. and Shawe-Taylor J. (2000). *An Introduction to Support Vector Machines*. Cambridge University Press, Cambridge, U.K.

[13] Dasarathy B. (1991). *Nearest Neighbor Pattern Classification Techniques*. IEEE Computer Society Press, Los Alamitos, CA.

[14] Duda R., Hart P. and Stork, D. (2001). *Pattern Classification (2nd edition)*. Wiley, New York.

[15] Fan J. and Li R. (2001). Variable selection via nonconcave penalized likelihood and its oracle properties. *Journal of American Statistical Association* **96**. 1348-1360.

[16] Friedman J. H., Hastie T. and Tibshirani R. (2000). Additive lobistic regression: a statistical view of boosting. *The Annals of Statistics* **28**, 337-407.

[17] Freund Y. (1995). Boosting a weak learning algorithm by majority. *Information and Computation* **121**, 256-285.

[18] Freund Y. and Schapire R. E. (1996). Experiments with a new boosting algorithm. In *Machine Learning: Proceedings of the Thirteenth International Conference*, 148-156. Morgan Kaufman, San Francisco.

[19] Fung G. and Mangasarian O. L. (2001). Proximal support vector machine classifiers. In *Proceedings KDD-2001: Knowledge Discovery and Data Mining*(Provost, F. and Srikant, F. eds.), 77-86. Asscociation for Computing Machinery.

[20] Guermeur Y. (2002). Combining discriminant models with new multiclass SVMS. *Pattern Analysis and Applications* **5**, 168-179.

[21] Hastie T., Tibshirani R. and Friedman J. H. (2001). *The Elements of Statistical Learning*. Springer, New York.

[22] Hsu C. and Lin C. (2002). A comparison of methods for multi-class support vector machines. *IEEE Transations on Neural Network* **13**, 415-425.

[23] Hunter D. and Li R. (2005). Variable selection using MM algorithms. *Annals of Statistics* **33**, 1617-1642.

[24] Kearns M. (1988). Thoughts on hypothesis boosting. Unpublished manuscript.

[25] Kimeldorf G. and Wahba G. (1971). Some results on Tchebycheffian spline functions. *Journal of Mathematical Analysis and Applications* **33**, 82-95.

[26] Lee Y., Lin Y. and Wahba G. (2004). Multicategory support vector machines, theory, and application to the classification of microarray data and satellite radiance data. *Journal of the American Statistical Association* **99**, 67-81.

[27] Lee Y., Kim Y., Lee S. and Koo J.-Y. (2006). Structured multicategory support vector machine with ANOVA decomposition. *Biometrika* **93**, 555-571.

[28] Lin X., Wahba G., Xiang D., Gao F., Klein R. and Klein B. (2000). Smoothing spline ANOVA models for large data sets with bernoulli observations and the randomized GACV. *The Annals of Statistics* **28**, 1570-1600.

[29] Lin Y. (2000). Some asymptotic properties of the support vector machine. *Technical Report*, **1029**. Department of Statistics, University of Wisconsin-Madison.

[30] Lin Y. (2002). Support vector machines and the Bayes rule in classification. *Data Mining and Knowledge Discovery* **6**, 259-275.

[31] Liu Y. (2007). Fisher consistency of multicategory support vector machines. *Eleventh International Conference on Artificial Intelligence and Statistics*, 289-296.

[32] Liu Y. and Shen X. (2006). Multicategory ψ-learning. *Journal of the American Statistical Association* **101**, 500-509.

[33] Liu Y., Shen X. and Doss H. (2005). Multicategory ψ-learning and support

vector machine: Computational tools. *Journal of Computational Graphical Statistics* **14**, 219-236.

[34] Liu Y. and Wu Y. (2006). Optimizing psi-learning via mixed integer programming. *Statistica Sinica* **16**, 441-457.

[35] Liu Y. and Wu Y. (2007). Variable selection via a combination of the L_0 and L_1 penalties. *Journal of Computational and Graphical Statistics* **16** (4), 782-798.

[36] Liu and Yuan (2009). Multicategory reinforced Support Vector Machines. Preprint.

[37] Liu Y., Zhang H. H., Park C. and Ahn J (2007). Support vector machines with adaptive Lq penalties. *Computational Statistics and Data Analysis* **51**, 6380-6394.

[38] Liu Y., Zhang H. H. and Wu Y. (2009). Hard or soft classification? Large margin unified machines. Preprint.

[39] Marron J. S., Todd M. and Ahn J. (2007). Distance-weighted discrimination. *Journal of the American Statistical Association* **102** (480), 1267-1271.

[40] McCullagh P. and Nelder J. A. (1989). *Generalized Linear Models*. Chapman & Hall, London.

[41] Park S. Y. and Liu Y. (2009). From the support vector machine to the bounded constraint machine. *Statistics and Its Interface*. **2**, 285-298.

[42] Paul D., Bair E., Hastie T. and Tibshirani R. (2008). Pre-conditioning for feature selection and regression in high-dimensional problems. *Annals of Statistics* **36** (4), 1595-1618.

[43] Platt J. C. (1999). Probabilistic outputs for support vector machines and comparisons to regularized likelihood methods. In *Advances in Large Margin Classifiers*(A. Smola, P. Bartlett, B. Schölkopf and D. Schuurmans, eds.), 61-74. MIT Press, Cambridge, MA.

[44] Rifkin R. and Klautau A. (2004). In defence of one-versus-all classificaiton. *Journal of Machine Learning Research* **5**, 101-141.

[45] Schapire R. E. (1990). The strength of weak learnability. *Machine Learning* **5**, 197-227.

[46] Shen X., Tseng G. C., Zhang X. and Wong W. H. (2003). On ψ-learning. *Journal of the American Statistical Association* **98**, 724-734.

[47] Schölkopf B. and Smola A. J.(2002). *Learning with Kernels*. MIT Press, Cambridge, MA.

[48] Shen X. and Wang L. (2007). Generalization error for multi-class margin classification. *Electronic Journal of Statistics* **1**, 307-330.

[49] Steinwart I. and Scovel C. (2006). Fast rates for support vector machines using Gaussian kernels. *The Annals of Statistics* **35**, 575-607.

[50] Suykens J. A. K. and Vandewalle J. (1999). Least squares support vector machine classifiers. *Neural Processing Letters* **9**, 293-300.

[51] Tarigan B. and van de Geer S. A. (2004). Adaptivity of support vector machines with L1 penalty. *Technical Report MI 2004-14*. University of Leiden.

[52] Tewari A. and Bartlett P. L. (2005). On the consistency of multiclass classification methods. In *Proceedings of the 18th Annual Conference on Learning Theory* **LNAI 3559**, 143-157. Springer.

[53] Tibshirani R. (1996). Regression shrinkage and selection via the LASSO. *Journal of Royal Statistical Society Series B* **58**, 267-288.

[54] Vapnik V. (1995). *The Nature of Statistical Learning Theory.* Springer, New York.

[55] Vapnik V. (1998). *Statistical Learning Theory.* Wiley, New York.

[56] Wahba G. (1990). *Spline Models for Observational Data.* CBMS-NSF Regional Conference Series in Applied Mathematics, 59. SIAM, Philadelphia.

[57] Wahba G. (1998). Support vector machines, reproducing Kernel Hilbert spaces, and randomized GACV. In *Advances in Kernel Methods: Support Vector Learning* (Schökopf, Burges, and Smola. eds.), 125-143. MIT Press, Cambridge, MA.

[58] Wang H., Li G. and Jiang G. (2007). Robust regression shrinkage and consistent variable selection through the lad-lasso. *Journal of Business & Economic Statistics* **25**, 347-355.

[59] Wang J., Shen X. and Liu Y. (2008). Probability estimation for large margin classifiers. *Biometrika* **95**, 149-167.

[60] Wang J., Shen X. and Pan W. (2009). On efficient large margin semisupervised learning: methodology and theory. *Journal of Machine Learning Research* **10**, 719-742.

[61] Wang J., Shen X. and Pan W. (2009b). On large margin hierarchical classification with multiple paths. *Journal of the American Statistical Association.* To appear.

[62] Wang L. and Shen X. (2007). On L1-norm multiclass support vector machines: methodology and theory. *Journal of the American Statistical Association* **102**, 583-594.

[63] Wang L. and Shen X. (2007). On L1-norm multi-class support vector machines: methodology and theory. *Journal of the American Statistical Association* **102**, 595-602.

[64] Wang L., Zhu J. and Zou H. (2006). The doubly regularized support vector machine. *Statistica Sinica* **16**.

[65] Weston J., Elisseeff A., Scholkopf B. and Tipping M. (2003). Use of the zero-norm with linear models and kernel methods. *Journal of Machine Learning Research* **3**, 1439-1461.

[66] Weston J. and Watkins C. (1999). Support vector machines for multi-class pattern recognition. In *Proceedings of the 7th European Symposium on Artificial Neural Networks (ESANN-99)*, 219-224.

[67] Wu Y. and Liu Y. (2007a). Robust truncated-hinge-loss support vector machines. *Journal of the American Statistical Association* **102** (479), 974-983.

[68] Wu Y. and Liu Y. (2007b). On multicategory truncated-hinge-loss support vector machines. *Contemporary Mathematics* **443**, 49-58.

[69] Wu Y. and Liu Y. (2009a). Variable selection in quantile regression. *Statistica Sinica* **19**, 801-817.

[70] Wu Y. and Liu Y. (2009b). Non-crossing large-margin probability estimation and its application to robust SVM via preconditioning. *Statistical Methodology.* To appear.

[71] Wu Y., Zhang H. H. and Liu Y. (2009). Robust model-free multiclass proba-

bility estimation. *Journal of the American Statistical Association.* To appear.

[72] Zhang T. (2004). Statistical analysis of some multi-category large margin classification methods. *Journal of Machine Learning Research* **5**, 1225-1251.

[73] Zhang H. H., Ahn J., Lin X. and Park C. (2006). Gene selection using support vector machines with nonconvex penalty. *Bioinformatics* **22**, 88-95.

[74] Zhang H. H. and Lu W. (2007). Adaptive-LASSO for Cox's proportional hazard model. *Biometrika* **94**, 691-703.

[75] Zhang T. and Oles F. J. (2001). Text categorization based on regularized linear classification methods. *Information Retrieval* **4**, 5-31.

[76] Zhao P., Rocha G. and Yu B. (2006). Grouped and hierarchical model selection through composite absolute penalties. *Annals of Statistics.* To appear.

[77] Zhu J. (2008). Semi-supervised learning literature survey. *Computer Sciences TR 1530* University of Wisconsin Madison.

[78] Zhu J., Rosset S., Hastie T. and Tibshirani R. (2003). 1-norm support vector machines. *Neural Information Processing Systems* **16**.

[79] Zou H. (2006). The adaptive lasso and its oracle properties. *Journal of the American Statistical Association* **101**, 1418-1429.

[80] Zou H. and Yuan M. (2008). The F-infinity-norm support vector machine. *Statistica Sinica* **18**, 379-398.

Part II

Large-Scale Multiple Testing

Chapter 3

A Compound Decision-Theoretic Approach to Large-Scale Multiple Testing

T. Tony Cai [*]   and   Wenguang Sun [†]

Abstract

In this chapter, we discuss topics in large-scale multiple testing and present a compound decision theoretical framework for false discovery rate (FDR) analysis. It is shown that conventional multiple testing procedures that threshold p-values can be much improved by a class of powerful data-driven procedures that exploit relevant information of the sample, including the proportion of non-nulls, the null and alternative distributions, the correlation structures as well as possible external information. Our discussion reveals the special features of large-scale inference problems and provides additional insights into the classic statistical decision theory. Both simulated and real data examples are presented for illustration of ideas and comparison of different procedures. Some important open problems for future research are also discussed.

Keywords: Compound decision theory, dependence, false discovery rate, grouped hypotheses, hidden Markov models, large-scale multiple testing.

1   Introduction

Large-scale multiple testing is an important area in modern statistics with a wide range of applications including genome-wide association studies, DNA microarray analysis, brain imaging studies and astronomical surveys. In these applications, one often tests thousands or even millions of hypotheses simultaneously. The analysis of these large-scale problems poses many statistical challenges not present in smaller scale studies. We discuss in this chapter the challenging statistical issues and new methodological developments in this field.

---

[*]Department of Statistics, Wharton School, University of Pennsylvania, Philadelphia, PA 19104, USA. Research supported in part by NSF Grant DMS-0604954 and NSF FRG Grant DMS-0854973.

[†]Department of Statistics, North Carolina State University, Raleigh, NC 27695, USA.

1.1 Setting and notation

We shall use DNA microarray studies to illustrate the typical setting and notation in large-scale multiple testing problems. In DNA microarray experiments, a standard technique for comparison of genes across two conditions is *differential analysis* (Dudoit et al. 2002; Sebastiani et al. 2003), where gene expression data are first collected on the same m genes for the two groups of subjects $\{X_1, \ldots, X_{n1}\}$ and $\{Y_1, \ldots, Y_{n_2}\}$, and then a two sample t-statistic t_i is calculated for each gene. The p-value and z-value of each test can be obtained using appropriate transformations: $p_i = 2F(-|t_i|)$ and $z_i = \Phi^{-1}\{F(t_i)\}$, where F and Φ are respectively the cdf's of the t-variable and standard normal variable. The data notation is summarized in Table 1.1. For example, Hedenfalk et al. (2001) analyzed a breast cancer (BRC) study where gene expression data were measured on the same $m = 3226$ genes for 15 breast cancer patients, 7 with BRCA1 mutation and 8 with BRCA2 mutation. Van't Wout et al. (2003) carried out a human immunodeficiency virus (HIV) study where gene expression levels are measured on the same $m = 7680$ genes for 4 HIV positive cases and 4 HIV negative controls. In both the BRC and HIV analyses, the m genes can be assumed to come from a mixture of two populations (null and non-null): in the null population, the gene expression levels are not associated with the disease condition, while in the non-null population, genes are differentially expressed between the two conditions. The goal is to test for differential gene expression and hence to identify genes associated with the disease status.

Table 1.1   Summary of the data notation

		Cond. I			Cond. II		t	z	p
	X_1	$\cdots$	X_{n_1}	Y_1	$\cdots$	Y_{n_2}	T	Z	P
Gene 1	x_{11}	$\cdots$	$x_{n_1 1}$	y_{11}	$\cdots$	$y_{n_2 1}$	t_1	z_1	p_1
Gene 2	x_{12}	$\cdots$	$x_{n_1 2}$	y_{12}	$\cdots$	$y_{n_2 2}$	t_2	z_2	p_2
$\vdots$	$\vdots$		$\vdots$	$\vdots$		$\vdots$	$\vdots$	$\vdots$	$\vdots$
Gene m	x_{1m}	$\cdots$	$x_{n_1 m}$	y_{1m}	$\cdots$	$y_{n_2 m}$	t_m	z_m	p_m

1.2 Type I error rates

When performing a single test, two types of errors may be committed: rejecting a hypothesis when it is a true null (type I error or false positive) or accepting it when it is a non-null (type II error or false negative). The outcomes of m simultaneous tests can be summarized in Table 1.2.

Table 1.2   Classification of tested hypotheses

	Claimed non-significant	Claimed significant	Total
Null	N_{00}	N_{10}	m_0
Non-null	N_{01}	N_{11}	m_1
Total	S	R	m

The consequences of the two types of errors are often different. In single hypothesis testing, it is desirable to control the Type I error rate at a prespecified level α and minimize the type II error rate. When multiple tests are considered jointly, a procedure that tests each hypothesis at level α, referred to as the pre-comparison error rate (PCER) procedure, in general leads to the inflation of type I errors. The family wise error rate (FWER), defined as FWER $= P(N_{10} \geqslant 1)$, has been widely used to avoid misleading inferences caused by the multiplicity in simultaneous testing. Here "family" refers to the collection of all tests being conducted. Instead of controlling the PCER at level α, an α-level FWER controlling procedure guarantees that the probability of making one or more type I errors in the family will not exceed the pre-specified level α. A well-known FWER procedure is the Bonferroni method that tests each hypothesis at level α/m. We refer to Hochberg and Tamhane (1987) and Shaffer (1995) for a review of FWER procedures. A natural extension of the FWER is the k-FWER, defined as the probability of making k or more false rejections in the family. Recently, some k-FWER procedures have been discussed in Lehmann and Romano (2005), Romano and Shaikh (2006) and Sarkar (2007).

However, the power to reject a non-null hypothesis while controlling for the FWER is greatly reduced as the number of tests increases. In situations where both the number of hypotheses and the number of true non-nulls are large, it is cost-effective to tolerate some type I errors, provided that the number is small compared to the total number of rejections. These considerations lead to a more powerful approach which calls for controlling the false discovery rate (FDR, Benjamini and Hochberg 1995). The FDR, defined as the expected proportion of false rejections among all rejections, provides a novel way to combine the errors in multiple comparisons. Using the notation in Table 1.2, the FDR can be defined as

$$\text{FDR} = E\left(\frac{N_{10}}{R} \,\middle|\, R > 0\right) P(R > 0) = E\left(\frac{N_{10}}{R \vee 1}\right). \tag{1.1}$$

According to the definition, the FDR is zero when no hypotheses are rejected. Other similar measures include the positive false discovery rate (Storey 2002) and the marginal false discovery rate (Genovese and Wasserman 2002; Sun and Cai 2007), respectively defined as

$$\text{pFDR} = E\left(\frac{N_{10}}{R} \,\middle|\, R > 0\right) \quad \text{and} \quad \text{mFDR} = \frac{E(N_{10})}{E(R)}.$$

The pFDR and mFDR are equivalent when test statistics come from a random mixture of the null and non-null distributions (Storey 2003). Genovese and Wasserman (2004) showed that, under mild conditions, mFDR $=$ FDR $+ O(m^{-1/2})$, where m is the number of hypotheses.

The FDR controlling procedures are more appropriate in large-scale multiple comparison problems and have been successfully applied in different scientific areas such as multistage clinical trials, microarray experiments, genome-wide association studies, brain imaging studies and astronomical surveys, among others (Weller et al. 1998; Efron et al. 2001; Miller et al. 2001; Tusher et al. 2001; Storey and

Tibshirani 2003; Dudoit et al. 2002; Sabatti et al. 2003; Menshausen and Rice 2006; Schwartzman et al. 2008). This chapter discusses the recent theoretical and methodological developments in the FDR field. The main goal is to develop a compound decision theoretical framework for large-scale multiple testing and to introduce a new class of powerful data-driven FDR procedures.

1.3   Optimality

In single hypothesis testing, the power is defined as the probability of correctly rejecting a non-null hypothesis. In the Neyman-Pearson testing framework, the *most powerful test* maximizes the power subject to a constraint on the type I error rate. The power can be generalized in different ways as we move from single hypothesis testing to multiple hypothesis testing. For FDR control, the most widely used measure is the false negative (or non-discovery) rate (FNR; Genovese and Wasserman 2002), the expected proportion of non-nulls among all non-rejections. Using the notation in Table 1.2, we have

$$\text{FNR} = E\left(\frac{N_{01}}{S}\bigg| S > 0\right) P(S > 0) = E\left(\frac{N_{01}}{S \vee 1}\right). \tag{1.2}$$

The FNR is a natural dual quantity to the FDR and will be used in this paper. Other quantities, including the missed discovery rate (MDR), the expected true positives (ETP) and the average power (AP), have also been considered in the literature (Spjøtvoll 1972; Storey 2007; Efron 2007). An FDR procedure is said to be *valid* if it controls the FDR at a prespecified level α and *optimal* if it has the smallest FNR among all valid FDR procedures at level α.

1.4   *P*-values and adjusted *p*-values

The p-value is a measure of how strongly the observed data contradict the null hypothesis. In single hypothesis testing, the p-value of a test can be interpreted as the probability under the null of observing a test statistic that is as extreme as or more extreme than the observed value in the direction of rejection. Therefore a statistical decision can be made by simply comparing the p-value with a given test level α. A similar concept, the adjusted p-value, was developed in the multiple testing context (Rosenthal and Rubin 1983; Wright 1992; Westfall and Young 1993). Given a testing procedure, the adjusted p-value for hypothesis i is the level of the entire testing procedure at which H_i would *just* be rejected, given the values of all test statistics. For example, if the interest is in controlling the FWER, the Bonferroni adjusted p-value for hypothesis i is $\tilde{p}_i = mp_i$, where p_i is the unadjusted p-value and m is the number of tests. Then H_i is rejected at nominal FWER α if $\tilde{p}_i \leqslant \alpha$.

1.5   Stepwise FWER procedures

The Bonferroni method is a *single-step* procedure, which evaluates each hypothesis using a common critical value that is essentially independent of other test statistics. Other single-step procedures include the Šidák procedure and the minP procedure (cf. Westfall and Young 1993; Dudoit et al. 2003). The single-step procedures are conservative and can be improved by procedures that have a more complicated structure. Two class of stepwise procedures, namely step-down and step-up procedures, have been developed in the literature. Due to their data-dependent nature, the stepwise procedures are capable of improving the power of a single-step procedure at the same FWER level without making additional assumptions. The most well known stepwise procedures for FWER control include the Holm procedure, the Simes procedure and the Hochberg procedure.

Holm procedure. Let $p_{(1)} \leqslant \cdots \leqslant p_{(m)}$ be the ordered p-values and $H_{(1)}, \ldots, H_{(m)}$ the corresponding hypotheses. The Holm procedure (Holm 1979) operates as follows: if $p_{(1)} \geqslant \alpha/m$, then accept all hypotheses and stop. Otherwise reject $H_{(1)}$ and test the remaining $m-1$ hypotheses at level $\alpha/(m-1)$. If $p_{(2)} \geqslant \alpha/(m-1)$, accept the remaining hypotheses and stop. Otherwise reject $H_{(2)}$ and test the remaining $m-2$ hypotheses at level $\alpha/(s-2)$. And so on. The adjusted p-value for the Holm procedure is

$$\tilde{p}_{(i)} = \max_{k=1,\ldots,i} [\min\{(m-k+1)p_{(k)}, 1\}],$$

for $i = 1, \ldots, m$. It can be shown that the Holm procedure controls the FWER at level α for all possible constellations of true and false hypotheses. This is referred to as the *strong control* of FWER. In addition, the Holm procedure starts with the most significant p-value and continues rejecting hypotheses as long as their p-values are small. This is called a *step-down* procedure.

Simes-Hochberg procedure. In contrast to step-down procedures, step-up procedures begin by looking at the least significant p-value and then move to the more significant ones. A well known step-up procedure is Simes-Hochberg procedure (Hochberg 1988), which operates as follows. Let $k = \max\{i : p_{(i)} \leqslant \alpha/(m-i+1)\}$, then reject hypotheses $H_1, \ldots, H_{(k)}$. The adjusted p-value for the Simes-Hochberg procedure is

$$\tilde{p}_{(i)} = \min_{k=i,\ldots,m} [\min\{(m-k+1)p_{(k)}, 1\}], \tag{1.3}$$

for $i = 1, \ldots, m$. We refer to Hochberg and Tamhane (1987) and Shaffer (1995) and for a comprehensive review of other issues in FWER control.

The rest of this chapter is organized as follows. Traditional FDR procedures are introduced in Section 2. In Section 3 we formulate the multiple testing problem in a compound decision theoretic framework and discuss an oracle and an adaptive procedure for FDR control. The simultaneous testing of grouped hypotheses and multiple testing under dependence are considered in Sections 4 and 5, respectively. We conclude the chapter with a discussion of some open problems and future directions.

2   FDR controlling procedures based on p-values

In contrast to the restrictive FWER criterion, an FDR procedure allows making more than one Type I errors as long as the total number of false rejections is small relative to the total number of rejections. Since the seminar work of Benjamini and Hochberg (BH 1995), the FDR procedures have been widely used in large-scale multiple comparison problems.

In this section, we first introduce the well known BH step-up procedure and then present several improvements over the BH procedure, including a q-value procedure (Storey 2002), an adaptive p-value procedure (Benjamini and Hochberg 2000), an oracle p-value procedure and the corresponding plug-in p-value procedure (Genovese and Wasserman 2004).

2.1   BH step-up procedure

Let α be an FDR level. Denote by $p_{(1)}, \ldots, p_{(m)}$ the ordered individual p-values and $H_{(1)}, \ldots, H_{(m)}$ the corresponding hypotheses. Benjamini and Hochberg (1995), designated by BH hereinafter, proposed the following step-up procedure:

$$\text{Let } k = \max\{i : P_{(i)} \leqslant i\alpha/m\}, \text{ then rejects all } H_{(i)}, i \leqslant k. \qquad (2.1)$$

BH showed that the step-up procedure (2.1) controls the FDR at the nominal level α when the tests statistics are independent. The BH procedure is more powerful than FWER controlling procedures at the same level.

The BH step-up procedure is distribution-free, which guarantees that the FDR is controlled at level α regardless of the p-value distribution. Thus it provides *strong control* of the FDR. However, with the gain of robustness, we pay a price for ignoring the distributional information in the sample. This issue was first raised in BH (2000), where it is argued that the BH step-up procedure is conservative when some of the hypotheses are in fact non-nulls. Specifically, the BH step-up procedure controls the FDR at level $(1 - p)\alpha$, instead of the nominal level α, where p is the proportion of non-nulls. Next we shall discuss several approaches that exploit the information of p to improve the classical BH procedure .

2.2   Adaptive p-value procedure

BH (2000) proposed a modified procedure for FDR control that is *adaptive* to the unknown non-null proportion. Let $\hat{p}$ be a conservative estimate of p. The adaptive p-value procedure operates as follows.

$$\text{Let } k = \max\{i : P_{(i)} \leqslant i\alpha/[(1 - \hat{p})m]\}, \text{ then rejects all } H_{(i)}, i \leqslant k, \qquad (2.2)$$

A graphical implementation of the BH adaptive procedure was presented in BH (2000) and it was shown that the adaptive procedure is more powerful than the BH step-up procedure.

2.3   Storey's q-value procedure

The conservativeness of the BH step-up procedure is also noted by Storey (2002), who suggested a direct approach to the FDR control. Instead of fixing the FDR level α and estimate the cutoff, Storey (2002) proposed to estimate the FDR conservatively for a given cutoff. Storey's approach also indicates that the efficiency of an FDR procedure can be improved by exploiting the information of p.

Let G denote the marginal distribution of the p-value. It can be shown that for the random mixture model, the pFDR for a given p-value cutoff λ is

$$\text{pFDR}(\lambda) = E\left(\frac{N_{10}}{R} \middle| R > 0\right) = \frac{(1-p)\lambda}{G(\lambda)}. \tag{2.3}$$

Consider a set of tests conducted with independent p-values. For an observed p-value, its q-value is defined as

$$q(p_i) = \inf_{\gamma \geq p_i} \{\text{pFDR}(\gamma)\} = \inf_{\gamma \geq p_i} \left\{\frac{(1-p)\gamma}{G(\gamma)}\right\}. \tag{2.4}$$

The q-value can be explained as the minimum pFDR level such that a hypothesis with the p-value of p_i is *just* rejected. Thus the q-value in multiple testing can be viewed as the pFDR analogue of the p-value in single hypothesis testing.

In practice, the non-null proportion p and G are unknown. Given a cutoff λ, a conservative estimate for p is proposed by Storey (2002): $\hat{p}(\lambda) = 1 - \frac{\#\{p_i > \lambda\}}{(1-\lambda)m}$. The estimate $\hat{p}(\lambda)$ is conservative because the largest p-values are most likely to come from the null. Next, an empirical estimate of G is $\hat{G}(\lambda) = \frac{\#\{p_i < \lambda\}}{m}$. Therefore the q-value can be estimated as

$$\hat{q}\left(p_{(i)}\right) = \widehat{\text{pFDR}}\left(p_{(i)}\right) = \frac{(1-\hat{p})p_{(i)}}{\hat{G}\left(p_{(i)}\right)}. \tag{2.5}$$

In practice, we can calculate the q-values for all individual p-values and determine the rejection region. The q-value procedure is equivalent to the BH procedure when $\hat{p}$ is estimated as zero, and is equivalent to the adaptive p-value procedure when the same $\hat{p}$ is used.

2.4   Oracle and plug-in p-value procedures

Let $G_1(t)$ denote the non-null distribution of p-value and p the proportion of non-nulls. We assume that G_1 is concave, then according to Genovese and Wasserman (2002), the *oracle p-value procedure* rejects all hypotheses whose p-value is less than u^*, the solution to the equation $\{(1-p)u\}/\{(1-p)u + pG_1(u)\} = \alpha$ or equivalently,

$$G_1(u)/u = (1/p - 1)(1/\alpha - 1). \tag{2.6}$$

The cutoff u^* is *optimal* in the sense that it has the smallest FNR among all p-value based procedures at FDR level α.

The idea that a testing problem is connected to an estimation problem is further developed in Genovese and Wasserman (2004), designated by GW hereinafter. Let $\hat{G}$ and $\hat{p}$ be estimates of the p-value cdf and the proportion of non-nulls, respectively. The FDR for a given cutoff t can be estimated as $\hat{Q}(t) = (1-\hat{p})t/\hat{G}(t)$. A class of plug-in FDR procedures were constructed (GW 2004) based on the p-value cutoff

$$t(\hat{p}, \hat{G}) = \sup\{t : \hat{Q}(t) \leqslant \alpha\}. \tag{2.7}$$

The BH step-up procedure, BH adaptive procedure and Storey's FDR approach can be identified as special cases when different estimates for p and G are chosen.

Although numerical results are promising, no theoretical supports for the uses of BH adaptive p-value procedure or Storey's q-value procedure were provided in the original works of BH (2000) and Storey (2002). GW (2004) developed a stochastic process framework for multiple testing and showed that, when consistent estimates of G and p are chosen, the class of plug-in procedures (2.7) controls the FDR at level $\alpha + o(1)$. Therefore the validity of BH adaptive procedure, GW plug-in procedure and Storey's FDR procedure, which can be viewed as special cases of (2.7), are established in an asymptotic sense.

2.5   Other issues

Resampling approaches (e.g., bootstrap, permutation) can be used to estimate the p-values and adjusted p-values without making any parametric assumptions on the joint distribution of the test statistics, and the correlation structure and distributional characteristics of the gene expression can be preserved. Algorithms for computing adjusted p-values are introduced, for example, in Westfall and Young (1993) and Dudoit et al. (2003). Permutation based methods have been applied to the significance analysis of microarrays (SAM). A widely used SAM procedure was proposed in Tusher et al. (2001).

The theoretical properties and operating characteristics of p-value based FDR procedures have been studied extensively in the literature. Here we mention a few references, Finner and Roters (2002), Sarkar (2002, 2006), Lehmann and Romano (2005), Finner et al. (2009), for interested readers.

3   Oracle and adaptive compound decision rules for FDR control

Developing the optimal FDR procedure is important from both theoretical and practical perspectives. The construction of an optimal multiple testing procedure involves two important steps: deriving an optimal test statistic T and setting a cutoff for T for a given FDR level. However, the focus of the FDR literature has been exclusively on the second step on how to threshold the p-values so that the FDR can be controlled, while the more fundamental problem on how to choose T is ignored.

In single hypothesis testing, p-value is a fundamental statistic for deciding whether a hypothesis should be rejected or accepted. This p-value testing framework is almost universally used in the FDR literature. For example, the FDR procedures reviewed in the previous section essentially involve first ranking the p-values from individual tests and then choosing a cutoff along the rankings. However, testing procedures built on ranked p-values fail to exploit all important distributional information in the sample (e.g., the symmetry of distribution and correlation in the sample), and hence are inefficient. Sun and Cai (2007) showed that p-value is not a fundamental building block in large-scale multiple testing, and proposed an adaptive data-driven procedure based on z-values that uniformly improves all p-value based FDR procedures.

In this section, we study the optimality issue in a compound decision theoretic framework. Our strategy for deriving an *optimal* testing procedure essentially involves three steps.

1. The first step is to derive an *oracle* test statistic T_{OR} that gives the optimal significance rankings of all tests. "Oracle" is used here to reflect the fact that we have assumed in this step that all distributional information of the sample is known. A useful technique in the derivation is to make connections between the multiple testing and *weighted classification* problems, then solve the former problem via finding the optimal classification rule.

2. The second step is to choose an optimal cutoff c_{OR} along the rankings produced by T_{OR} so that the FDR is controlled at the nominal level α. The essential idea is to first evaluate the distributions of T_{OR}, then calculate the FDR level for a given cutoff c, and finally choose the largest cutoff c_{OR} that controls the FDR. The resulting testing procedure $\delta(T_{OR}, c_{OR}\mathbf{1}) = I(T_{OR} < c_{OR}\mathbf{1})$ is referred to as the *oracle procedure*, which has the smallest FNR or the largest power to detect non-null cases among all valid FDR procedures.

3. The third step is to develop a *data-driven* procedure that mimics the oracle procedure by plugging-in estimates of the unknown parameters. To achieve this goal, we need to (i) construct good estimates of the unknown parameters from the sample, and (ii) establish the *asymptotic validity and optimality* of the data-driven procedure by showing that it achieves the performance of the oracle procedure asymptotically.

3.1   Two-component random mixture model

A two-component random mixture model provides a convenient and efficient framework for large-scale multiple testing and has been widely used in the FDR literature (Efron et al. 2001; Storey 2002; Newton et al. 2004; Sun and Cai 2007). Let $\boldsymbol{\theta} = (\theta_1, \ldots, \theta_m)$ be independent Bernoulli(p) variables, where $\theta_i = 1$ indicates that hypothesis i is a non-null and $\theta_i = 0$ otherwise. It is assumed that $\boldsymbol{x} = (x_1, \ldots, x_m)$ are observations generated conditional on $\boldsymbol{\theta}$:

$$X_i|\theta_i \sim (1 - \theta_i)F_0 + \theta_i F_1, \tag{3.1}$$

where F_0 and F_1 are the conditional cumulative distribution functions (cdf) of X_i under the null and alternative, respectively. Let $F(x) = (1 - p)F_0(x) + pF_1(x)$ be the mixture cdf and $f(x) = (1 - p)f_0(x) + pf_1(x)$ the mixture density, with f_0 and f_1 the corresponding conditional probability distribution functions (pdf).

3.2   Compound decision problem

Consider the random mixture model (3.1). In a multiple testing problem, we are interested in separating the non-null cases ($\theta_i = 1$) from the null cases ($\theta_i = 0$). A solution to this problem can be represented by a general decision rule $\boldsymbol{\delta} = (\delta_1, \ldots, \delta_m) \in \{0, 1\}^m$, where $\delta_i = 1$ indicates that we claim case i is a non-null and $\delta_i = 0$ otherwise. In an FDR analysis, the m decisions are combined and evaluated integrally; this is referred to as a *compound decision problem* (Robbins 1951). The decision rule $\boldsymbol{\delta}$ is *simple* if δ_i is only a function of x_i, i.e., $\delta_i(\mathbf{x}) = \delta_i(x_i)$. The simple rules correspond to solving the m component problems separately. In contrast, $\boldsymbol{\delta}$ is *compound* if δ_i depends on other x_j's, $j \neq i$. A decision rule $\boldsymbol{\delta}$ is *symmetric* if $\boldsymbol{\delta}(\pi(\mathbf{x})) = \pi(\boldsymbol{\delta}(\mathbf{x}))$ for all permutation operators π.

Robbins (1951) considered the following compound decision problem. Let X_i, $i = 1, \ldots, n$, be independent normal random variables with unit variance and mean θ_i, where each θ_i is 1 or -1. It is desired to classify each θ_i according to its sign, and the risk of a classification procedure is taken to be the expected number of errors, i.e., the risk for a classification rule $\boldsymbol{\delta} \in \{-1, 1\}^m$ is

$$R(\boldsymbol{\theta}, \boldsymbol{\delta}) = E \left\{ \sum_{i=1}^{m} I(\theta_i = 1)I(\delta_i = -1) + I(\theta_i = -1)I(\delta_i = 1) \right\}, \tag{3.2}$$

Robbins showed that the *simple rule*

$$\boldsymbol{\delta}^S = [\text{sgn}(x_i) : i = 1, \ldots, n] \tag{3.3}$$

is the unique minimax decision rule with constant risk, say r_n. At the same time, he argued that minimax did not equal *best* by exhibiting another decision rule

$$\boldsymbol{\delta}^C = \left[\text{sgn}\left(x_i - \frac{1}{2}\log \frac{1 - \bar{x}}{1 + \bar{x}} \right) : i = 1, \ldots, n \right], \tag{3.4}$$

which has a much lower risk r_n^* when p_0 approaches 0 or 1, and only exceeds r_n slightly near 0.5. As $n \to \infty$, $r_n^* - r_n \to 0$ at point 0.5, so it is *subminimax*. We shall give a graphical illustration of this interesting result shortly. An important feature of the decision rule $\boldsymbol{\delta}^C$ is that it classifies θ_i depending on the whole vector $x = (x_1, \ldots, x_n)$, not on x_i alone. Thus $\boldsymbol{\delta}^C$ is a compound rule.

The subminimax rule can be motivated as follows. Assume that an oracle knows the proportion of $\theta = 1$, say p, then it can be shown that the optimal classification rule (Bayes oracle rule) is

$$\boldsymbol{\delta}^* = \left[\text{sgn}\left\{ x_i - \log\left(\frac{1-p}{p} \right) \right\} : i = 1, \ldots, n \right]. \tag{3.5}$$

The subminimax rule can be obtained by replacing the unknown p using its unbiased estimate

$$\hat{p} = \frac{1}{n}(\text{no. of } i \text{ for which } x_i > 0) = \frac{1 + \bar{x}}{2} \tag{3.6}$$

The subminimax rule is a compound decision rule, which improves the efficiency of simple rules by utilizing the information of $\hat{p}$ that is estimated from the entire sample x. Some calculations show that the classification risks for the minimax rule, the Bayes oracle rule, and the subminimax rule are

$$R_{\text{Minimax}} = \Phi(-1),$$

$$R_{\text{Oracle}} = p\Phi\left(-1 + \frac{1}{2}\log\frac{1-p}{p}\right) + (1-p)\Phi\left(-1 - \frac{1}{2}\log\frac{1-p}{p}\right), \text{ and}$$

$$R_{\text{Subminimax}} = r(p) + \frac{1}{n}p\left[\left\{1 + \frac{1}{4p(1-p)}\right\}^2 - 1\right]\Phi\left(-1 + \frac{1}{2}\log\frac{1-p}{p}\right),$$

respectively. We plot the classification risks as functions of the proportion p. The results are shown in Figure 3.1. It is easy to see that the risk of the subminimax rule approaches the risk of the Bayes oracle rule as $n \to \infty$. For large n, the subminimax rule is better than the minimax rule for most values of p except the values near 0.5. The efficiency gain is the largest as p approaches 0 and 1.

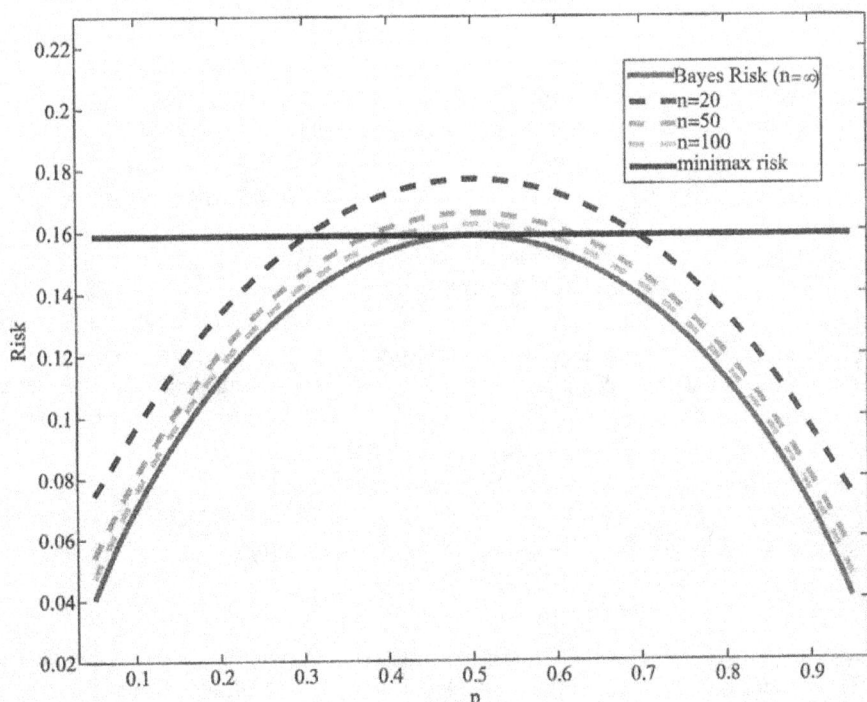

Figure 3.1   The classification risks of the minimax, subminimax and Bayesian oracle rules.

An important implication of Robbin's results is that the precision of individual decisions can be much increased by pooling information from different samples. The distributional information among all hypotheses, such as the proportion of non-nulls, is important for construction of efficient testing procedures. Therefore we anticipate that, in large-scale multiple testing, conventional testing procedures may be much improved by an estimation-testing combined procedure.

3.3   Derivation of the optimal test statistic

In the loss function (3.2) that was considered by Robbins (1951), it is assumed that a false positive and a false negative have the same cost. However, in practice, a false positive is often considered to be more serious than a false negative. Therefore it is desirable to treat the two types of errors differently. Let λ be the known relative cost of a false positive to a false negative, then a weighted classification rule $\boldsymbol{\delta} = (\delta_1, \cdots, \delta_m) \in \{0,1\}^m$, where $\delta_i = 1$ indicates that we classify θ_i as a non-null and $\delta_i = 0$ otherwise, can be used to separate the non-nulls from the nulls. The false positives and false negatives of the m simultaneous decisions can be combined using the following loss function

$$L_\lambda(\boldsymbol{\theta}, \boldsymbol{\delta}) = \frac{1}{m} \sum_i \left[\lambda(1 - \theta_i)\delta_i + \theta_i(1 - \delta_i) \right]. \tag{3.7}$$

The goal of a weighted classification problem is to find a decision rule $\boldsymbol{\delta}^\lambda$ that minimizes the weighted classification risk $R_\lambda = E[L_\lambda(\boldsymbol{\theta}, \boldsymbol{\delta})]$.

However, it may be difficult to prespecify the relative cost of a false positive to a false negative, especially in many large-scale studies where the non-nulls are very sparse. We can use the FDR and FNR to combine the respective false positives and false negatives instead, and apply a multiple testing procedure to select non-null cases from the nulls. In a multiple testing problem, the two types of errors are also treated differently, where a higher penalty on false positives is achieved by prespecifying a smaller FDR level α. The goal of multiple testing is to find a decision rule $\boldsymbol{\delta}^\alpha \in \{0,1\}^m$ that has the smallest FNR among all FDR procedures at level α. We define a multiple testing procedure in terms of a function T and a constant c such that $\boldsymbol{\delta}(T, c) = [I(T(x_i) < c) : i = 1, \ldots, m]$. In conventional p-value based FDR procedures, $T(x_i)$ is taken as $F_0(-|x_i|)$ or $2F_0(-|x_i|)$, where F_0 is the null cdf of X_i. We consider a wider class of testing procedures in which T is allowed to also depend on other quantities. We are interested in finding the optimal choices of T and its corresponding cutoff c.

A monotone ratio condition. The goal of both the multiple testing and weighted classification problems is to separate the non-null cases from the null cases, and the solution to both problems can be represented by a decision rule of the form

$$\boldsymbol{\delta}(\boldsymbol{T}, c\mathbf{1}) = \{I(T_i < c) : i = 1, \ldots, m\}, \tag{3.8}$$

where $\boldsymbol{T}$ is a classifier or a test statistic and c is a cutoff. In the multiple testing literature, the following assumption has been used (e.g., Genovese and Wasserman 2004; Storey 2004):

the FDR level yielded by $\boldsymbol{\delta}(\boldsymbol{T}, c\mathbf{1})$ is increasing in c; \qquad (3.9)

the FNR level yielded by $\boldsymbol{\delta}(\boldsymbol{T}, c\mathbf{1})$ is decreasing in c. \qquad (3.10)

Assumptions (3.9) and (3.10) are desirable for developing multiple testing procedures since it implies that in order to minimize the FNR, we should choose the largest cutoff c that satisfies FDR$\leqslant \alpha$. Let G_0 and G_1 be the null and non-null cdfs of T, respectively. Denote by $G = (1 - \pi)G_0 + \pi G_1$ the marginal cdf of T. A general condition that guarantees (3.9) and (3.10) is the monotone ratio condition (MRC), which assumes that

$$g_1(t)/g_0(t) \text{ is monotonically decreasing in } t. \tag{3.11}$$

See Sun and Cai (2007) for a proof. Denote by $\mathcal{T}$ the collection of all test statistics that satisfy (3.11). The MRC class $\mathcal{T}$ is fairly general. Let $T_i = p_i$, the p-value of an individual test based on the observation x_i. Assume that $p(x_i) \sim G = (1-p)G_0 + pG_1$, where G_0 and G_1 are the p-value distributions under the null and the alternative, respectively. Next assume that $G_1(t)$ is twice differentiable. Note $g_0(t) = G_0'(t) = 1$, the assumption $p(.) \in \mathcal{T}$ implies that $G_1''(t) = g_1'(t) < 0$, i.e., $G_1(t)$ is concave. Therefore the MRC condition (3.11) can be viewed as a generalized version of the concavity assumption on G_1 for p-value (Storey 2002, Genovese and Wasserman 2004). In addition, other test statistics, including the local false discovery rate (Lfdr, Efron et al. 2001) and the local index of significance (Sun and Cai 2009) also belong to $\mathcal{T}$. See Sun and Cai (2007) for more discussions of the MRC.

Connection between multiple testing and weighted classification. An important step for our derivation is to show that the multiple testing and weighted classification problems are "equivalent" when the MRC holds (Sun and Cai 2007). Specifically, let $\mathcal{D}_\alpha$ be the collection of all α-level FDR procedures of the form $\boldsymbol{\delta} = I(\boldsymbol{T} < c\boldsymbol{1})$. Suppose that the classification risk with the loss function defined in (3.7) is minimized by $\delta^\lambda\{\boldsymbol{T}, c(\lambda)\}$, so that $\boldsymbol{T}$ is optimal in the weighted classification problem. If $\boldsymbol{T} \in \mathcal{T}$, then $\boldsymbol{T}$ is also optimal in the multiple testing problem, in the sense that for each FDR level α, there exists a unique $\lambda(\alpha)$, and hence $c\{\lambda(\alpha)\} = c(\alpha)$, such that $\delta^{\lambda(\alpha)}\{\boldsymbol{T}, c(\alpha)\}$ controls the FDR at level α with the smallest FNR level among all testing rules in $\mathcal{D}_\alpha$. This implies that the more complicated multiple testing problem can be solved by studying an equivalent weighted classification problem.

The next step is to derive the optimal classification rule. We first study an ideal setup where there is an oracle that knows p, f_0 and f_1. Then the oracle rule in this weighted classification problem gives the optimal choice of δ.

Theorem 3.1. (The oracle rule for weighted classification.) *Consider the random mixture model (3.1). Suppose p, f_0, f_1 are known. Then the classification risk $E[L_\lambda(\theta, \delta)]$ with loss function (3.7) is minimized by $\delta^\lambda(\Lambda, 1/\lambda) = \{\delta_1, \ldots, \delta_m\}$, where*

$$\delta_i = I\left\{\Lambda(x_i) = \frac{(1-p)f_0(x_i)}{pf_1(x_i)} < \frac{1}{\lambda}\right\}, i = 1, \ldots, m. \tag{3.12}$$

3.4 Oracle testing procedure

We have shown that $\delta^\lambda(\Lambda, 1/\lambda) = [I\{\Lambda(x_1) < 1/\lambda\}, \ldots, I\{\Lambda(x_m) < 1/\lambda\}]$ is the oracle rule in the weighted classification problem. The equivalence between multiple testing and weighted classification implies the optimal testing rule is also of the form $\delta^{\lambda(\alpha)}[\Lambda, 1/\lambda(\alpha)]$ if $\Lambda \in \mathcal{T}$, although the cutoff $1/\lambda(\alpha)$ is not obvious. Note that $\Lambda(x) = \text{Lfdr}(x)/[1 - \text{Lfdr}(x)]$ is monotonically increasing in $\text{Lfdr}(x)$, where

Lfdr$(\cdot) = (1 - p)f_0(\cdot)/f(\cdot)$ is the local false discovery rate (Lfdr) introduced by Efron et al. (2001) and Efron (2004), so the optimal rule for mFDR control is of the form $\delta(\mathrm{Lfdr}(\cdot), c) = \{I[\mathrm{Lfdr}(x_i) < c] : i = 1, \ldots, m\}$. The Lfdr has been widely used in the FDR literature to provide a Bayesian version of the frequentist FDR measure and interpret results for individual cases (Efron 2004). We rediscover it here as the optimal (oracle) statistic in the multiple testing problem in the sense that the thresholding rule based on Lfdr(X) controls the mFDR at the nominal level with the smallest mFNR.

The MRC implies that in order to minimize the mFNR level, we should choose the largest threshold for the Lfdr statistic. Therefore the *oracle testing procedure* is

$$\delta(\mathrm{Lfdr}, c_{OR}) = \{I[\mathrm{Lfdr}(x_i) < c_{OR}] : i = 1, \ldots, m\}, \tag{3.13}$$

where the oracle threshold $c_{OR} = \sup\{c \in (0, 1) : \mathrm{mFDR}(c) \leqslant \alpha\}$. The oracle procedure (3.13) provides an ideal target for evaluating different multiple testing procedures. In particular, it is more efficient than the p-value oracle procedure proposed in Genovese and Wasserman (2002). Hence the z-value oracle procedure is more efficient than all p-value based FDR procedures.

The Lfdr statistic is defined in terms of the *z-values*, which can be converted from other test statistics including the t-statistic and χ^2-statistic using appropriate transformations. Note that the non-null proportion p is a global parameter. The expression Lfdr$(z) = (1 - p)f_0(z)/f(z)$ therefore implies that we actually rank the relative importance of the observations according to their likelihood ratios, and that the rankings are generally different from the rankings of p-values. An interesting consequence of using the Lfdr statistic in multiple testing is that an observation located farther from the null may have a lower significance level. It is therefore possible that the test accepts a more extreme observation while rejecting a less extreme observation, which implies that the rejection region is asymmetric. This is not possible for a testing procedure based on the individual p-values, whose rejection region is always symmetric about the null.

3.5   A data-driven procedure

The oracle procedure is not applicable in practice because the distributional information is usually unknown. This section first discusses the estimation of the null distribution and the non-null proportion in large-scale multiple comparisons. Then we introduce a data-driven procedure that mimics the oracle procedure.

Efron (2004) raised an important issue that in many large-scale studies the usual assumption that the null distribution is known is incorrect, and seemingly negligible differences in the null may result in large differences in subsequent studies. It was demonstrated that the null distribution should be estimated from data instead of being assumed known. Besides the null distribution, the proportion of non-null effects p is also an important quantity. The implementation of many FDR procedures requires the knowledge of p (BH 2000; Storey 2002; GW 2004). Developing good estimators for the proportion of non-nulls is a challenging task. Recent work includes that of Genovese and Wasserman (2004), Langaas, Lindqvist and Ferkingstad (2005), Meinshausen and Rice (2006), Cai, Jin and Low (2007), and Jin and Cai (2007).

Jin and Cai (2007) developed an approach based on the empirical characteristic function and Fourier analysis for simultaneous estimation of both the null distribution f_0 and proportion of non-null effects p. The estimators are shown to be uniformly consistent over a wide class of parameters. Numerical results also showed that the estimators perform favorably in comparison to other existing methods. This method will be used in our data-driven procedure.

Next we outline the steps for an intuitive derivation of the adaptive z-value based procedure. The derivation essentially involves mimicking the operation of the z-value oracle procedure and evaluating the distribution of $T_{OR}(z)$ empirically. Let $z_1, \ldots, z_m$ be a random sample from the mixture model (3.1) with the CDF $F = (1 - p)F_0 + pF_1$ and PDF $f = (1 - p)f_0 + pf_1$. Let $\hat{p}$, $\hat{f}_0$ and $\hat{f}$ be consistent estimates of p, f_0 and f. Such estimates are provided, for example, in Jin and Cai (2007). Define $\hat{T}_{OR}(z_i) = [(1 - \hat{p})\hat{f}_0(z_i)/\hat{f}(z_i)] \wedge 1$. The mFDR of decision rule $\boldsymbol{\delta}(T_{OR}, \lambda) = \{I[T_{OR}(z_i) < \lambda] : i = 1, \ldots, m\}$ is given by $Q_{OR}(\lambda) = (1 - p)G^0_{OR}(\lambda)/G_{OR}(\lambda)$, where $G_{OR}(t)$ and $G^0_{OR}(t)$ are the marginal cdf and null cdf of T_{OR}, respectively. Let $S_\lambda = \{z : T_{OR}(z) < \lambda\}$ be the rejection region. Then $G_{OR}(\lambda) = \int_{S_\lambda} f(z)dz = \int 1\{T_{OR}(z) < \lambda\}f(z)dz$. We estimate $G_{OR}(\lambda)$ by $\hat{G}_{OR}(\lambda) = \frac{1}{m}\sum_{i=1}^m 1\{\hat{T}_{OR}(z_i) < \lambda\}$. The numerator of $Q_{OR}(\lambda)$ can be written as $(1 - p)G^0_{OR}(\lambda) = (1 - p)\int_{S_\lambda} f_0(z)dz = \int 1\{T_{OR}(z) < \lambda\}T_{OR}(z)f(z)dz$ and we estimate this quantity by $\frac{1}{m}\sum_{i=1}^m 1\{\hat{T}_{OR}(z_i) < \lambda\}\hat{T}_{OR}(z_i)$. Then $Q_{OR}(\lambda)$ can be estimated as

$$\hat{Q}_{OR}(\lambda) = [\sum_{i=1}^m 1\{\hat{T}_{OR}(z_i) < \lambda\}\hat{T}_{OR}(z_i)]/[\sum_{i=1}^m 1\{\hat{T}_{OR}(z_i) < \lambda\}].$$

Set the estimated threshold as $\hat{\lambda}_{OR} = \sup\{t \in (0, 1) : \hat{Q}_{OR}(t) \leqslant \alpha\}$ and let R be the set of the ranked $\hat{T}_{OR}(z_i)$: $R = \{\text{L}\hat{\text{f}}\text{dr}_{(1)}, \ldots, \text{L}\hat{\text{f}}\text{dr}_{(m)}\}$. We only consider the discrete cutoffs in set R, where the estimated mFDR is reduced to $\hat{Q}_{OR}(\text{L}\hat{\text{f}}\text{dr}_{(k)}) = \frac{1}{k}\sum_{i=1}^k \text{L}\hat{\text{f}}\text{dr}_{(i)}$. We propose the following adaptive step-up procedure:

$$\text{Let } k = \max\{i : \frac{1}{i}\sum_{j=1}^i \text{L}\hat{\text{f}}\text{dr}_{(j)} \leqslant \alpha\}, \text{ then reject all } H^{(i)}, i = 1, \ldots, k. \quad (3.14)$$

In the FDR literature, z-value based methods such as the Lfdr procedure (Efron, 2004) are only used to calculate individual significance levels whereas the p-value based procedures are used for global FDR control to identify non-null cases. It is also notable that the goals of global error control and individual case interpretation are naturally unified in the adaptive procedure. The procedure (3.14) is more *adaptive* than the BH adaptive procedure in the sense that it adapts both to the global feature (p) and local feature (f_0/f). In contrast, the BH method only adapts to the global feature p. Suppose we use the theoretical null $N(0, 1)$ in the expression of $\text{L}\hat{\text{f}}\text{dr} = (1 - \hat{p})f_0/\hat{f}$. The p-value approaches treat points $-z$ and z equally, whereas the z-value approaches evaluate the relative importance of

$-z$ and z according to their estimated densities. For example, if there is evidence in the data that there are more non-nulls around $-z$ (i.e., $\hat{f}(-z)$ is larger), then observation $-z$ will be correspondingly ranked higher than observation z.

The following theorem shows that the adaptive procedure (3.14) asymptotically attains the performance of the oracle procedure based on the z-values in the sense that both the mFDR and mFNR levels achieved by the oracle procedure are also asymptotically achieved by the adaptive z-value procedure.

Theorem 3.2 (Asymptotic validity and optimality of the adaptive procedure). *Consider the random mixture model* (3.1). *Suppose f is continuous and positive on the real line. Assume $T_{OR}(z_i) = (1-p)f_0(z_i)/f(z_i)$ is distributed with the marginal PDF $g = (1-p)g_0 + pg_1$ and $T_{OR} \in T$ satisfies the MRC assumption. Let $\hat{p}$, $\hat{f}_0$, $\hat{f}$ be estimates of p, f_0 and f such that $\hat{p} \xrightarrow{qm} p$, $E\|\hat{f} - f\|^2 \to 0$ and $E\|\hat{f}_0 - f_0\|^2 \to 0$. Define test statistic $\hat{T}_{OR}(z_i) = (1-\hat{p})\hat{f}_0(z_i)/\hat{f}(z_i)$. Let $\hat{Lfdr}_{(1)}, \ldots, \hat{Lfdr}_{(m)}$ be the ranked values of $\hat{T}_{OR}(z_i)$, then the FDR level of the adaptive procedure* (3.14) *is $\alpha + o(1)$, and the mFNR level of the adaptive procedure* (3.14) *is $\tilde{Q}_{OR}(\lambda_{OR}) + o(1)$, where $\tilde{Q}_{OR}(\lambda_{OR})$ is the mFNR level achieved by the oracle oracle procedure.*

3.6   Numerical results

We now turn to the numerical performance of our adaptive z-value procedure. When the Lfdr statistic is needed to be estimated, f_0 is chosen to to be the theoretical null density $N(0,1)$, p is estimated consistently using the approach of Jin and Cai (2007), and f is estimated using the kernel density estimator. The new data-driven procedure is compared with the BH step-up procedure and the adaptive p-value procedure (BH 2000; GW 2004). These three procedures are designated respectively by SC, BH and AP hereinafter.

Example 3.3.   We generate $m = 3000$ observations from the normal mixture model $0.8N(0,1) + p_1 N(\theta_{1i}, 1) + (0.2 - p_1)N(\theta_{2i}, 1)$, where θ_{1i} and θ_{2i} are randomly generated from uniform distributions $U(\mu_1 - \epsilon_1, \mu_1 + \epsilon_1)$ and $U(\mu_2 - \epsilon_2, \mu_2 + \epsilon_2)$. We apply the BH, AP and SC with FDR = 0.10, $\mu_1 = -3$, $\epsilon_1 = 0.4$ and $\epsilon_2 = 0.2$. The comparison results are displayed in Figure 3.2. In panel (a), we set $\mu_2 = 3$ and plot the FNR's by BH, GW and SC as functions of p_1. In panel (b), we set $p_1 = 0.18$ and plot the FNR's by BH, AP and SC as functions of μ_2. We can see that the BH is dominated by AP, which is again dominated by SC. The efficiency gain of SC becomes more prominent when the alternative distribution is more asymmetric.   □

Next we illustrate our method in the analysis of the microarray data from an HIV study. The goal of the HIV study (van't Wout et al., 2003) is to discover differentially expressed genes between HIV positive patients and HIV negative controls. Gene expression levels were measured for four HIV positive patients and four HIV negative controls on the same $m = 7680$ genes. A total of m two sample t-tests were performed and the corresponding two-sided p-values were obtained. The z-values were then converted from the t-statistics using the transformation $z_i = \Phi^{-1}[G_0(t_i)]$, where Φ and G_0 are the CDFs of a standard normal and a t variable with six degrees of freedom, respectively. The histograms of the z-values

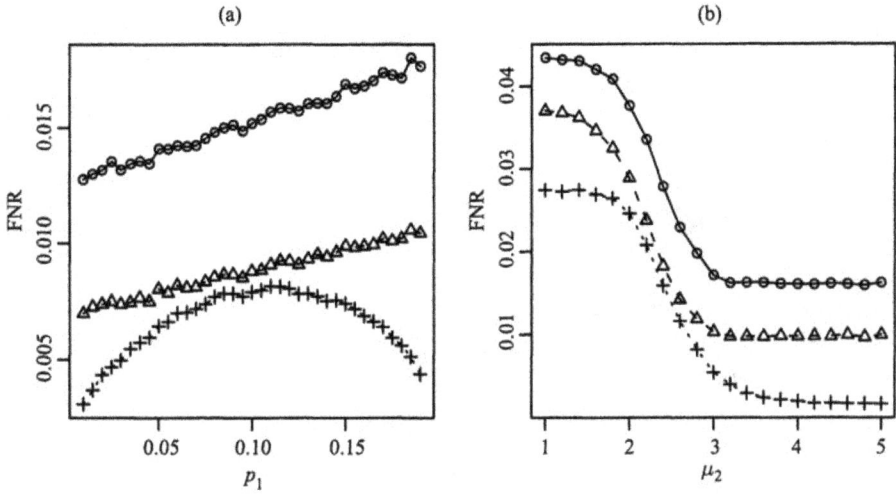

Figure 3.2   The comparison of BH, AP and SC when the alternative is concentrated ('o': BH; '△': AP; '+': SC). The FDR level is set at 0.1. (a) The FNR versus p_1; (b) The FNR versus μ_2. Both the BH and AP are dominated by SC.

and p-values were presented in Figure 3.3. An important feature in this data set is that the z-value distribution is asymmetric about the null. The distribution is skewed to the right.

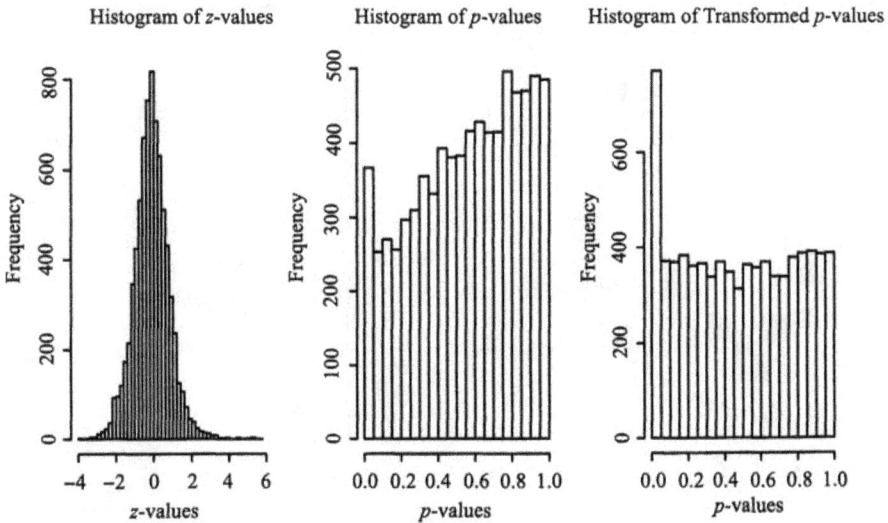

Figure 3.3   The histograms of the HIV data: p-values and z-values. The transformed p-values are approximately distributed as uniform $(0, 1)$ for the null cases.

When the null hypothesis is true, the p-values and z-values should follow their *theoretical null distributions*, which are uniform and standard normal, respectively. However, the theoretical nulls are usually quite different from the empirical nulls for the data arising from microarray experiments. We take the approach in Jin and Cai (2006) to estimate the null distribution as $N(\hat{\mu}_0, \hat{\sigma}_0^2)$. The estimates $\hat{\mu}_0$ and $\hat{\sigma}_0^2$ are consistent. We then proceed to estimate the proportion of the non-nulls $\hat{p}$ based on $\hat{\mu}_0$ and $\hat{\sigma}_0^2$. The marginal density f is estimated by a kernel density estimate $\hat{f}$ with the bandwidth chosen by cross validation. The Lfdr statistics are then calculated as $\text{Lfdr}(z_i) = (1 - \hat{p})\hat{f}_0(z_i)/\hat{f}(z_i)$. The transformed p-values are obtained as $\hat{F}_0(z_i)$, where $\hat{F}_0$ is the estimated null CDF $\Phi(\frac{x - \hat{\mu}_0}{\hat{\sigma}_0})$. As we can see from the right panel of Figure 3.3, after transformation, the distribution of the transformed p-values is approximately uniform when the null is true.

We compare the BH, AP and SC using both the theoretical nulls and estimated nulls. We calculate the number of rejections for each mFDR level and the results are shown in Figure 3.4. For the left panel, f_0 is chosen to be the theoretical null $N(0,1)$ and the estimate for the proportion of nulls is 1. The BH and AP procedures therefore yield the same number of rejections. For the right panel, the estimated null distribution is $N(-0.08, 0.77^2)$ with estimated proportion of nulls $\hat{p}_0 = 0.94$. Transformed p-values as well as the Lfdr statistics are calculated according to the estimated null. The following observations can be made from the results displayed. (i) The number of rejections is increasing as a function of the mFDR. (ii) For both the p-value and z-value based approaches, more hypotheses are rejected by using the estimated null. (iii) Both comparisons show that SC is more powerful than the BH and AP that are based on p-values.

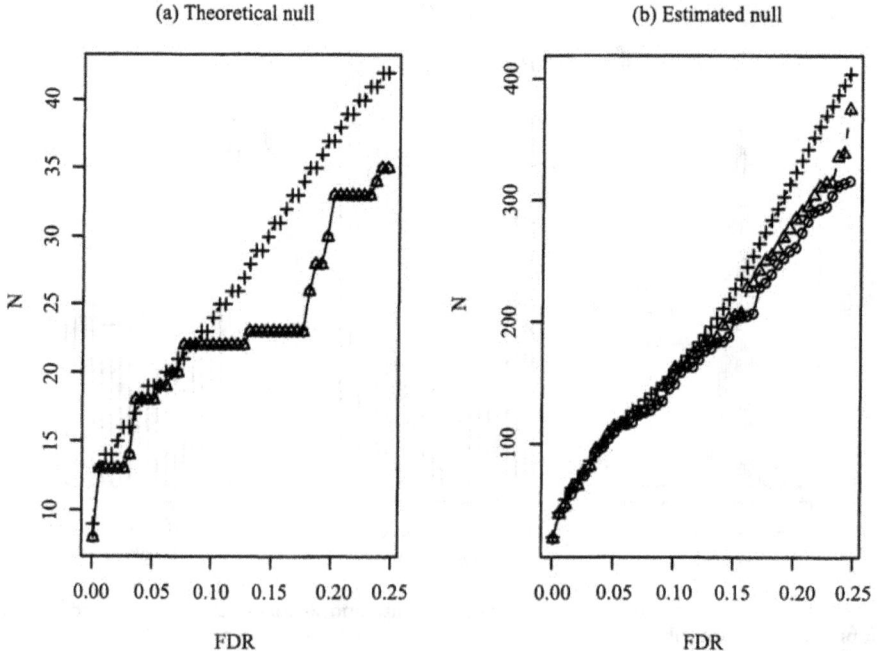

Figure 3.4   Analysis of the HIV data: Number of rejections versus FDR levels: 'o': BH; '△': AP; '+': SC.

4 Simultaneous testing of grouped hypotheses

So far we have made two important assumptions: all hypotheses are independent and all observations come from a homogeneous distribution. Next we shall deal with more complicated situations where these assumptions do not hold. In this section, we consider the multiple testing problem when hypotheses come from heterogeneous groups. In Section 5, we discuss the multiple testing problem under dependence. For both problems, we will focus on the motivation and the main ideas of our solution. More technical details are given in Cai and Sun (2009) and Sun and Cai (2009).

4.1 Motivating examples

Conventional multiple testing procedures, such as the false discovery rate analyses (Benjamini and Hochberg 1995; Efron et al. 2001; Storey 2002; Genovese and Wasserman 2002; van der laan et al. 2004), implicitly assume that data are collected from repeated or identical experimental conditions, and hence the hypotheses are exchangeable. However, in many applications, data are known to be collected from heterogeneous sources and hypotheses intrinsically form into different groups.

Consider the following two examples. The adequate yearly progress (AYP) study compares the academic performances of social-economically advantaged (SEA) versus social-economically disadvantaged (SED) students of California high schools (Rogosa 2003). Standard tests in mathematics were administered to 7867 schools and a z-value for comparing SEA and SED students was obtained for each school. The estimated null densities of the z-values for small, medium and large schools are plotted on the left panel of Figure 4.1. It is interesting to see that the null density of the large group is much wider than those of the other two densities. The differences in the null distributions have significant effects on the outcomes of a multiple testing procedure. Another example is the brain imaging study analyzed in Schwartzman et al. (2005). In this study, 6 dyslexic children and 6 normal children received diffusion tensor imaging brain scans on the same 15443 brain locations (voxels). A z-value (converted from a two-sample t-statistic) for comparing dyslexic versus normal children was obtained for each voxel. The right panel in Figure 4.1 plots the estimated null densities of the z-values for the front and back halves of the brain. We can see that the null cases from two groups centered on different means, and the density of the back half is narrower. There are many other examples where the hypotheses are naturally grouped. For instance, in analysis of geographical survey data, individual locations are aggregated into several large clusters; and in meta-analysis of large biomedical studies, the data are collected from different clinical centers. An important common feature of these examples is that data are collected from heterogeneous sources and the hypotheses being considered are grouped and no longer exchangeable. We shall see that incorporating the grouping information is important for optimal simultaneous inference with samples collected from different groups.

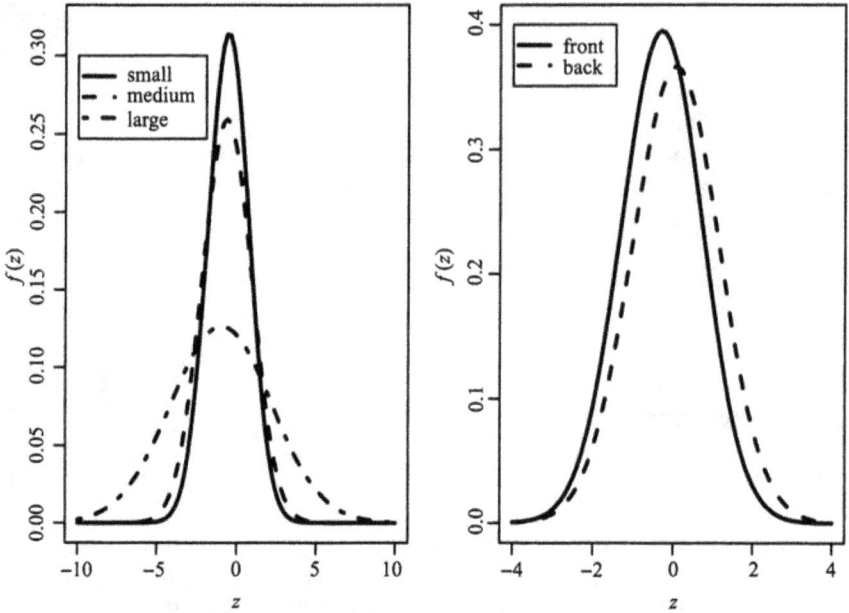

Figure 4.1   Empirical null densities of the AYP study and the brain imaging study. The null density of the large group is much wider than those of the other two densities. In the right panel, the null densities of the front and back halves of the brain are $N(0.06, 1.09^2)$ and $N(-0.29, 1.01^2)$, respectively, which are centered at different means.

4.2   The multiple-group model

The multiple-group random mixture model (Efron 2008a; see Figure 4.2) extends the previous random mixture model (3.1) (for a single group) to cover the situation where the m cases can be divided into K groups. It is assumed that within each group, the random mixture model (3.1) holds separately.

Let $g = (g_1, \ldots, g_K)$ be a multinomial variable with probabilities $\{\pi_1, \ldots, \pi_K\}$, where $g_i = k$ indicates that case i belongs to group k. We assume that prior to analysis, the group labels g have been determined by external information derived from other data or *a priori* knowledge. Let $\theta = (\theta_1, \ldots, \theta_m)$ be Bernoulli variables, where $\theta_i = 1$ indicates that case i is a non-null and $\theta_i = 0$ otherwise. Given g, θ can be grouped as $\theta = (\theta_1, \ldots, \theta_K) = \{(\theta_{k1}, \ldots, \theta_{km_k}) : k = 1, \ldots, K\}$, where m_k is the number of hypotheses in group k. Different from g, θ are unknown and need to be inferred from observations x. Let $\theta_{ki}, i = 1, \ldots, m_k$, be independent Bernoulli (p_k) variables and $X = (X_{ki})$ be generated conditional on θ:

$$X_{ki}|\theta_{ki} \sim (1 - \theta_{ki})F_{k0} + \theta_{ki}F_{k1}, \ i = 1, \ldots, m_k, \ k = 1, \ldots, K. \quad (4.1)$$

Hence within group k, the X_{ki}'s, $i = 1, \ldots, m_k$, are i.i.d. observations with mixture distribution $F_k = (1 - p_k)F_{k0} + p_k F_{k1}$. Denote by f_k the mixture density of group k, the null and non-null densities by f_{k0} and f_{k1}, respectively. Then $f_k = (1 - p_k)f_{k0} + p_k f_{k1}$.

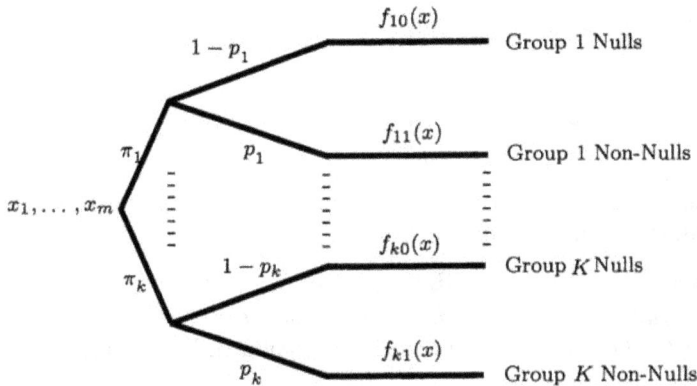

Figure 4.2   The multiple group model: the m hypotheses are divided into K groups with prior probability π_k; the random mixture model (3.1) holds separately within each group, with possibly different p_k, f_{k0} and f_{k1}.

4.3   Conventional FDR procedures

We first consider the problem in an ideal setting where all distributional information is assumed to be known. This section considers two conventional FDR approaches: pooled and separate analyses.

Pooled FDR analysis. A naive approach to testing grouped hypotheses is to simply ignore the group labels and combine all cases into a pooled sample. Denote by f the mixture density,

$$f = \sum_k \pi_k[(1-p_k)f_{k0} + p_k f_{k1}] = (1-p)f_0^* + pf_1^*,$$

where $p = \sum_k \pi_k p_k$ is the non-null proportion, $f_0^* = \sum_k [(\pi_k - \pi_k p_k)/(1-p)]f_{k0}$ and $f_1^* = \sum_k (\pi_k p_k/p) f_{k1}$ are the pooled or global null and non-null densities, respectively. Denote the pooled null distribution by $F_0^* = \sum_k [(\pi_k - \pi_k p_k)/(1-p)]F_{k0}$. In a pooled analysis, the group labels are ignored and one tests against the common pooled null distribution F_0^* in all individual tests. Define the pooled Lfdr statistic (PLfdr) by

$$\mathrm{PLfdr}(x_i) = \frac{(1-p)f_0^*(x_i)}{f(x_i)}, \quad i = 1, \ldots, m. \tag{4.2}$$

The results in Section 3 imply that among all testing procedures that adopt the pooled-analysis strategy, the optimal one is

$$\boldsymbol{\delta}(\mathbf{PLfdr}, c_{OR}(\alpha)\mathbf{1}) = [I\{\mathrm{PLfdr}(x_i) < c_{OR}(\alpha)\} : i = 1, \ldots, m], \tag{4.3}$$

where $c_{OR}(\alpha)$ is the largest cutoff for the PLfdr statistic that controls the overall FDR at level α. Let $\mathrm{PLfdr}_{(1)}, \ldots, \mathrm{PLfdr}_{(m)}$ be the ranked PLfdr values and $H_{(1)}, \ldots, H_{(m)}$ the corresponding hypotheses. An asymptotically equivalent version of (4.3) is the *PLfdr procedure*:

$$\text{Reject all } H_{(i)}, i = 1,\ldots,l, \quad \text{where } l = \max\left\{i : (1/i)\sum_{j=1}^{i}\text{PLfdr}_{(j)} \leqslant \alpha\right\}. \quad (4.4)$$

The following result shows that the PLfdr procedure is valid for FDR control when testing against the pooled null distribution F_0^*.

Theorem 4.1. *Consider the mixture model (4.1). Let $PLfdr_{(i)}, i = 1,\ldots,m$ be the ranked PLfdr values defined in (4.2). Then the PLfdr procedure (4.4) controls the FDR at level α when testing against the pooled null distribution F_0^*.*

We should emphasize here that a pooled analysis makes sense only when the null distributions F_{k0} are the same for all groups, in which case F_0^* coincides with the common group null. When F_{k0} are different across groups, in general the pooled null distribution F_0^* differs from any of the group null F_{k0}. In this case a pooled analysis is not appropriate at all because for each individual case a rejection against F_0^* does not imply rejection against a null distribution F_{k0} for a given group. To further illustrate this important point, let us take the most extreme case. Consider two groups where the null distribution of the first group is the alternative distribution of the second, and vice versa. It is then impossible to decide whether a case is a null or non-null without knowing the grouping information. In this case F_0^* is not the right null distribution to test against for any individual tests and therefore it is entirely inappropriate to perform a pooled analysis.

Separate FDR analysis. Another natural approach to testing grouped hypotheses is the *separate analysis* where each group is analyzed separately at the same FDR level α. Define the conditional Lfdr for group k as

$$\text{CLfdr}^k(x_{ki}) = \frac{(1 - p_k)f_{k0}(x_{ki})}{f_k(x_{ki})}, i = 1,\ldots,m_k; \ k = 1,\ldots,K. \quad (4.5)$$

Again implied by the results in Sun and Cai (2007), the optimal procedure for testing hypotheses from group k is of the form

$$\delta^k(\mathbf{CLfdr}^k, c_{OR}^k(\alpha)\mathbf{1}) = [I\{\text{CLfdr}^k(x_{ki}) < c_{OR}^k(\alpha)\} : i = 1,\ldots,m_k], k = 1,\ldots,K, \quad (4.6)$$

where $c_{OR}^k(\alpha)$ is the largest cutoff for CLfdr statistic that controls the FDR of group k at level α. By combining testing results from separate groups together, we have $\delta = (\delta^1,\ldots,\delta^K)$.

Similarly we can propose the separated Lfdr *(SLfdr) procedure* that is asymptotically equivalent to (4.6). Denote by $\text{CLfdr}_{(1)}^k,\ldots,\text{CLfdr}_{(m_k)}^k$ the ranked CLfdr values in group k and $H_{(1)}^k,\ldots,H_{(m_k)}^k$ the corresponding hypotheses. The testing procedure for group k is:

$$\text{Reject all } H_{(i)}^k, i = 1,\ldots,l_k, \quad \text{where } l_k = \max\left\{i : (1/i)\sum_{j=1}^{i}\text{CLfdr}_{(j)}^k \leqslant \alpha\right\}. \quad (4.7)$$

The final rejection set of the SLfdr procedure is obtained by combining the K rejection sets from all separate analyses: $\mathcal{R}_{\text{SLfdr}} = \cup_{k=1}^{K} \{H_{(i)}^k : i = 1, \ldots, l_k\}$. The next theorem shows that the SLfdr procedure is also valid for global FDR control.

Theorem 4.2. *Consider the random mixture model (4.1). Let $CLfdr_{(i)}^k, i = 1, \ldots, m_k$, $k = 1, \ldots, K$, be the ranked CLfdr values defined by (4.5) for group k. Then the SLfdr procedure (4.7) controls the global FDR at level α.*

4.4   Optimal FDR procedures for grouped tests

The pooled and separate analyses are inefficient in reducing the overall FNR. In this section, we begin by considering an ideal setting where all distributional information is known and propose an optimal (oracle) FDR procedure that uniformly outperforms both the pooled and separate procedures. We then turn to the situation where the distributions are unknown and introduce a data-driven procedure that is asymptotically valid and optimal.

Consider a weighted classification problem with loss function

$$L(\boldsymbol{\theta}, \boldsymbol{\delta}) = (1/m) \sum_{k=1}^{K} \sum_{i=1}^{m_k} \lambda(1 - \theta_{ki})\delta_{ki} + \theta_{ki}(1 - \delta_{ki}). \tag{4.8}$$

The goal in a weighted classification problem is to find $\boldsymbol{\delta} \in \{0, 1\}^m$ that minimizes the classification risk $E[L_\lambda(\boldsymbol{\theta}, \boldsymbol{\delta})]$. Cai and Sun (2009) showed that the multiple testing and weighted classification problems are "equivalent" under mild conditions for model (4.1). Consider an ideal setting where an oracle knows p_k, f_{k0} and f_{k1}, $k = 1, \ldots, K$. The optimal classification rule is given by the next theorem.

Theorem 4.3. *Consider the random mixture model (4.1). Suppose p_k, f_{k0}, f_{k1} are known. Then the classification risk with loss function (4.8) is minimized by $\boldsymbol{\delta}^\lambda = (\delta_{ki})$, where*

$$\delta_{ki} = I\left\{ \Lambda^k(x_{ki}) = \frac{(1 - p_k)f_{k0}(x_{ki})}{p_k f_{k1}(x_{ki})} < \frac{1}{\lambda} \right\}. \tag{4.9}$$

Note that $\Lambda^k(x) = \text{CLfdr}^k(x)/[1 - \text{CLfdr}^k(x)]$ is strictly increasing in $\text{CLfdr}^k(x)$, where $\text{CLfdr}^k(x)$ is the conditional local false discovery rate defined in (4.5), an equivalent optimal test statistic is $\textbf{CLfdr} = [\text{CLfdr}^k(x_{ki}) : i = 1, \ldots, m_k, k = 1, \ldots, K]$. Therefore the optimal testing procedure is of the form $\boldsymbol{\delta}[\textbf{CLfdr} < c(\alpha)]$. The MRC implies the cutoff should be chosen as $c_{OR}(\alpha) = \sup\{c \in (0, 1) : \text{mFDR}(c) \leqslant \alpha\}$. Therefore the optimal (oracle) procedure for multiple group hypothesis testing is the following *CLfdr oracle procedure:*

$$\boldsymbol{\delta}[\textbf{CLfdr}, c_{OR}(\alpha)\textbf{1}] = [I\{\text{CLfdr}^k(x_{ki}) < c_{OR}(\alpha)\} : i = 1, \ldots, m_k, k = 1, \ldots, K], \tag{4.10}$$

Note that different from (4.6), the oracle procedure (4.10) suggests using a universal cutoff for all CLfdr statistics regardless of their group identities.

For a given FDR level, it is difficult to calculate the optimal cutoff $c_{OR}(\alpha)$ directly. Also, the CLfdr oracle procedure requires the distributional information of all individual groups, which is usually unknown in practice. Cai and Sun (2009) derived the following CLfdr procedure that is asymptotically equivalent to the oracle procedure (4.10). Let $\hat{p}_k$, $\hat{f}_{k0}$ and $\hat{f}_k$ be estimates obtained for separate groups. The CLfdr procedure involves the following three steps:

1. Calculate the plug-in CLfdr statistic $\widehat{\text{CLfdr}}^k(x_{ki}) = (1 - \hat{p}_k)\hat{f}_{k0}(x_{ki})/\hat{f}_k(x_{ki})$.
2. Combine and rank the plug-in CLfdr values from all groups. Denote by $\widehat{\text{CLfdr}}_{(1)}, \ldots, \widehat{\text{CLfdr}}_{(m)}$ the ranked values and $H_{(1)}, \cdots, H_{(m)}$ the corresponding hypotheses.
3. Reject all $H_{(i)}$, $i = 1, \ldots, l$, where $l = \max\left\{i : (1/i)\sum_{j=1}^{i} \widehat{\text{CLfdr}}_{(j)} \leqslant \alpha\right\}$.

The next theorem shows that the data-driven procedure is *asymptotically valid and optimal* in the sense that both the FDR and FNR levels of the oracle procedure are asymptotically achieved by the data-driven procedure.

Theorem 4.4. (Cai and Sun 2009). *Consider the multiple group model* (4.1). *Let* $\hat{p}_k$, $\hat{f}_{k0}$ *and* $\hat{f}_k$ *be consistent estimates of* p_k, f_{k0} *and* f_k *such that* $\hat{p}_k \xrightarrow{P} p_k$, $E\|\hat{f}_{k0} - f_{k0}\|^2 \to 0$, $E\|\hat{f}_k - f_k\|^2 \to 0$, $k = 1, \ldots, K$. *Let*

$$\widehat{CLfdr}^k(x_{ki}) = (1 - \hat{p})\hat{f}_0(x_{ki})/\hat{f}(x_{ki}),$$

for $i = 1, \ldots, m_k, k = 1, \ldots, K$. *Combine all test statistics from separate groups and let* $\widehat{CLfdr}_{(1)}, \ldots, \widehat{CLfdr}_{(m)}$ *be the ranked values. Then*

(i). *The FDR and FNR levels of the data-driven procedure are respectively* $\alpha + o(1)$ *and* $FNR_{OR} + o(1)$, *where* FNR_{OR} *is the FNR level of the oracle procedure* (4.10).

(ii). *The FDR level of the data driven procedure in group* k *can be consistently estimated as* $\widehat{FDR}^k = (1/R_k)\sum_{i=1}^{R_k} \widehat{CLfdr}_{(i)}^k$. *In addition,* $\widehat{FDR}^k = FDR_{OR}^k + o(1)$, *where* $FDR_{OR}^k + o(1)$ *is the FDR level of the oracle procedure* (4.10) *in group* k.

It is important to note that in Step 1, the external information of group labels is utilized to calculate the CLfdr statistic; this is the feature from a separate analysis. However, in Steps 2 and 3, the group labels are dropped and the rankings of all hypotheses are determined globally; this is the feature from a pooled analysis. Therefore the CLfdr procedure is a hybrid strategy that enjoys features from both pooled and separate analyses.

Unlike for the separate analysis, the group-wise FDR levels of the CLfdr procedure are in general different from α. In addition to its validity, one may be interested in knowing the actual group-wise FDR levels FDR^k yielded by the CLfdr procedure; this can be conveniently obtained based on the quantities that we have already calculated. Specifically, let R_k be the number of rejections in group k. The actual FDR^k's can be consistently estimated by $\widehat{FDR}^k = \frac{1}{R_k}\sum_{i=1}^{R_k} \text{CLfdr}_{(i)}^k$.

4.5   Simulation studies

Consider the following two-group normal mixture model:

$$X_{ki} \sim (1 - p_k)N(\mu_{k0}, \sigma_{k0}^2) + p_k N(\mu_k, \sigma_k^2), \quad k = 1, 2. \tag{4.11}$$

The numerical performances of the PLfdr, SLfdr and CLfdr procedures are investigated in the next simulation study. The nominal global FDR level is 0.10.

Example 4.5. The null distributions of both groups are fixed as $N(0, 1)$. Three simulation settings are considered: (i) The group sizes are $m_1 = 3000$ and $m_2 = 1500$; the group mixture pdf's are $f_1 = (1 - p_1)N(0, 1) + p_1 N(-2, 1)$ and $f_2 = 0.9N(0, 1) + 0.1N(4, 1)$. We vary p_1, the proportion of non-nulls in group 1, and plot the FDR and FNR levels as functions of p_1. (ii) The groups sizes are also $m_1 = 3000$ and $m_2 = 1500$; the group mixture pdf's are $f_1 = 0.8N(0, 1) + 0.2N(\mu_1, 1)$ and $f_2 = 0.9N(0, 1) + 0.1N(2, 0.5^2)$. The FDR and FNR levels are plotted as functions of μ_1. (iii) The marginal pdf's are $f_1 = 0.8N(0, 1) + 0.2N(-2, 0.5^2)$ and $f_2 = 0.9N(0, 1) + 0.1N(4, 1)$. The sample size of group 2 is fixed at $m_2 = 1500$, the FDR and FNR levels are plotted as functions of m_1. The simulation results with 500 replications are given in Figure 4.3. The top row compares the actual FDR levels of the three procedures; the results for setting (i), (ii) and (iii) are shown in Panels (a), (b) and (c), respectively. The group-wise FDR levels of the CLfdr procedure are also provided (the dashed line for group 1 and dotted line for group 2). The bottom row compares the FNR levels of the three procedures; the results for setting (i), (ii) and (iii) are shown in Panels (d), (e) and (f), respectively. □

We can see that all three procedures control the global FDR level at the nominal level 0.10, indicating that all three procedures are valid. It is important to note that the CLfdr procedure chooses group-wise FDR levels automatically (dashed and dotted lines in Panels (a)-(c)), and the levels are in general different from the nominal level 0.10. The relative efficiency of PLfdr versus SLfdr is inconclusive (depends on simulation settings). For example, the SLfdr procedure yields lower FNR levels in Panel (d), but higher FNR levels in Panel (f). However, all simulations show that both the PLfdr and SLfdr procedures are uniformly dominated by the CLfdr procedure.

4.6   A case study

We now return to the adequate yearly progress (AYP) study mentioned in the introduction. In this section, we analyze the data collected from $m = 7867$ of California high schools (Rogosa 2003) by using the PLfdr, SLfdr and CLfdr procedures.

One goal of the AYP study is to compare the success rates in Math exams of social-economically advantaged (SEA) versus social-economically disadvantaged (SED) students. Since the average success rates of the SEA students are in general (7370 out of 7867 schools) higher that the SED students, it is of interest to identify a subset of schools in which the advantaged-disadvantaged performance differences are unusually small or large. Denote by X_i and Y_i the success rates, and n_i and n_i' the numbers of scores reported for SEA and SED students in school i,

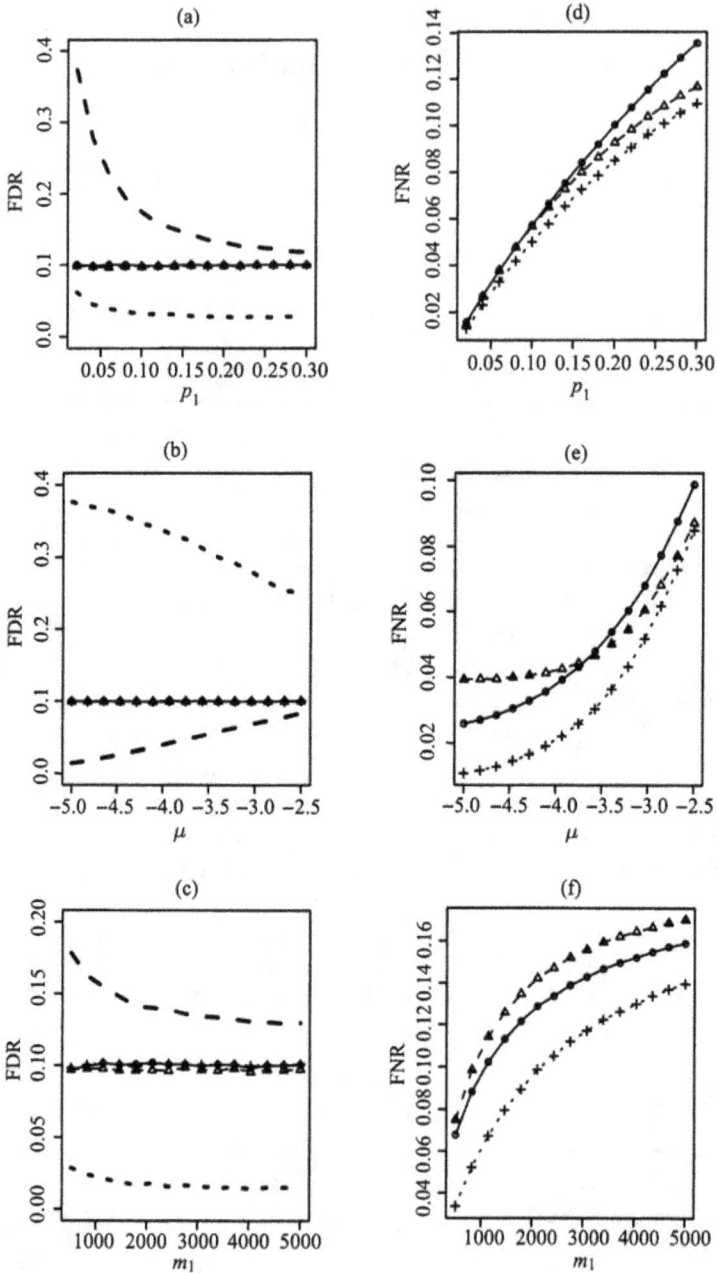

Figure 4.3   Results for Simulation Study 4.5: the top row compares the FDR levels and the bottom row compares the FNR levels (○, PLfdr; △, SLfdr; +, CLfdr). The optimal group-wise FDR levels suggested by the CLfdr procedure are provided together with the global FDR levels (dashed line, Group1; dotted line, Group2).

$i = 1, \ldots, m$. Define the centering constant $\Delta = \text{median}(X_i) - \text{median}(Y_i)$. A z-value for comparing the SEA students versus the SED students can be computed for each school:

$$z_i = \frac{X_i - Y_i - \Delta}{\sqrt{X_i(1 - X_i)/n_i + Y_i(1 - Y_i)/n_i'}}, \tag{4.12}$$

for $i = 1, \ldots, m$. We claim school i is "interesting" if the observed $|z_i|$ is large.

The AYP data has been analyzed by Efron (2007 and 2008b), where he first estimated the global null density $\hat{f}_0$, then searched for interesting cases in the tail areas of $\hat{f}_0$. This pooled-analysis strategy ignores the fact that the hypotheses formed for different schools are not exchangeable. In particular, the number of scores reported by each school varies from less than a hundred to more than ten thousands. A pooled analysis tends to over-select too many large schools, which often express themselves as "very significant" in the tail areas due to small denominators in (4.12). In contrast, small schools are likely to be hidden in the central area of $\hat{f}_0$ and appear "uninteresting". This is not desirable because, in practice, investigators are interested in identifying significant differences from all schools, not only from large schools. As we shall see, an important feature of the AYP data is that the empirical null distributions of the z-values are substantially different for small and large schools, therefore a pooled analysis is inappropriate and one should perform a separate analysis to take into account the effect of school size. Based on a preliminary cluster analysis, we divide all schools into three groups according to the number of scores reported ($n_i + n_i'$): small schools ($n_i + n_i' \leqslant 120$), medium schools ($120 < n_i + n_i' \leqslant 900$) and large schools ($n_i + n_i' > 900$). The group characteristics are summarized in Table 4.1, where the empirical null distributions are estimated using Jin and Cai (2007)'s method. Note that the variance of the empirical null distribution for the scores from the large schools is more than four times than those for the scores from the other two groups. See also Figure 4.1 in the introduction.

Table 4.1   Group characteristics in the AYP data: 7867 schools in total. The global null density is $\hat{f}_0 = N(-0.59, 1.59^2)$

Group	Group Definition	Group Size	Proportion	Empirical Null
Small	$n_i + n_i' \leqslant 120$	516	6.6%	$\hat{f}_{10} = N(-0.51, 1.27^2)$
Medium	$120 < n_i + n_i' \leqslant 900$	6514	80.6%	$\hat{f}_{20} = N(-0.61, 1.54^2)$
Large	$n_i + n_i' > 900$	837	12.8%	$\hat{f}_{30} = N(-0.95, 3.16^2)$

We then apply the PLfdr, SLfdr and CLfdr procedures to the AYP data at different FDR levels. The PLfdr procedure claims the most discoveries, followed by the CLfdr and then SLfdr procedure. It is important to emphasize that the PLfdr procedure is inappropriate here because the pooled null distribution is not the correct null to test against. The PLfdr procedure is too liberal for the large group yet too conservative for the small group: around 50%-70% significant schools come from the large group, although its population proportion is only 13%; in contrast, only around 1% interesting cases come from the small group, although its population proportion is more than 6%. The SLfdr procedure considers the groups separately; large schools are no longer over-selected and more small schools are identified. The CLfdr procedure further improves the SLfdr procedure by efficiently exploiting the important grouping information and weighting the numbers of discoveries among groups. The optimal group-wise FDR levels estimated by the CLfdr procedure at different nominal FDR levels are plotted in Figure 4.4, suggesting that we should choose higher FDR levels for the medium group and lower FDR level for the large group. Note that the SLfdr procedure uses the same FDR level for all groups, the CLfdr procedure usually identifies more cases from the medium group, but fewer cases from the large group.

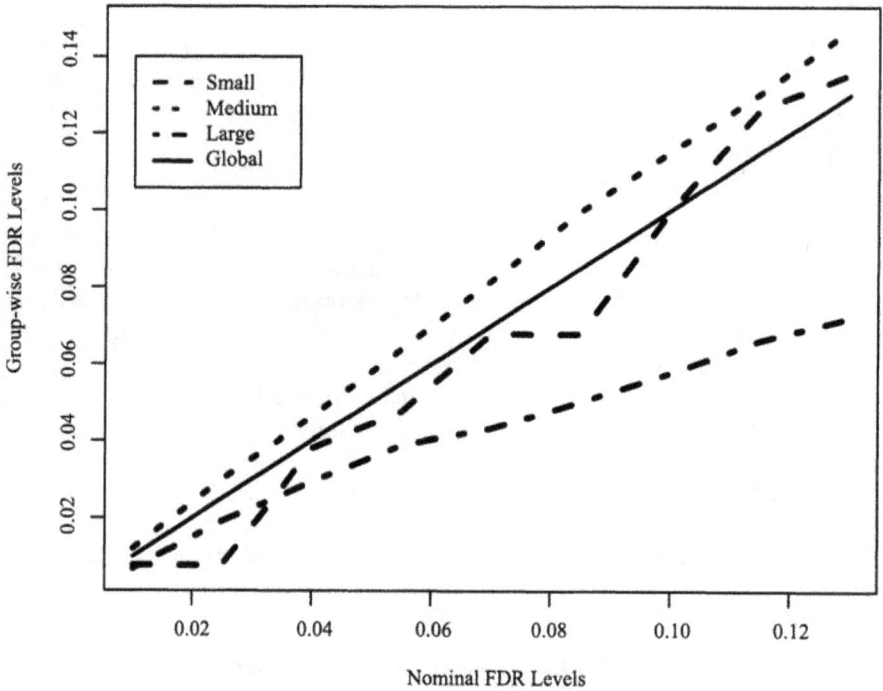

Figure 4.4   AYP study. Optimal group-wise FDR levels estimated by the CLfdr procedure.

5   Large-scale multiple testing under dependence

Observations arising from large scale multiple comparison problems are often dependent. For example, in microarray experiments, different genes may cluster into groups along biological pathways and exhibit high correlation. In public health surveillance studies, the observed data from different time periods and locations are often serially or spatially correlated. Correlation has big effects on a multiple testing procedure. Finner and Roters (2002) and Owen (2005) showed that both the expectation and variance of the number of Type I errors are greatly affected by the correlation among the hypotheses. Qiu et al. (2005) noted that the correlation effects can substantially deteriorate the performance of many FDR procedures. The correlation effects on the z-value null distribution is studied by Efron (2007), who suggested that an adjusted FDR estimate should be combined with the use of an Lfdr procedure to remove the bias caused by the correlation. Nevertheless, the works by Benjamini and Yekutieli (2001), Farcomeni (2006) and Wu (2009) show that the FDR is controlled at the nominal level by the BH step-up and adaptive p-value procedure under different dependence assumptions, supporting the "do nothing" approach.

Among the suggestions with respect to the correlation effects on an FDR procedure, the validity issue is overemphasized, and the efficiency issue is ignored. The FDR procedures developed under the independence assumption, even valid, may suffer from substantial efficiency loss when the dependence structure is highly informative. These situations include the geographical disease mapping studies, multiple-stage clinical trials, functional Magnetic Resonance Imaging analyses and comparative microarray experiments, where the non-null cases are often structured in some way, e.g., correlated temporally, spatially or functionally. Benjamini and Heller (2007) and Genovese et al. (2005) suggested incorporating scientific or spatial information into a multiple testing procedure to improve the efficiency. However, their approaches essentially rely on prior information, such as well defined clusters or prespecified weights, and the correlation structure among the hypotheses is not modeled.

We study multiple testing under dependency in a compound decision-theoretic framework. An important dependency structure, the hidden Markov model (HMM), is considered. The HMM is an effective tool for modeling the dependency structure and has been widely used in areas such as speech recognition, signal processing (Rabiner 1989; Ephraim and Merhav 2002). Also see Churchill (1992), Krogh et al. (1994) for its applications in analyzing biological sequences and processes.

In this section, we first propose an oracle testing procedure in an ideal setting where the HMM parameters are assumed to be known. Under mild conditions, the oracle procedure is shown to be optimal in the sense that it minimizes the FNR subject to a constraint on the FDR. This approach is distinguished from the conventional methods in that the proposed procedure is built on a new test statistic (local index of significance, LIS) instead of the p-values. Unlike p-values, the LIS takes into account the observations in adjacent locations by exploiting the local dependency structure in the HMM. The precision of individual tests is hence improved by utilizing the dependency information.

We then introduce a data-driven procedure that mimics the oracle procedure by plugging in consistent estimates of the unknown HMM parameters. The data-driven procedure is shown to be *asymptotically optimal* in the sense that it attains both the FDR and FNR levels of the oracle procedure asymptotically. Simulation studies conducted in Section 5.4 indicate the favorable performance of the LIS procedure. Our findings show that the correlation among hypotheses is highly informative in simultaneous inference and can be exploited to construct more efficient testing procedures. The LIS procedure is illustrated in analyzing the SNP data from a genome-wide association study of T1D disease in Section 5.5.

5.1   The hidden Markov model

Let $\boldsymbol{\theta} = (\theta)_1^m = (\theta_1, \cdots, \theta_m)$ be a sequence of Bernoulli random variables and distributed as a stationary Markov chain, where $\theta_i = 1$ indicates that case i is a non-null and $\theta_i = 0$ otherwise. Assume that observations $\boldsymbol{x} = (x_1, \ldots, x_m)$ are generated according to the following conditional probability model:

$$P(\boldsymbol{x}|\boldsymbol{\theta}, \mathcal{F}) = \prod_{i=1}^m P(x_i|\theta_i, \mathcal{F}), \tag{5.1}$$

where $P(x_i < x|\theta_i = j) = F_j(x)$, $j = 0, 1$ and $\mathcal{F} = (F_0, F_1)$. Denote by f_0 and f_1 the corresponding pdf's. The HMM can be illustrated in Figure 5.1.

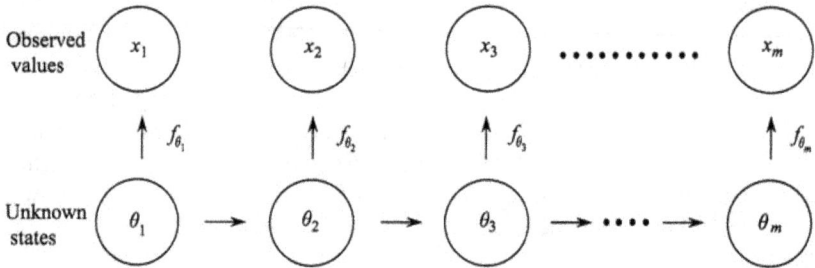

Figure 5.1   Graphical representation of an HMM.

We assume that the Markov chain $(\theta_i)_1^m = (\theta_1, \ldots, \theta_m)$ is stationary, irreducible and aperiodic. Specifically, the transition probabilities are homogeneous and bounded away from 0 and 1. That is, $a_{jk} = P(\theta_i = k|\theta_{i-1} = j), 0 \leqslant j, k \leqslant 1$, do not depend on i, with the standard stochastic constraints $0 < a_{jk} < 1, a_{j0} + a_{j1} = 1$. The convergence theorem of a Markov chain implies that $(1/m)\sum_{i=1}^m I(\theta_i = j) \to \pi_j$ almost surely as $m \to \infty$. The Bernoulli variables $\theta_1, \ldots, \theta_m$ are identically distributed (but correlated) with $P(\theta_i = j) = \pi_j$. Denote by $\mathcal{A} = \{a_{jk}\}$ the transition matrix, $\boldsymbol{\pi} = (\pi_0, \pi_1)$ the stationary distribution, $\mathcal{F} = \{F_0, F_1\}$ the observation distribution, and $\vartheta = (\mathcal{A}, \boldsymbol{\pi}, \mathcal{F})$ the collection of all HMM parameters.

5.2   The oracle procedure

Let $\boldsymbol{\delta} \in \{0, 1\}^m$ be a general decision rule defined as before. Sun and Cai (2009) showed that under the HMM dependency and a monotone ratio condition, the multiple testing problem is equivalent to a weighted classification problem with loss function

$$L_\lambda(\boldsymbol{\theta}, \boldsymbol{\delta}) = \frac{1}{m}\sum_i [\lambda(1 - \theta_i)\delta_i + \theta_i(1 - \delta_i)]. \tag{5.2}$$

It can be shown that the optimal solution to the weighted classification problem is $\boldsymbol{\delta}(\boldsymbol{\Lambda}, 1/\lambda) = (\delta_1, \ldots, \delta_m)$, where

$$\Lambda_i(\boldsymbol{x}) = \frac{P_\vartheta(\theta_i = 0|\boldsymbol{x})}{P_\vartheta(\theta_i = 1|\boldsymbol{x})} \tag{5.3}$$

and $\delta_i = I\{\Lambda_i(\boldsymbol{x}) < 1/\lambda\}$ for $i = 1, \ldots, m$.

Remark 5.1. Given ϑ, the oracle classification statistic $\Lambda_i(\boldsymbol{x})$ can be expressed in terms of the forward and backward density variables, which are defined as $\alpha_i(j) = f_\vartheta[(x_t)_1^i, \theta_i = j]$ and $\beta_i(j) = f_\vartheta[(x_t)_{i+1}^m | \theta_i = j]$, respectively (note that the dependence of $\alpha_i(j)$ on $(x_t)_1^i$ has been suppressed, similarly for $\beta_i(j)$). It can be shown that $P_\vartheta(\boldsymbol{x}, \theta_i = j) = \alpha_i(j)\beta_i(j)$ and hence $\Lambda_i(\boldsymbol{x}) = [\alpha_i(0)\beta_i(0)]/[\alpha_i(1)\beta_i(1)]$. The forward variable $\alpha_i(j)$ and backward variable $\beta_i(j)$ can be calculated recursively using the *forward-backward procedure* (Baum et al. 1970 and Rabiner 1989). Specifically, we initialize $\alpha_1(j) = \pi_j f_j(x_1)$, $\beta_m(j) = 1$, then by induction we have $\alpha_{i+1}(j) = \left[\sum_{k=0}^1 \alpha_i(k)a_{kj}\right] f_j(x_{i+1})$ and $\beta_i(j) = \sum_{k=0}^1 a_{jk} f_k(x_{i+1})\beta_{i+1}(k)$.

Since $\Lambda_i(\boldsymbol{x})$ is increasing in $P_\vartheta(\theta_i = 0|\boldsymbol{x})$, an optimal multiple-testing rule in an HMM can be written in the form of $\boldsymbol{\delta} = [I\{P_\vartheta(\theta_i = 0|\boldsymbol{x}) < t\} : i = 1, \ldots, m]$. Define the *local index of significance* (LIS) for hypothesis i by

$$\mathrm{LIS}_i = P_\vartheta(\theta_i = 0|\boldsymbol{x}). \tag{5.4}$$

The LIS depends only on x_i and reduces to Efron's local false discovery rate (Lfdr) in the independent case, i.e., $\mathrm{LIS}_i(\boldsymbol{x})$ simplifies to $\mathrm{Lfdr}(x_i) = (1 - \pi)f_0(x_i)/f(x_i)$, where π is the proportion of non-nulls and f is the marginal pdf. The oracle testing procedure is

$$\boldsymbol{\delta} = [I\{\mathbf{LIS} < c_{OR}\mathbf{1}\} : i = 1, \ldots, m],$$

where c_{OR} is the largest cutoff for LIS that controls the FDR at level α.

Remark 5.2. It is important to note that the conventional testing procedures essentially involve ranking and thresholding p-values, whereas under our framework the optimal statistic is the LIS. Now we compare p-value and LIS from a compound decision theoretic view. Let $\boldsymbol{\delta}$ be a general decision rule, then $\boldsymbol{\delta}$ is *symmetric* if $\boldsymbol{\delta}(T(\boldsymbol{x})) = T(\boldsymbol{\delta}(\boldsymbol{x}))$ for all permutation operators T. In situations where one expects the non-null hypotheses appear in clusters, it is natural to treat differently a hypothesis surrounded by non-nulls from one surrounded by nulls. However, these two hypotheses are exchangeable when a symmetric rule is applied. The FDR procedures that threshold the p-value or Lfdr are symmetric rules; so are not desirable when hypotheses are correlated. In contrast, we consider decision rule $\boldsymbol{\delta}(\mathrm{LIS}, \lambda) = \{I(\mathrm{LIS}_i(\boldsymbol{x}) < \lambda) : i = 1, \ldots, m\}$. It is easy to see that $\boldsymbol{\delta}(\mathrm{LIS}, \lambda)$ is asymmetric, and the order of the sequence $(x_i)_1^m$ is accounted for in deciding the significance level of hypothesis i. In particular, the local dependency structure is captured by the HMM, and the operation of the forward-backward procedure implies that a large (small) observation will increase (decrease) the significance level of its neighbors. The performance of the testing procedure is hence improved by pooling information from adjacent locations. In addition, the signal to noise ratio is increased since the information from the whole sequence is integrated to calculate the LIS value of a single hypothesis. Therefore, the LIS is more robust against local disturbance, which further increases the efficiency of our testing procedure.

5.3   A data-driven procedure for dependent tests in an HMM

The oracle procedure is difficult to implement since c_{OR} is difficult to calculate. In addition, the HMM parameters ϑ are usually unknown. Sun and Cai (2009) derived a data-driven procedure that mimics the oralce procedure. We first estimate the unknown quantities by $\hat{\vartheta}$, then plug-in $\hat{\vartheta}$ to obtain $\hat{\text{LIS}}_i$. The maximum likelihood estimate (MLE) is commonly used and is strongly consistent and asymptotically normal under certain regularity conditions (Baum and Petrie, 1966; Leroux, 1992; Bickel et al., 1998). The MLE can be computed using the EM algorithm or other standard numerical optimization schemes, such as the gradient search, or downhill simplex algorithm. These methods are reviewed by Ephraim and Merhav (2002). In many practical applications, the number of components in the non-null mixture L is unknown, yet the information is needed by the algorithms used to maximize the likelihood function. Consistent estimates of L can be obtained using the method proposed by Kiefer (1993) and Liu and Narayan (1994), among others. Alternately, one can use likelihood based criteria, such as Akaike or Bayesian information criterion (BIC) to select the number of components in the normal mixture.

Let $\hat{\vartheta}$ be an estimate of the HMM parameter ϑ. Define the plug-in test statistic $\hat{\text{LIS}}_i(\boldsymbol{x}) = P_{\hat{\vartheta}}(\theta_i = 0|\boldsymbol{x})$. For given $\hat{\vartheta}$, $\hat{\text{LIS}}_i$ can be computed via the forward-backward procedure. Denote by $\hat{\text{LIS}}_{(1)}(\boldsymbol{x}), \ldots, \hat{\text{LIS}}_{(m)}(\boldsymbol{x})$ the ranked plug-in test statistics and $H_{(1)}, \ldots, H_{(m)}$ the corresponding hypotheses. In light of the oracle procedure, we propose the following data-driven procedure:

$$\text{Let } k = \max\left\{ i : \frac{1}{i}\sum_{j=1}^{i} \hat{\text{LIS}}_{(j)}(\boldsymbol{x}) \leqslant \alpha \right\}, \text{ then reject all } H_{(i)}, i = 1, \ldots, k. \quad (5.5)$$

The testing procedure given in (5.5) is referred to as the *LIS procedure*. We shall show that the performance of OR is asymptotically attained by LIS under some standard assumptions on the HMM. The asymptotic properties of the LIS procedure are studied by the following theorems. Theorem 5.3 shows that the rejection sets yielded by OR and LIS are asymptotically equivalent in the sense that the ratio of the number of rejections and the ratio of the number of true positives yielded by the two procedures approach 1 as $m \to \infty$.

Theorem 5.3. *Consider an HMM defined as in (5.1). Let R and $\hat{R}$, V and $\hat{V}$ be the number of rejections and number of false positives yielded by OR and LIS procedures, respectively. Under some regularity conditions (see Sun and Cai 2009), we have $\hat{R}/R \xrightarrow{P} 1$, $\hat{V}/V \xrightarrow{P} 1$.*

Theorem 5.4 below, together with the validity of the oracle procedure, implies that the FDR is controlled at level $\alpha + o(1)$ by LIS, so the LIS procedure is asymptotically valid. Theorem 5.4 also shows that the performance of OR is attained by the LIS procedure asymptotically in the sense that the FNR level yielded by LIS approaches that of OR as $m \to \infty$, therefore the LIS procedure is asymptotically efficient.

Theorem 5.4. *Consider an HMM defined as in* (5.1). *Let FDR$_{OR}$ and FDR$_{LIS}$, FNR$_{OR}$ and FNR$_{LIS}$ be the FDR levels and FNR levels yielded by OR and LIS, respectively. Under some regularity conditions (see Sun and Cai 2009) FDR$_{OR}$ – FDR$_{LIS}$ → 0. In addition, if at least a fixed proportion of hypotheses are not rejected, then FNR$_{OR}$ – FNR$_{LIS}$ → 0 as m → ∞.*

5.4   Simulation studies

We first assume that L, the number of components in non-null mixture, is known or estimated correctly from the data. The situation where L is misspecified is considered in Sun and Cai (2009). In all simulations, we choose the number of hypotheses $m = 3000$ and the number of replications $N = 500$.

Example 5.5. The Markov chain $(\theta_i)_1^m$ is generated with the initial state distribution $\pi^0 = (\pi_0, \pi_1) = (1, 0)$ and transition matrix $\mathcal{A} = [0.95, 0.05; 1 - a_{11}, a_{11}]$. The observations $(x_i)_1^m$ are generated conditional on $(\theta_i)_1^m$: $x_i|\theta_i = 0 \sim N(0, 1)$, $x_i|\theta_i = 1 \sim N(\mu, 1)$. Figure 5.2 compares the performance of BH, AP, OR and LIS. In the top row we choose $\mu = 2$ and plot the FDR, FNR and average number of true positives (ATP) yielded by BH, AP, OR and LIS as functions of a_{11}. In the bottom row, we choose $a_{11} = 0.8$ and plot the FDR, FNR and ATP as functions of μ. The nominal FDR in all simulations is set at level 0.10.

From Panel (a), we can see that the FDR levels of all four procedures are controlled at 0.10 asymptotically, and the BH procedure is conservative. From Panels (b) and (c), we can see that the two lines of the oracle procedure and LIS procedure are almost overlapped, indicating that the performance of the oracle procedure is attained by the LIS procedure asymptotically. In addition, the two p-value based procedures are dominated by the LIS procedure and the difference in FNR and ATP levels becomes larger as a_{11} increases. Note that a_{11} is the transition probability from a non-null case to a non-null case, therefore it controls how likely the non-null cases cluster together. It is interesting to observe that the p-value procedures have higher FNR levels as the non-nulls cluster in larger groups. In contrast, the FNR levels of the LIS procedure decreases as a_{11} increases. This observation shows that if modeled appropriately, the positive dependency is a blessing (the FNR level decreases in a_{11}); but if it is ignored, the positive dependency may become a disadvantage. In situations where the non-null cases are prevented from forming into clusters ($a_{11} < 0.5$), the LIS procedure is still more efficient than BH and AP, although the gain in efficiency is not as much as the situation where $a_{11} > 0.5$.

Panel (d) similarly shows that all procedures are valid and BH is conservative. On Panels (e) and (f), we plot the FNR and ATP levels as functions of the non-null mean μ. We can see that BH and AP are dominated by LIS, and the difference is large when μ is small to moderate. This is due to the fact that the LIS procedure can integrate information from adjacent locations, so is still very efficient even when the signals are weak.

The superiority of LIS is achieved by incorporating the informative dependency structure; hence more efficient rankings of the SNPs are produced. The LIS rankings are fundamentally different from the rankings by BH. Table 5.1 compares

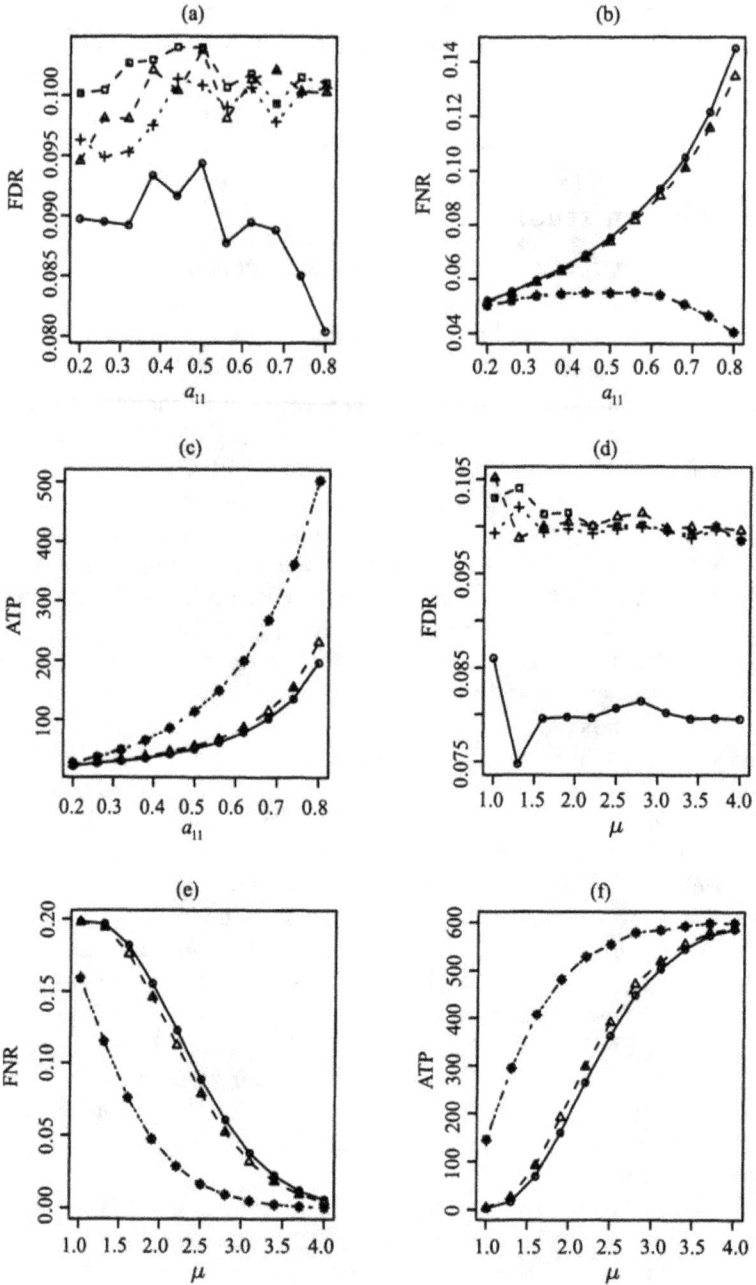

Figure 5.2   A comparison of the BH, AP, OR and LIS in an HMM with simple alternative (○, BH; △, AP; +, OR; □, LIS). (a)-(c) The FDR, FNR and ATP versus a_{11}. (d)-(f) The FDR, FNR and ATP versus μ. The FDR level is set at 0.10.

the outcomes of the LIS and BH procedures for testing two clusters of non-null SNPs, where ○ denotes a null hypothesis or an acceptance and ● denotes a non-null hypothesis or a rejection. It is interesting to note that BH suggests site 2720 be less significant than site 2723. In contrast, LIS suggests that site 2720, being surrounded by significant neighbors, be more significant than site 2723. LIS tends to identify disease-associated SNPs in clusters, while the BH procedure only identifies sporadic suspicious SNPs. Using HMM, LIS efficiently increases the signal to noise ratio by integrating information from adjacent locations. The precision is greatly improved in the sense that (51) the number of false positives is greatly reduced and (52) the statistical power to reject a non-null is substantially increased. This indicates that dependence can make the testing problem "easier" and is a *blessing* if incorporated properly in a testing procedure.

Table 5.1   **Significance levels suggested by BH & LIS**

Sequence	States	p-values	LIS Values	BH *Procedure*	LIS *Procedure*
2694	●	3.62e-01	1.59e-03	○	●
2695	●	1.52e-03	6.85e-05	●	●
2696	●	6.62e-04	9.29e-04	●	●
2697	●	7.19e-01	1.00e-01	○	●
⋮	⋮	⋮	⋮	⋮	⋮
2718	●	1.23e-02	1.65e-04	○	●
2719	●	3.37e-03	8.19e-05	●	●
2720	●	5.59e-01	2.19e-03	○	●
2721	●	2.07e-04	2.67e-05	●	●
2722	●	3.42e-02	7.42e-04	○	●
2723	●	2.88e-01	6.19e-03	○	●

5.5   A case study

Because of the recent advancements in comprehensive genomic information and cost-effective genotyping technologies, genome-wide association studies (GWAS) have become a popular tool to detect genetic variants that contribute to complex diseases. However, the established genetic associations with T1D only explain around 50% of the genetic risk for T1D. Many more genes with small to moderate effects remain to be discovered (Todd et al., 2007 and Hakonarson et al., 2007). Finding these unknown "weak" genes is important for improving the understanding of the pathology of T1D disease.

It has been appreciated that genomic dependency information can significantly improve the efficiency in analysis of large-scale genomic data. We expect that the information of the SNP dependency can be exploited to construct more efficient tests. From a biological point of view, the SNP dependency is informative in constructing more efficient association tests because, when a SNP is associated with a disease, it is likely that the neighboring SNPs are also disease-associated (due to the co-segregation). Therefore, when deciding the significance level of a SNP, the neighboring SNPs should be taken into account. The dependency of adjacent SNPs is captured using an HMM.

To search systematically for these unknown loci, Hakonarson et al. (2007) performed a genome-wide association study, where a discovery cohort, including 563 cases and 1146 controls, was collected. All participants were of European ancestry and recruited through paediatric diabetes clinics in Philadelphia, Montreal, Toronto, Ottawa and Winnipeg. The study subjects were genotyped using the Illumina HumanHap550 BeadChip at the Children's Hospital of Philadelphia (CHOP). A replication cohort, consisting of 483 parents-offspring trios with affected children, was also collected and genotyped (Hakonarson et al., 2007; Grant et al., 2008). A series of standard quality control procedures was performed to eliminate markers with minor allele frequency less than 1%, with hardy-Weinberg Equilibrium p-values lower than 1e-6, or with genotype nocall rate higher than 5%. After the quality control, 534213 markers on 23 chromosomes in the discovery cohort are eligible for further analysis. The same quality control procedure was applied to the replication cohort. Additional markers showing excessive (>families) Mendelian inconsistencies were eliminated. After the screening, 532662 markers over 23 chromosomes in the replication cohort are eligible for further analysis.

We first conducted a χ^2-test for each SNP to assess the association between the allele frequencies and the disease status, then obtain p-values and z-values using appropriate transformations. Each chromosome is modeled separately to obtain chromosome-specific HMM parameters Ψ_k and LIS values $\widehat{\text{LIS}}_{ki}$. We assume that the null distribution is standard normal $N(0,1)$ and the non-null distribution is a normal mixture $\sum_{l=1}^{L} c_l N(\mu_l, \sigma_l^2)$. The number of components L in the non-null distribution is determined by the BIC criterion, $\text{BIC} = \log\{P(\hat{\Psi}_L | z)\} - \frac{|\Psi_L|}{2} \log(m)$, where $P(\Psi_L | z)$ is the likelihood function, $\hat{\Psi}_L$ is the MLE of HMM parameters, and $|\Psi_L|$ is the number of HMM parameters. We vary L and evaluate different choices of L for each chromosome. The high transition probabilities (a_{00} and a_{11}) indicate that the genomic dependency is strong.

Table 5.2   The 7 known T1D susceptibility loci identified by PLIS

Chr.	T1D Loci	SNP	Dist.	D'	LIS Stat.	p Value
1	rs2476601	rs2476601	0		3.19e-09	1.22e-12
6	rs3129871	rs3129871	0		2.42e-89	2.91e-90
16	rs725613	rs725613	0		2.73e-08	4.92e-09
11	rs4244808	rs4320932	8	1	8.40e-05	5.07e-04
12	rs1701704	rs10876864	11	0.955	3.26e-04	4.80e-07
10	rs12251307	rs4147359	15	0.610	8.48e-05	1.55e-04
6	rs3757247	rs10498965	42	1	1.96e-04	8.25e-04

Three loci are identified directly; four loci are identified by nearby SNPs (within 50Kb) in LD with them. Chr., chromosome; Dist., distance of the significant SNPs to the known T1D loci, in Kb; D', disequilibrium scores of the significant SNPs to the known T1D loci, derived from HapMap CEPH Utah (CEU) data.

A meta-analysis based on recent GWAS has confirmed 15 T1D susceptibility loci, among which four are identified by BH and seven are identified by the LIS procedure. Detailed results are provided in Table 5.2. In contrast, LIS does not claim two loci, namely the gene for collagen type 1 a2 (COL1A2; rs10255021) and rs672797, in the vicinity of latrophilin 2 (LPHN2), are disease-associated, while the p-value based approaches even claimed their significant association under the Bonferroni correction. However, these two loci failed to replicate in follow-up studies. A closer look at the nearby regions shows that these two SNPs are both surrounded by insignificant SNPs; hence the "significance" is more likely to be "noise". By borrowing information from nearby locations, LIS successfully classifies them as non-disease associated SNPs.

6   Open problems

We conclude with discussions on some open problems and possible directions for future research in large-scale multiple testing.

It is of great importance from both theoretical and practical perspectives to develop efficient multiple testing procedures under general dependency. The problem is challenging. The asymptotic optimality of a data-driven procedure requires the estimates of the unknown model parameters to be consistent. However, to the best of our knowledge, such theoretical results for other dependency structures, such as for a higher dimensional random field, have not been developed in the literature. Hence the optimality of the LIS procedure may be lost in the estimation step. In addition, the implementation of the data-driven procedure for other correlation structures may be very complicated. The forward-backward procedure and EM algorithm for an HMM is known to be efficient and relatively easy to program. However, such efficient algorithms may not exist for other dependency structures.

Another important direction is to extend the testing procedure in Section 4 to deal with the situation when the grouping information is unknown. One needs to take into account the interplay among grouping, estimation and testing in developing an efficient procedure. On the one hand, if we use many small groups instead of several large groups, the hypotheses within each small group will be more homogeneous, which will increase the precision of a testing procedure. On the other hand, the estimated test statistic will become less precise if a group consists of too few cases; hence the efficiency may be lost in the estimation step as the number of groups increases. A problem of particular interest in this direction is multiple testing with covariates.

The false discovery exceedance (FDX) control is an alternative approach to the FDR control. The false discovery proportion (FDP) is the proportion of false positives among all rejections and the FDX is the tail probability that the FDP exceeds a specified bound (Genovese and Wasserman 2006). In many applications, the FDP in each realization has very large deviations from the nominal FDR level. This unfavorable situation is allowed by FDR controlling procedures. Instead of controlling the average error rate, an FDX procedure aims to control the tail probability of FDP $> \alpha$ below a pre-specified level γ. The FDX procedures are argued to be more appropriate since the variability of FDP is taken into account. The procedures related to the FDX/FDP control are discussed in Pacifico et al.(2004), Lehmann and Romano (2005) and Genovese and Wasserman (2006). However, these procedures are essentially based on thresholding p-values.

We anticipate that more efficient testing procedures can be developed under the compound decision theoretic framework.

Finally, we mention a few other interesting directions that are worth pursuing for future research with related references: multiple testing with weights (Genovese et al 2006; Roeder et al. 2007); conjunction and partial conjunction analysis of sets of hypotheses (Friston et al. 2005; Pacifico et al. 2007; Benjamini and Heller 2007); multiple testing with a hierarchical structure (Meinshausen 2008; Yekutieli 2008; Bickel et al. 2009).

References

[1] Baum L. and Petrie T. (1966). Statistical inference for probalistic functions of finite Markov chains. *Ann. Math. Statis.* **37**, 1554-1563.

[2] Benjamini Y. and Heller R. (2007). False discovery rates for spatial signals. *Journal of the American Statistical Association* **102**, 1272-1281.

[3] Benjamini Y. and Heller R. (2008). Screening for partial conjunction hypotheses. *Biometrics* **64**, 1215-1222.

[4] Benjamini Y. and Hochberg Y. (1995). Controlling the false discovery rate: a practical and powerful approach to multiple testing. *Journal of the Royal Statistical Society Series B* **57**, 289-300.

[5] Benjamini Y. and Hochberg Y. (2000). On the adaptive control of the false discovery rate in multiple testing with independent statistics. *Journal of Educational and behavioral statistics* **25**, 60-83.

[6] Benjamini Y. and Yekutieli D. (2001). The control of false discovery rate in multiple testing under dependency. *The Annals of Statistics* **29**, 1165-1188.

[7] Bickel P., Ritov Y. and Rydèn T. (1998). Asymptotic normality of the maximum likelihood estimator for general hidden markov models. *The Annals of Statistics* **26**, 1614-1635.

[8] Cai T., Jin J. and Low M. (2007). Estimation and confidence sets for sparse normal mixtures. *The Annals of Statistics* **35**, 2421-2449.

[9] Cai T. and Sun W. (2009). Simultaneous testing of grouped hypotheses: finding needles in multiple haystacks. *Journal of the American Statistical Association* **104**, 1467-1481.

[10] Churchill G. (1992). Hidden Markov chains and the analysis of genome structure. *Computers in Chemistry* **16**, 107-115.

[11] Copas J. (1974). On symmetric compound decision rules for Dichotomies. *The Annals of Statistics* **2**, 199-204.

[12] Dudoit S., Shaffer J. and Boldrick J. (2003). Multiple hypothesis testing in microarray experiments. *Statistical Science* **18**, 71-103.

[13] Dudoit S., Yang Y., Callow M. and Speed, T. (2002). Statistical methods for identifying differentially expressed genes in replicated cDNA microarray experiments. *Statistica Sinica* **12**, 111-139.

[14] Efron B., Tibshirani R., Storey J. and Tusher V. (2001). Empirical Bayes analysis of a microarray experiment. *Journal of the American Statistical Association* **96**, 1151-1160.

[15] Efron B. (2004). Large-scale simultaneous hypothesis testing: the choice of a null hypothesis. *Journal of the American Statistical Association* **99**, 96-104.

[16] Efron B. (2007). Correlation and large-scale simultaneous testing. *Journal of the American Statistical Association* **102**,93-103.

[17] Efron B. (2008a). Simultaneous inference: When should hypothesis testing problems be combined? *The Annals of Applied Statistics* **1**, 197-223.

[18] Efron B. (2008b). Microarrays, empirical Bayes and the two-groups model. *Statistical Science* **23**, 1-22.

[19] Ephraim Y. and Merhav N. (2002). Hidden Markov processes. *Invited paper, IEEE transactions on Information Theory* **48**, 1518-1569.

[20] Farcomeni A. (2007). Some results on the control of the false discovery rate under Dependence. *Scandinavian Journal of Statistics*, **34**, 275-297.

[21] Ferkinstad E., Frigessi A., Thorleifsson G. and Kong A. (2008). Unsupervised empirical Bayesian multiple testing with external covariates. *The Annals of Applied Statistics* **2**, 714-735.

[22] Finner H. and Roters M. (2002). Multiple hypotheses testing and expected number of type I errors. *The Annals of Statistics* **30**, 220-238.

[23] Finner H., Dickhaus T. and Roters M. (2009). On the false discovery rate and an asymptotically optimal rejection curve. **37**, 596-618.

[24] Friston K., Penny W. and Glaser D. (2005). Conjunction revisited. *NeuroImage* **25**, 661-667.

[25] Genovese C., Roeder K. and Wasserman L. (2006). False discovery control with p-value weighting. *Biometrika* **93**, 509-524.

[26] Genovese C. and Wasserman L. (2002). Operating characteristic and extensions of the false discovery rate procedure. *Journal of the Royal Statistical Society Sries B* **64**, 499-517.

[27] Genovese C. and Wasserman L. (2004a). A Stochastic process approach to false discovery control. *Annals of Statistics* **32**, 1035-61.

[28] Genovese C. and Wasserman L. (2006). Exceedance control of the false discovery proportion. *Journal of the American Statistical Association* **101**, 1408-1417.

[29] Genovese C., Roeder K., Wasserman L. (2006). False discovery control with p value weighting. *Biometrika* **93**, 509-524.

[30] Hakonarson H., Grant S., Bradfield J. et al. (2007). A genome-wide association study identifies KIAA0350 as a type 1 diabetes gene. *Nature* **448**, 591-594.

[31] Hedenfalk I., Duggen D., Chen Y., et al. (2001). Gene expression profiles in hereditary breast cancer. *New England Journal of Medicine* **344**, 539-548.

[32] Holm S. (1979). A simple sequentially rejective multiple test procedure. *Scandinavian Journal of Statistics* **6**, 65-70.

[33] Hochberg Y. (1988). A sharper Bonferroni procedure for multiple tests of significance. *Biometrika* **75**, 800-803.

[34] Hochberg Y. and Tamhane A. (1987). *Multiple Comparison Procedures*. Wiley, New York.

[35] Jin J. and Cai T. (2007). Estimating the null and the proportion of non-null effects in large-scale multiple comparisons. *Journal of the American Statistical Association* **102**, 495-506.

[36] Kieffer J. (1993). Strongly consistent code-based identification and order estimation for constrained finite-state model classes, *IEEE Transactions on Information Theory* **39**, 893-902.

[37] Krogh A., Brown M., Mian I., Sjölander K. and Haussler D. (1994). Hidden Markov models in computational bilogy applications to protein modeling. *Journal of Molecular Biology* **235**, 1501-1531.

[38] Langaas M., Lindqvist B. and Ferkinstad E. (2005). Estimating the proportion of true null hypotheses, with application to DNA microarray data. *Journal of the Royal Statistical Society , Series B* **67**, 555-572.

[39] Lehmann E. and Romano J. (2005). Generalizations of the familywise error rate. *The Annals of Statistics* **33**, 1138-1154.

[40] Leroux B. (1992). Maximum-likelihood estimation for hidden Markov models. *Stochastic Processes and Their Applications* **40**, 127-143.

[41] Liu C. and Narayan P. (1994). Order estimation and sequential universal data compression of a hidden Markov source by the method of mixtures. *IEEE Transactions on Information Theory* **40**, 1167-1180.

[42] Meinshausen N. and Rice J. (2006). Estimating the proportion of false null hypotheses among a large number of independently tested hypotheses. *The Annals of Statistics* **34**, 373-393.

[43] Miller C., Genovese C., Nichol R., Wasserman L., Connolly A., Reichart D., Hopkins D., Schneider J. and Moore A. (2001). Controlling the false-discovery rate in astrophysical data analysis. *The Astronomical Journal* **122**, 3492-3505.

[44] Newton M., Noueiry A., Sarkar D., and Ahlquist P. (2004). Detecting differential gene expression with a semiparametric hierachical mixture method. *Biostatstics* **5**, 155-176.

[45] Owen A. (2005).Variance of the number of false discoveries. *Journal of the Royal Statistical Society Series B* **67**, 411-426.

[46] Pacifico M., Genovese C., Verdinelli I. and Wasserman L. (2004). False discovery control for random fields. *Journal of the American Statistical Association* **99**, 1002-1014.

[47] Qiu X., Klebanov L. and Yakovlev A. (2005). Correlation between gene expression levels and limitations of the empirical Bayes methodology for finding differentially expressed genes. *Statistical Applications in Genetics and Molecular Biology* **4**, Article 34. Available at: http://www.bepress.com/sagmb/vol4/iss1/art34.

[48] Rabiner L. (1989). A tutorial on hidden Markov models and selected applications in speech recognition. *Proceedings of the IEEE* **77**, 257-286.

[49] Robbins H. (1951).Asymptotically subminimax solutions of compound statistical decision problems. *Proceedings of Second Berkeley Symposium on Mathematical Statistics and Probability.* University of California Press, Berkeley.

[50] Roeder K., Devlin B. and Wasserman L. (2007). Improving power in genome-wide association studies: Weights tip the scale. *Genetic Epidemiology* **31**, 741-747.

[51] Romano J. and Shaikh A. (2006). Step-up procedures for control of generalizations of the familywise error rate. *The Annals of Statistics* **34**, 1850-1873.

[52] Rogasa D. (2003). Accuracy of API index and school base re-

port elements: 2003 Academic Performance Index, California Department of Education. Technical Report, Department of Statistics and School of Education, Stanford University, available at http://www.cde.ca.gov/ta/ac/ap/researchreports.asp.

[53] Rosenthal R. and Rubin D. (1983). Ensemble-adjusted p-values. *Psychological Bulletin* **94**, 540-541.

[54] Sabatti C., Service S. and Freimer N. (2003). False discovery rate in linkage and association genome screens for complex disorders. *Genetics* **164**, 829-833.

[55] Sarkar S. (2002). Some results on false discovery rate in stepwise multiple testing procedures. *Annals of statistics* **30**, 239-257.

[56] Sarkar S. (2006). False discovery and false nondiscovery rates in single-step multiple testing procedures. *Annals of statistics* **34**, 394-415.

[57] Sarkar S. K. (2007). Step-up procedures controlling generalized FWER and generalized FDR. *The Annals of Statistics* **35**, 2405-2420.

[58] Sebastiani P., Gussoni E., Kohane I. and Ramoni M. (2003). Statistical challenges in functional genomics. *Statistical Science* **18**, 33-70.

[59] Shaffer J.(1995).Multiple hypothesis testing. *Annual Review of Psychology* **46**, 561-584.

[60] Spjøtvoll E. (1972) On the optimality of some multiple comparison procedures. *The Annals of Mathematical Statistics* **43**, 398-411.

[61] Storey J. (2002). A Direct approach to false discovery rates. *Journal of the Royal Statistical Society, Series B* **64**, 479-498.

[62] Storey J. (2003). The positive false discovery rate: a Bayesian interpretation and the Q-value. *The Annals of Statistics* **31**, 2012-35.

[63] Storey J. and Tibshirani R. (2003) Statistical significance for genome-wide studies. *Proceedings of the National Academy of Sciences* **100**, 9440-9445.

[64] Storey J. (2007). The optimal discovery procedure: A new approach to simultaneous significance testing. *Journal of Royal Statistical Society, Series B* **69**, 347-368.

[65] Schwartzman A., Dougherty R., Taylor J. (2008). False discovery rate analysis of brain diffusion direction maps. *Annals of Applied Statistics* **2**, 153-175.

[66] Sun W. and Cai T. (2007). Oracle and adaptive compound decision rules for false discovery rate control. *Journal of the American Statistical Association* **102**, 901-912.

[67] Sun W. and Cai T. (2009). Large-scale multiple testing under dependence. *Journal of the Royal Statistical Society, Series B* **71**, 393-424.

[68] Todd J. A., Walker N. M., Cooper J. D., et al. (2007). Robust associations of four new chromosome regions from genome-wide analyses of type 1 diabetes. *Nature Genetics* **39**, 857-864.

[69] Tusher V., Tibshirani R. and Chu G. (2001). Significance analysis of microarrays applied to the ionizing radiation response. *Proceedings of National Academy of Science, USA* **98**, 5116-5121.

[70] van der Laan M., Dudoit S. and Pollard K. (2004). Augmentation procedures for control of the generalized family-wise error rate and tail probabilities for the proportion of false positives. *Statistical Applications in Genetics and Molecular Biology* **3**, Article 15.

[71] van't Wout A., Lehrman G., Mikheeva S., O'Keeffe G., Katze M., Bumgar-
    ner R., Geiss G. and Mullins J. (2003). Cellular gene expression upon human
    immunodeficiency virus type 1 infection of $CD4^+$-T-cell lines. *Journal of Vi-
    rology* **77**, 1392-1402.

[72] Weller J., Song J., Heyen D., Lewin H. and Ron M. (1998). A new approach
    to the problem of multiple comparisons in the genetic dissection of Complex
    traits. *Genetics* **150**, 1699-1706.

[73] Westfall P. and Young S. (1993). *Resamplingbased Multiple Testing.* Wiley,
    New York.

[74] Wright S. (1992). Adjusted p-values for simultaneous inference. *Biometrics*
    48, 1005-1013.

[75] Wu W. (2009) On false discovery control under dependence. *The Annals of
    statistics* **36**, 364-380.

Part III

Model Building with Variable Selection

Chapter 4

Model Building with Variable Selection [*]

Abstract

In this chapter, we give a selective review of several popular variable selection techniques ranging from the classical stepwise regression to the more recent regularization based techniques.

Keywords: Information criteria, linear model, model selection, regularization, variable selection.

1 Introduction

Variable selection is a classical problem in statistics. It is an essential tool to statistical model building as it results in more interpretable and therefore in practice more useful models. Variable selection has been studied and used extensively. As of May 2009, a search of google scholar gives nearly four million items. It is clearly impossible to give an exhaustive review of the vast literature. Instead, we shall focus here on surveying several common ideas behind some of the most popular approaches.

To fix ideas, we shall restrict ourselves to the context of linear regression although the techniques discussed are also applicable to more general predictive modeling framework. In the common normal linear regression model, we have n observations $(\mathbf{x}_1, y_1), \ldots, (\mathbf{x}_n, y_n)$ on a dependent variable Y and p predictors $(X_1, X_2, \ldots, X_p)' =: X$, and

$$Y = X'\beta + \epsilon, \tag{1.1}$$

where $\epsilon \sim N(0, \sigma^2)$, and $\beta = (\beta_1, \ldots, \beta_p)'$. The underlying notion behind variable selection is that some of the predictors are redundant and therefore only an unknown subset of the β coefficients are nonzero. By effectively identifying the subset of important predictors, variable selection can improve estimation accuracy and enhance model interpretability.

The purpose of variable selection can be viewed as identifying a subset $\mathcal{M} \subset \{1, \ldots, p\}$, such that $X_\mathcal{M} := \{X_j : j \in \mathcal{M}\}$ is sufficient to predict Y. In other

---

[*]This research was supported in part by NSF grants DMS-0706724 and DMS-0846234, and a grant from Georgia Cancer Coalition.

[†]H. Milton Stewart School of Industrial and Systems Engineering, Georgia Institute of Technology, 755 Ferst Dr NW, Atlanta, GA 30332-0205, USA, E-mail: myuan@isye.gatech.edu

words $\beta_j = 0$ for $j \notin \mathcal{M}$. With slight abuse of notation, in what follows, we shall use $\mathcal{M}$ to denote both a subset of $\{1, \ldots, p\}$ and the linear model with only variables $\{X_j : j \in \mathcal{M}\}$ as predictors. The meaning of the notation should be clear given its context.

2   Why variable selection

To illustrate the importance and effect of variable selection, we begin with a simple data example from Anderson, Sweeney and Williams (2003). The data, reproduced in Table 2.1, consist of six performance measures on twenty five players in the PGA golf tour. The goal is to examine the importance of a golfer's shot-making skills measured by average length of a drive in yards (Drive), percentage of drives that land in the fairway (Fair), percentage of green hit in regulation (Green), average number of putts for greens that have been hit in regulation (Putt), as well as percentage of sand saves (landing in a sand trap and still scoring par or better; Sand), to the players overall performance measured as the average score per round of golf (Score).

Table 2.1   PGA Data

Drive	Fair	Green	Putt	Sand	Score
277.6	.681	.667	1.768	.550	69.10
259.6	.691	.665	1.810	.536	71.09
269.1	.657	.649	1.747	.472	70.12
267.0	.689	.673	1.763	.672	69.88
267.3	.581	.637	1.781	.521	70.71
255.6	.778	.674	1.791	.455	69.76
272.9	.615	.667	1.780	.476	70.19
265.4	.718	.699	1.790	.551	69.73
272.6	.660	.672	1.803	.431	69.97
263.9	.668	.669	1.774	.493	70.33
267.0	.686	.687	1.809	.492	70.32
266.0	.681	.670	1.765	.599	70.09
258.1	.695	.641	1.784	.500	70.46
255.6	.792	.672	1.752	.603	69.49
261.3	.740	.702	1.813	.529	69.88
262.2	.721	.662	1.754	.576	70.27
260.5	.703	.623	1.782	.567	70.72
271.3	.671	.666	1.783	.492	70.30
263.3	.714	.687	1.796	.468	69.91
276.6	.634	.643	1.776	.541	70.69
252.1	.726	.639	1.788	.493	70.59
263.0	.687	.675	1.786	.486	70.20
263.0	.639	.647	1.760	.374	70.81
253.5	.732	.693	1.797	.518	70.26
266.2	.681	.657	1.812	.472	70.96

The following regression equation emerges from the usual least squares estimate:

$$Score = 74.157 - .0407(Drive) - 6.931(Fair) - 10.357(Green)$$
$$+ 10.267(Putt) + 0.407(Sand).$$

The corresponding R^2 is 72.6% suggesting a reasonable fit. The only problem with this model, however, is how to interpret the sign of the coefficient for Sand, which should be negative by intuition. This puzzle turns out to be a possible artifact as most variable selection methods suggest that Sand is actually not an important predictor. Removing Sand from the regression will result in the following regression equation

$$\mathsf{Score} = 74.678 - .0398(\mathsf{Drive}) - 6.686(\mathsf{Fair}) - 10.342(\mathsf{Green}) + 9.858(\mathsf{Putt}) \quad (2.1)$$

with R^2 72.2%. Therefore, removing the dubious effect of Sand yields a more interpretable model.

In addition to interpretability, variable selection can also lead to improved predictive performance. As Box famously put, "All models are wrong, but some are useful". A statistical model only serves as an approximation to the truth. Clearly with more variables in a model, the quality of approximation improves and bias reduces. On the other hand, with more variables come more coefficients to be estimated and subsequently more variation in terms of the estimation accuracy. From this perspective, variable selection serves as a tool to balance the trade-off between bias and variance. Figure 2.1 shows qualitatively how squared bias, variance and prediction ability measured by mean squared error typically vary with the model complexity, often measured by the number of predictors.

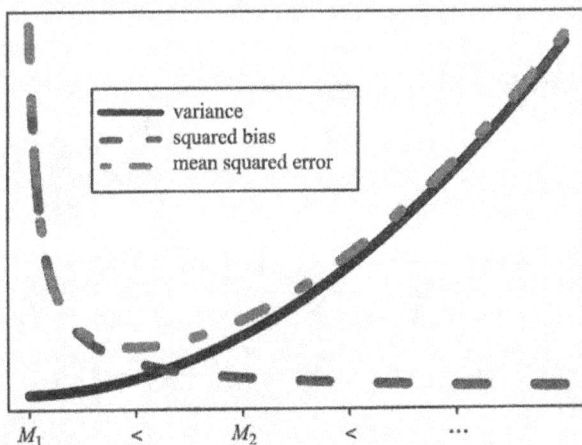

Figure 2.1   An illustration of possible effect of model complexity on the squared bias, variance and mean squared error.

3   Classical approaches

3.1   Model selection via hypothesis testing

Among the first approaches to model selection are those based on hypothesis testing. To fix ideas, we consider first a set of nested models $\mathcal{M}_1 \subset \mathcal{M}_2 \subset \cdots$. This situation naturally arises in the context of linear time series modeling when we are interested in choosing an appropriate lag for predicting future observations. More specifically, let $Z(t)$ be a time series, then we are interested in the following regression model:

$$Z(t) = \beta_0 + \beta_1 Z(t-1) + \cdots + \beta_q Z(t-q) + \epsilon_t, \qquad (3.1)$$

denoted by $\mathcal{M}_{q+1}$. The choice of the lag q can then be cast as a variable selection problem.

With these candidate models $\{\mathcal{M}_1, \mathcal{M}_2, \ldots\}$, a sequence of hypotheses can be formulated:

$$H_{j0} : \mathcal{M}_j \text{ is true} \qquad \text{vs} \qquad H_{ja} : \mathcal{M}_{j+1} \text{ is true.} \qquad (3.2)$$

There are two ways to proceed from here. In *forward selection*, we start by testing H_{10} vs H_{1a}. If the null is rejected, we continue testing H_{20} vs H_{2a}. The procedure stops when a null hypothesis is accepted or the full model with all variables is reached. Alternatively, in *backward elimination*, we start with the last hypothesis in the sequence, say H_{p0} vs H_{pa}. If the null is accepted, we then test $H_{p-1,0}$ vs $H_{p-1,a}$. The procedure continues until a null hypothesis is rejected or the null model with only intercept is obtained. The significance level of the tests are often set to either 5% or 10%

Coming back to the more general linear regression setting, similar approach can be adopted. As in forward selection, we follow an iterative procedure, starting with the null model, i.e., $M^{[0]}$ is empty set. We then look at all possible subsets $M^{[1]}$ such that it contains exactly one more element that $\mathcal{M}^{[0]}$, i.e., $\mathcal{M}^{[1]}$ can be $\{j\}$ for any $j = 1, \ldots, p$. For each possible choice of $\mathcal{M}^{[1]}$, the following hypotheses are tested:

$$H_{10} : \mathcal{M}^{[0]} \text{ is true} \qquad \text{vs} \qquad H_{1a} : \mathcal{M}^{[1]} \text{ is true.} \qquad (3.3)$$

The choice with the smallest p-value is then taken, if the smallest p-value is smaller than a pre-specified threshold, say 5%, we continue examining all possible subsets $M^{[2]}$ such that it contains exactly one more element thn $\mathcal{M}^{[1]}$ and test the following hypotheses

$$H_{20} : \mathcal{M}^{[1]} \text{ is true} \qquad \text{vs} \qquad H_{2a} : \mathcal{M}^{[2]} \text{ is true.} \qquad (3.4)$$

As before, we continue if the smallest p-value is smaller than 5%. The procedure stops when the smallest p-value is larger than 5% or the full model is reached. The backward elimination procedure can also be generalized accordingly. Further details can be found for example in Miller (1990).

To see the two methods in action, we apply both forward selection and backward elimination to the PGA data. The p-values at each iteration are reported in Table 3.1 along with the candidate models that are entertained. In particular, forward selection chooses Green in the first iteration and then Putt. But after two iteration, adding any other variable yields a p-value at least 20%. Therefore, the final model selected is $\mathcal{M} = \{$Green, Putt$\}$. In contrast, backward elimination starts with all five variables and removes Sand in the first step. Afterwards, removing any other variable results in a p-values smaller than 5% and therefore the method stops with a model $\mathcal{M} = \{$Drive, Fair, Green, Putt$\}$.

Table 3.1   **Application of forward selection and backward elimination to PGA data – p-values are given for each iteration of the methods. The last row gives the candidate models that are entertained at each iteration**

	Forward Selection			Backward Elimination		
	1st Iteration	2nd Iteration	3rd Iteration		1st Iteration	2nd Iteration
Adding Drive	0.46	0.31	0.43	Dropping Drive	0.00	0.00
Fair	0.03	0.23	0.21	Fair	0.00	0.00
Green	**0.00**			Green	0.01	0.01
Putt	0.21	**0.00**		Putt	0.01	0.01
Sand	0.18	0.19	0.65	Sand	**0.70**	
$\mathcal{M}$	{}	{G}	{G,P}		{D,F,G,P,S}	{D,F,G,P}

3.2   Best subset and selection criteria

Another classical idea for variable selection is to directly compare all 2^p possible candidate models and select the best model according to a certain selection criteron. Several popular selection criteria are given below.

- Final prediction error (Akaike, 1969)

$$\mathrm{FPE}(\mathcal{M}) = S_{\mathcal{M}} \frac{n + |\mathcal{M}|}{n - |\mathcal{M}|} \tag{3.5}$$

  where $S_{\mathcal{M}}$ is the residual sum of squares for $\mathcal{M}$ and $|\mathcal{M}|$ represents the cardinality of the set $\mathcal{M}$.
- Mallow's C_p (Mallows, 1973)

$$C_p(\mathcal{M}) = S_{\mathcal{M}} + 2\hat{\sigma}^2 |\mathcal{M}| \tag{3.6}$$

  where $\hat{\sigma}^2 = S_{\text{full}}/(n - p)$ is the unbiased estimate of σ^2 with the full model and S_{full} is the residual sum of squares for the full model.
- Akaike's information criterion (Akaike, 1973)

$$\mathrm{AIC}(\mathcal{M}) = n \log(S_{\mathcal{M}}/n) + 2|\mathcal{M}|. \tag{3.7}$$

- Bayesian information criterion (Akaike, 1978; Schwartz, 1978)

$$\mathrm{BIC}(\mathcal{M}) = n \log(S_{\mathcal{M}}/n) + |\mathcal{M}| \log n. \tag{3.8}$$

- Generalized cross validation (Golub, Heath and Wahba, 1979)

$$\mathrm{GCV}(\mathcal{M}) = S_{\mathcal{M}}/(n - |\mathcal{M}|)^2. \tag{3.9}$$

- Risk inflation criterion (Foster and George, 1994)

$$\mathrm{RIC}(\mathcal{M}) = S_{\mathcal{M}} + 2|\mathcal{M}|\hat{\sigma}^2 \log p. \tag{3.10}$$

The list can go much longer and the readers are referred to textbooks such as Linhart and Zucchini (1986), Miller (1990), and McQuarrie and Tsai (1998) for a more thorough review. These criteria are often devised to reflect the prediction error, e.g., FPE, C_p, GCV and RIC; or motivated by information theoretic considerations, e.g., AIC and BIC. For each of the selection criterion, the minimizing

Table 3.2	Various selection criteria evaluated for the PGA data. The GCV scores have been multiplied by 100

$\mathcal{M}$	FPE	Cp	AIC	BIC	GCV	RIC	$\mathcal{M}$	FPE	Cp	AIC	BIC	GCV	RIC
{}	5.24	5.24	−39.05	−39.05	0.84	5.24	{S}	5.24	4.99	−39.05	−37.83	0.84	5.08
{P}	5.30	5.04	−38.78	−37.56	0.85	5.14	{P,S}	5.47	4.96	−37.99	−35.55	0.88	5.15
{G}	3.92	3.77	−46.29	−45.08	0.63	3.87	{G,S}	3.92	3.64	−46.33	−43.89	0.63	3.83
{G,P}	2.79	2.68	−54.79	−52.36	0.45	2.87	{G,P,S}	3.00	2.81	−53.04	−49.38	0.49	3.09
{F}	4.65	4.44	−42.07	−40.85	0.74	4.53	{F,S}	4.86	4.44	−40.97	−38.54	0.78	4.62
{F,P}	4.47	4.11	−43.05	−40.61	0.72	4.29	{F,P,S}	4.81	4.23	−41.23	−37.57	0.78	4.51
{F,G}	3.98	3.69	−45.98	−43.54	0.64	3.87	{F,G,S}	4.08	3.66	−45.33	−41.67	0.66	3.94
{F,G,P}	2.80	2.65	−54.75	−51.09	0.45	2.93	{F,G,P,S}	3.04	2.80	−52.77	−47.90	0.50	3.17
{D}	5.55	5.27	−37.64	−36.43	0.89	5.36	{D,S}	5.52	5.01	−37.76	−35.32	0.89	5.19
{D,P}	5.66	5.12	−37.16	−34.72	0.91	5.31	{D,P,S}	5.81	5.02	−36.50	−32.84	0.94	5.30
{D,G}	4.06	3.76	−45.48	−43.04	0.65	3.94	{D,G,S}	4.02	3.62	−45.69	−42.03	0.65	3.89
{D,G,P}	2.94	2.76	−53.54	−49.89	0.48	3.04	{D,G,P,S}	3.15	2.89	−51.87	−46.99	0.52	3.25
{D,F}	2.79	2.68	−54.82	−52.38	0.45	2.86	{D,F,S}	3.02	2.82	−52.90	−49.25	0.49	3.10
{D,F,P}	2.62	2.51	−56.41	−52.75	0.43	2.79	{D,F,P,S}	2.83	2.65	−54.53	−49.66	0.46	3.02
{D,F,G}	2.73	2.60	−55.40	−51.74	0.44	2.87	{D,F,G,S}	2.93	2.73	−53.66	−48.78	0.48	3.10
{D,F,G,P}	2.00	2.05	−63.21	−58.33	0.33	2.42	{D,F,G,P,S}	2.16	2.19	−61.41	−55.32	0.36	2.65

model $\mathcal{M}$ is chosen as the final model. To illustrate, we again examine the PGA data. Table 3.2 gives the values of the aforementioned criteria evaluated on all $2^5 = 32$ possible models. All criteria identify $\mathcal{M} = \{\text{Drive, Fair, Green, Putt}\}$ as the most plausible model.

Many of these criteria have been shown to enjoy various nice theoretical properties. However, the application of these methods are increasingly limited by its inability to handle high dimensional problems. Since the number of candidate models increases exponentially with the number of variables, it is clearly computationally infeasible to enumerate all possible candidate models in order to find the "best" model even for a moderate number of variables. Therefore, these criteria are most commonly used in conjunction with stepwise greedy algorithms, such as forwards selection or backward elimination. The myopic nature of these greedy algorithm, however, often leads to suboptimal solutions (Chen, Donoho and Saunders, 1999). Another subtle problem with the subset selection procedures is their instability. Since the final estimate is based upon the selected model, which is not a continuous function of the data, a small change in the data may result in significant changes in statistical inferences (Breiman, 1996).

3.3   Resampling methods

Resampling methods such as cross validation and bootstrap have also been used to compare the performance of candidate models. Both target at constructing proxies to the prediction error. In the multi-fold cross validation (Stone, 1974), the data $\mathcal{L} = \{(y_i, \mathbf{x}_i) : i = 1, \ldots, n\}$ are first equally split into V subsets $\mathcal{L}_1, \ldots, \mathcal{L}_V$. All candidate models are fitted with data $L^{(v)} = \mathcal{L} - \mathcal{L}_v$. Denote the corresponding coefficient estimate $\widehat{\beta}^{(v)}(\mathcal{M})$. The prediction error associated with model $\mathcal{M}$ can then be naturally estimated by

$$\widehat{PE}(\widehat{\beta}(\mathcal{M})) = \sum_v \sum_{(y_i, \mathbf{x}_i) \in \mathcal{L}_v} \left(y_i - \mathbf{x}_i \widehat{\beta}^{(v)}(\mathcal{M}) \right)^2 .$$

The models that minimizes $\widehat{PE}(\widehat{\beta}(\mathcal{M}))$ is then selected. It is often suggested to use $V = 10$ in practice (Breiman, 1995). A special case of the multi-fold cross validation discussed here is the so-called leave-one-out cross validation or full cross validation. In fact, the GCV criterion is an approximation to the full cross-validation. Other versions of cross validation can also be found in Shao (1993).

Similarly, Breiman (1992) suggests to estimate the prediction error, more precisely the model error, through the so-called little bootstrap. Other bootstrap related approaches can be found in Linhart and Zucchini (1986), Breiman (1992), Shao (1996), and Shibata (1997), and references therein.

4   Bayesian and stochastic search

Variable selection can be addressed coherently in a Bayesian framework. A large number of Bayesian approaches have been introduced in recent years. See, e.g.,

Berger and Pericchi (1996), Brown, Vannucci and Fearn (1998), Carlin and Chib (1995), Clyde, Parmigiani and Vidakovic (1998), George and McCulloch (1993; 1997), Green (1995), Ishwaran and Rao (2005), Kass and Raftery (1995), O'Hagan (1995), Speigelhalter et al. (2002), Smith and Kohn (1997), and Wood et al. (2002), among others.

4.1 Bayesian model selection

Treating the model itself as an unknown parameter and assigning a prior to it, variable selection can be performed in the same fashion as the usual Bayesian inference. More specifically, a hierarchical Bayesian formulation for variable selection in linear models consists of the following three main ingredients:

- a prior probability $P(\mathcal{M})$ for each candidate model $\mathcal{M}$;
- a prior $P(\beta_{\mathcal{M}}|\mathcal{M})$ for regression coefficient $\beta_{\mathcal{M}}$ associated with model $\mathcal{M}$;
- a data generating mechanism conditional on $(\mathcal{M}, \beta_{\mathcal{M}})$, $P(\text{data}|\mathcal{M}, \beta_{\mathcal{M}})$.

Once these three components are specified, one can combine data and priors to form the posterior

$$P(\mathcal{M}|\text{data}) = \frac{P(\mathcal{M})P(\text{data}|\mathcal{M})}{\sum_{\mathcal{M}'} P(\mathcal{M}')P(\text{data}|\mathcal{M}')} \qquad (4.1)$$

where

$$P(\text{data}|\mathcal{M}) = \int P(\text{data}|\mathcal{M}, \beta_{\mathcal{M}})P(\beta_{\mathcal{M}}|\mathcal{M})d\beta_{\mathcal{M}} \qquad (4.2)$$

is the marginal likelihood of $\mathcal{M}$. Pairwise model comparison can be directly made using these posterior probabilities, i.e., through the posterior odds ratio:

$$\frac{P(\mathcal{M}_1|\text{data})}{P(\mathcal{M}_2|\text{data})} = \frac{P(\text{data}|\mathcal{M}_1)}{P(\text{data}|\mathcal{M}_2)} \times \frac{P(\mathcal{M}_1)}{P(\mathcal{M}_2)}. \qquad (4.3)$$

In particular, if a noninformative prior $P(\mathcal{M}_1) = P(\mathcal{M}_2) = \cdots = 2^{-p}$ is adopted, the posterior odds $P(\mathcal{M}_1|\text{data})/P(\mathcal{M}_2|\text{data}) = P(\text{data}|\mathcal{M}_1)/P(\text{data}|\mathcal{M}_2)$ reduces to Bayes factor. Note however the "noninformativeness" of this prior is rather deceiving. Under this prior, models of size k appear with probability $2^{-p}\binom{k}{p}$, which can be vastly different with different k.

4.2 Spike and slab prior

A particularly popular class of prior is the so-called spike and slab prior (e.g., Mitchell and Beauchamp, 1988; George and McCulloch, 1993; Ishwaran and Rao, 2005). It begins by indexing each candidate model with a binary vector $\gamma = (\gamma_1, \ldots, \gamma_p)'$. An element γ_j takes value 0 or 1 depending on whether the j-th predictor is excluded from the model or not. The idea is then to put a prior to the coefficient β_j depending on the value of γ_j. When $\gamma_j = 1$, a flat, or in a certain sense noninformative prior is given to β_j. When $\gamma_j = 0$, a "spiky" prior highly concentrated around 0 is given to β_j.

In particular, the following hierarchical model is taken by the popular stochastic search variable selection procedure of George and McCulloch (1993):

$$y_i|\mathbf{x}_i, \beta, \sigma^2 \sim \mathcal{N}(\mathbf{x}_i'\beta, \sigma^2), \qquad i = 1, \ldots, n$$
$$\beta|\gamma \sim \mathcal{N}(0, D_\gamma R D_\gamma)$$
$$\gamma_j \sim Bernoulli(w_j), \qquad j = 1, \ldots, p$$
$$\sigma^2|\gamma \sim \pi(d\sigma^2|\gamma)$$

where R is a correlation matrix, i.e., $R = I$ or $(\mathbf{X}'\mathbf{X})^{-1}$ and

$$D_\gamma = \text{diag}(a_1 T_1, \ldots, a_p T_p) \qquad (4.4)$$

where $a_j = 1$ if $\gamma_j = 0$, and c_j if $\gamma_j = 1$. We generally choose a rather small T_j to give a spike around 0 when $\gamma_j = 0$, and a very large c_j to yield a flat prior for β_j when $\gamma_j = 1$. The prior π for σ^2 is often taken to be a conjugate prior, i.e., an inverse Gamma distribution, to ensure fast computation.

The choice of hyperparameters in Bayesian variable selection problems is often tricky. One common approach is the so-called empirical Bayes calibration where the hyperparameters are chosen to maximize the marginal likelihood:

$$P(\text{data}) = \sum_{\mathcal{M}} P(\mathcal{M})P(\text{data}|\mathcal{M}). \qquad (4.5)$$

The computation, however, can be rather difficult as it involves evaluating the marginal likelihood $P(\text{data}|\mathcal{M})$ for all candidate models. For this reason, one often resorts to various approximations of $P(\text{data})$ (e.g., George and Foster, 2000; Yuan and Lin, 2005).

4.3 Stochastic search

Similar to the best subset selection, direct calculation of all the posterior probabilities $P(\mathcal{M}|\text{data})$ is a daunting task when p is large. Sampling schemes are used to alleviate this problem. The idea to construct a Markov chain whose stationary distribution is $P(\mathcal{M}|\text{data})$. In particular for the stochastic search variable selection, a Gibbs sampling procedure can be devised. The procedure successively samples $\beta^{[j]}, \sigma^{[j]}$ and $\gamma^{[j]}$ from their respective conditional distributions. As suggested by George and McCulloch (1993), $\beta^{[0]}$ and $\sigma^{[0]}$ are initialized as the least squares estimate of the full model. Then, $\gamma^{[0]}$ is sampled from the conditional distribution $\gamma|\text{data}, \beta = \beta^{[0]}, \sigma = \sigma^{[0]}$. Subsequently, $\beta^{[1]}$ is sampled from $\beta|\text{data}, \gamma = \gamma^{[0]}, \sigma = \sigma^{[0]}$, and $\sigma^{[1]}$ from $\sigma|\text{data}, \gamma = \gamma^{[0]}, \beta = \beta^{[1]}$. We continue this iterative sampling scheme to generate an auxiliary "Gibbs sequence":

$$\beta^{[0]}, \sigma^{[0]}, \gamma^{[0]}, \beta^{[1]}, \sigma^{[1]}, \gamma^{[1]}, \ldots.$$

Some linear algebraic manipulation shows that all conditional distributions involved can be computed in closed form and boils down to common distributions: multivariate normal for β, inverse gamma for σ, and binomial for γ. Therefore, the sampling can be implemented fairly efficiently.

The Gibbs sequence forms a homogeneous Markov chain that converges geometrically to its unique stationary distribution $\gamma|$data. Therefore, after a certain burn-in period B, $\gamma^{[B+1]}, \gamma^{[B+2]}, \ldots$ can be treated as samples from the posterior distribution $\gamma|$data. This allows us to compute the corresponding posterior probabilities.

5 Regularization

To overcome the limits of the classical variable selection methods, a large number of new techniques have been introduced in recent years. Most of them can be formulated as regularization estimating procedures.

5.1 Nonnegative garrote

The nonnegative garrote estimator (Breimain, 1995) is a scaled version of the least square estimate. We consider shrinking the least squares estimate of the full model $\hat{\beta}^{LS} = (\hat{\beta}_1^{LS}, \ldots, \hat{\beta}_p^{LS})'$ by shrinking factor $d = (d_1, \ldots, d_p)'$, i.e., estimates of the form $\beta = (d_1 \hat{\beta}_1^{LS}, \ldots, d_p \hat{\beta}_p^{LS})$. To determine the shrinkage factors, we solve the following constrained least squares problem:

$$\min ||\mathbf{Y} - \mathbf{Z}d||^2, \qquad \text{subject to} \quad \sum_{j=1}^{p} d_j \leqslant M \qquad \text{and} \qquad d_j \geqslant 0, \forall j, \qquad (5.1)$$

where $\mathbf{Z} = \text{diag}(\hat{\beta}^{LS})\mathbf{X}$ and $M \geqslant 0$ is a tuning parameter. Clearly when $M \geqslant p$, the solution to (5.1) is $d_1 = \ldots = d_p = 1$ and the nonnegative garrote estimate reduces to the least squares. As M decreases to 0, the shrinking factors decrease and eventually, all of them will become zero.

The mechanism of the nonnegative garrote can be illustrated under orthogonal designs, where $\mathbf{X}'\mathbf{X} = nI$. In this case, the solution to (5.1) has an explicit form:

$$d_j(\lambda) = \left(1 - \frac{\lambda}{\left(\hat{\beta}_j^{LS} \right)^2} \right)_+ , \qquad j = 1, \ldots, p, \qquad (5.2)$$

where $\lambda \geqslant 0$ is chosen so that $\sum d_j = \min(M, p)$. Therefore, for those coefficients whose full least square estimate is large in magnitude, the shrinking factor will be close to 1. But for a redundant predictor, the least square estimate is likely to be small and consequently the shrinking factor will have a good chance to be exactly zero.

To demonstrate the nonnegative garrote, we apply it to the PGA data. Figure 5.1 gives the trajectory of the shrinking factors as M varies.

As in any other regularization method, the nonnegative garrote gives a continuum of estimates indexed by the tuning parameter M, and we need choose a final estimate among them. This is often done in a similar fashion as the classical variable selection approaches with a tuning parameter selected via a certain criterion. Such a criterion often reflects the prediction accuracy, which depends on the

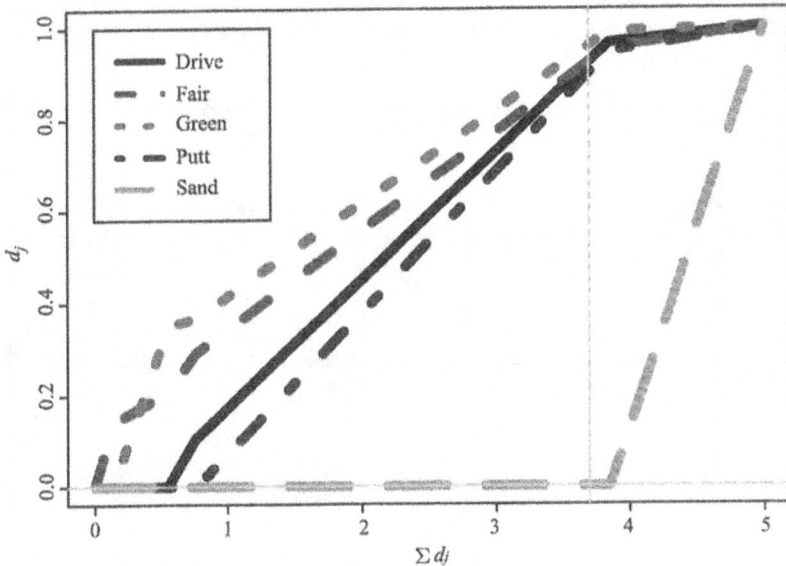

Figure 5.1   Nonnegative garrote applied to PGA data.

unknown parameters and needs to be estimated. This is often done either through cross validation or some data perturbation techniques (Shen and Ye, 2002). In particular, the vertical grey line in Figure 5.1 represents the tuning parameter selected by ten fold cross validation in a typical run.

One of the main drawbacks of the original nonnegative garrote is its explicit reliance on the full least square estimate. Obviously, with a small sample size, the least squares may perform poorly, and the nonnegative garrote is expected to suffer as well. In particular, the original nonnegative garrote, as proposed by Breiman (1995) can not be applied when the sample size is smaller than the number of predictors. But as demonstrated recently by Yuan and Lin (2006), other initial estimates can also be used in place of the least squares and therefore to alleviate the problem.

5.2   LASSO and bridge regression

Tibshirani (1996) proposed the popular LASSO which stands for "least absolute shrinkage and selection operator". The LASSO estimator is defined as the solution to

$$\min_{\beta} \|\mathbf{Y} - \mathbf{X}\beta\|^2 \qquad \text{subject to } \|\beta\|_{\ell_1} \leq M, \qquad (5.3)$$

where M is a tuning parameter, and $\|\cdot\|_{\ell_1}$ stands for the vector ℓ_1 norm. The ℓ_1 norm penalty induces sparsity in the solution meaning that some of the components of the estimated coefficient vector can be exactly zero with an appropriately chosen tuning parameter. Figure 5.2 shows the solution path of the LASSO estimate for the PGA data. Similar to before, the vertical grey line corresponds to the tuning parameter selected by ten fold cross validation. The corresponding model includes variable Fair, Green and Putt.

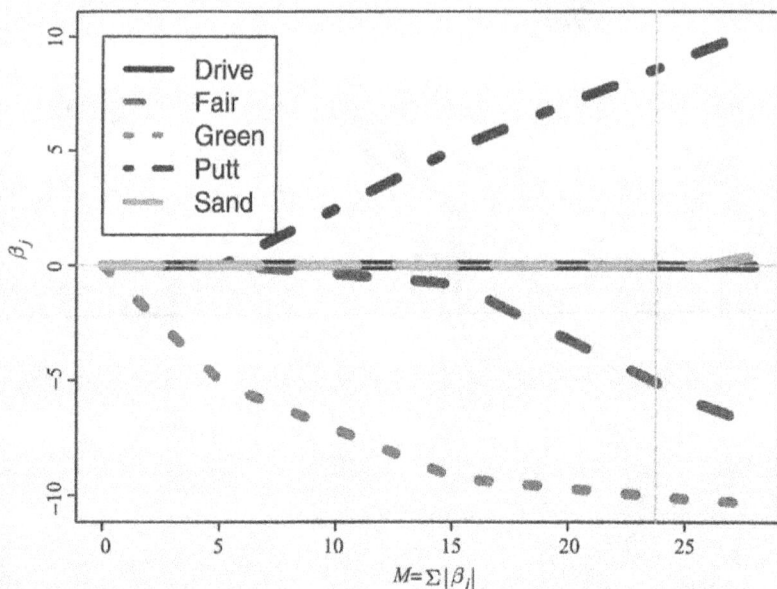

Figure 5.2   LASSO applied to PGA data.

The LASSO is a member of a large family of constrained least squares estimates, often referred to as the bridge estimates (Frank and Friedman, 1989):

$$\min_{\beta} \|\mathbf{Y} - \mathbf{X}\beta\|^2 \qquad \text{subject to} \sum_{j=1}^{p} |\beta_j|^{\gamma} \leqslant M, \qquad (5.4)$$

where $\gamma > 0$. Figure 5.3 shows the feasible region of bridge regression with four different values of γ in the case of $p = 2$. Notice that the feasible region is convex if and only if $\gamma \geqslant 1$. On the other hand the singularity of the feasible region suggests that the bridge estimate can induce sparsity if and only if $\gamma \leqslant 1$.

The shrinkage effect can be better illustrated by examining the orthogonal design case where the bridge estimate of each coefficient can be expressed in terms of the corresponding least squares estimate. In particular, when $\gamma = 2$,

$$\hat{\beta}_j = \frac{1}{1 + \lambda} \hat{\beta}^{\text{LS}}, \qquad (5.5)$$

where $\lambda \geqslant 0$ is such that

$$\sum_{j=1}^{p} \hat{\beta}_j^2 = M. \qquad (5.6)$$

When $\gamma = 1$,

$$\hat{\beta}_j = \left(|\hat{\beta}^{\text{LS}}| - \lambda/2 \right)_+ \text{sign} \left(\hat{\beta}^{\text{LS}} \right), \qquad (5.7)$$

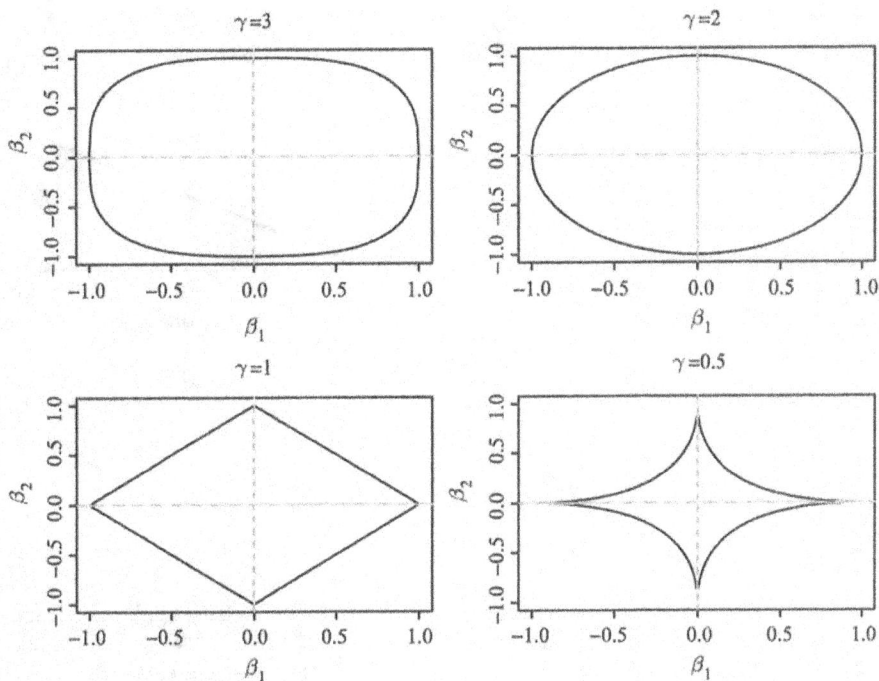

Figure 5.3   Feasible region of different constraints.

where $(x)_+$ stands for the positive part of x, i.e., $(x)_+ = \max(x, 0)$ and $\lambda \geqslant 0$ satisfies $\sum |\hat{\beta}_j| = M$. Closed form solutions for general γs are not always available. To compare the shrinkage effect with different values of γ, we plot in Figure 5.4 the bridge estimate as functions of the least squares estimate for $\gamma = 3, 2, 1$ and 0.5. It is evident that that the bridge estimate encourages sparsity when $\gamma \leqslant 1$.

5.3   LARS

A close relative of the LASSO is the so-called least angle regression (LARS) introduced by Efron et al. (2004). The LARS uses a variable selection strategy similar to the forward selection. As shown in Figure 5.5, we start with all coefficients equal to zero, the algorithm finds the variable that is most correlated with the response, in this case X_1, and proceeds in this direction. Instead of taking a full step towards the projection of Y on the variable, as would be done in a forward selection, the LARS only takes the largest step possible in this direction until some other variable has as much correlation with the current residual. Then this new variable is entered and the process is continued. The procedure is called the least angle regression because for any estimate visited by the procedure, the residual has the smallest angle with all variables that are included in the model.

The great computational advantage of the LARS comes from the fact that the LARS path is piecewise linear and all we need to do is to locate the change points. More specifically, the LARS algorithm can be described as follows.

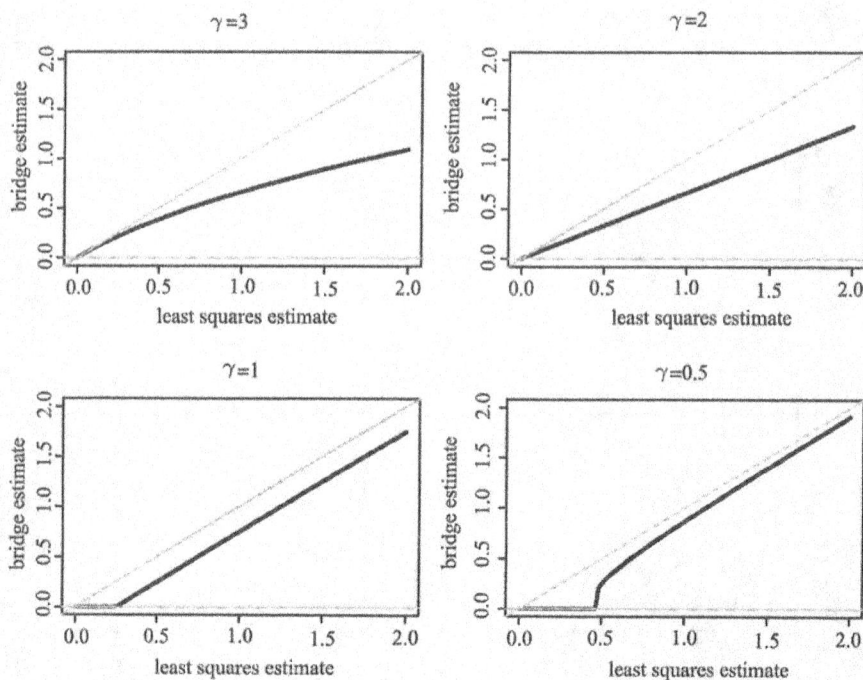

Figure 5.4   Shrinkage effect of different constraints.

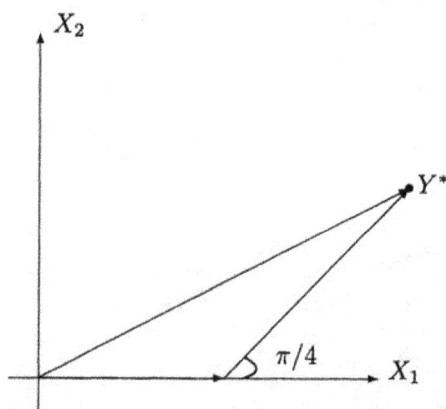

Figure 5.5   Lars algorithm.

Algorithm – LARS

1. Start from $\beta^{[0]} = 0$, $k = 1$ and $r^{[0]} = Y$
2. Find a variable X_j that is most correlated with $r^{[0]}$ and set $\mathcal{B}_k = \{j\}$.
3. Compute the current direction γ, which is a p dimensional vector with $\gamma_{\mathcal{B}_k^c} = $

0 and

$$\gamma_{\mathcal{B}_k} = \left(X'_{\mathcal{B}_k} X_{\mathcal{B}_k} \right)^{-} X'_{\mathcal{B}_k} r^{[k-1]}. \tag{5.8}$$

4. For every $i \notin \mathcal{B}_k$ compute how far the algorithm will march in direction γ before X_i has the same amount of correlation with the residual as the variables in $\mathcal{B}_k$. This can be measured by the smallest $\alpha_i \in [0,1]$ such that

$$|X'_i(r^{[k-1]} - \alpha_i X\gamma)| = |X'_{\mathcal{B}_1}(r^{[k-1]} - \alpha_i X\gamma)|. \tag{5.9}$$

5. If $\mathcal{B}_k \neq \{1, \ldots, p\}$, let $\alpha = \min_{i \notin \mathcal{B}_k} \alpha_i \equiv \alpha_{i^*}$ and update $\mathcal{B}_{k+1} = \mathcal{B}_k \cup \{i^*\}$. Otherwise, set $\alpha = 1$.
6. Update $\beta^{[k]} = \beta^{[k-1]} + \alpha\gamma$, $r^{[k]} = Y - X\beta^{[k]}$ and $k = k+1$. Go back to step 3 until $\alpha = 1$ when the full least squares estimate is reached.

Here $\mathcal{B}_k$ keeps track of the variables that are included in the model at the kth stage, γ determines the direction in which the coefficient estimate will move along, and α measures how far the algorithm will march along that direction. Note that (5.9) is equivalent to

$$X'_i(r^{[k-1]} - \alpha_i X\gamma) = \pm X'_{\mathcal{B}_1}(r^{[k-1]} - \alpha_i X\gamma), \tag{5.10}$$

which can be easily solved for α_i.

The LARS is closely related to the LASSO. In fact a simple modification of the LARS algorithm can be employed to produce the whole solution path of the LASSO estimator. The modification occurs when one of the coefficients marches from a nonzero value to zero. Once it hits zero, we simply remove it from the active set $\mathcal{B}_k$ and re-calibrate the direction the estimate moves towards, and continue as the usual LARS algorithm. More specifically, Steps 4 and 5 of the LARS algorithm will be changed to:

4'. For every $i \notin \mathcal{B}_k$, compute α_i as in Step 4 of the LARS algorithm. For every $i \in \mathcal{B}_k$, compute how far the algorithm will march before β_i hits zero, i.e., $\alpha_i = -\beta_i/\gamma_i$ if $\beta_i/\gamma_i < 0$ and $+\infty$ otherwise.
5'. If $\alpha_i \notin \mathcal{B}_k$, proceed as in Step 5 of the LARS algorithm. If $\alpha_i \in \mathcal{B}_k$, update $\mathcal{B}_{k+1} = \mathcal{B}_k \setminus \{i^*\}$. Otherwise, set $\alpha = 1$.

Similar algorithm for computing the solution path for the LASSO was first considered by Osborne, Presnell and Turlach (2000) and later formalized by Efron et al. (2004). Note that in the case of PGA data, the event that a nonzero coefficient becomes zero, i.e., $i^* \in \mathcal{B}_k$ never occurs. Hence the LARS and the LASSO give the same solution.

5.4   Other regularization techniques

In addition to the ℓ_1 regularization employed by the LASSO, there are also other alternative constraints that can be used to encourage sparsity. Most notable examples include the smoothly clipped absolute deviation (SCAD) penalty (Fan and Li, 2001), elastic net (Zou and Hastie, 2004), and adaptive LASSO (Zou, 2006).

These estimates are often expressed in terms of the penalized least squares:

$$\min_{\beta} \left\{ \|\mathbf{Y} - \mathbf{X}\beta\|^2 + \sum_{j=1}^{p} J_\lambda(\beta_j) \right\}, \tag{5.11}$$

where J_λ is a penalty depending on tuning parameter $\lambda \geqslant 0$.

The SCAD penalty is given as

$$J_\lambda(\beta_j; a) = \begin{cases} \lambda|\beta_j| & |\beta_j| \leqslant \lambda \\ -(\beta_j^2 - 2a\lambda|\beta_j| + \lambda^2)/(2(a-1)) & \lambda < |\beta_j| \leqslant a\lambda \\ (a+1)\lambda^2/2 & |\beta_j| > a\lambda \end{cases}, \tag{5.12}$$

where $a > 2$ is a prespecified parameter.

The elastic net uses a mixture of ℓ_1 and ridge penalty:

$$J_\lambda(\beta_j; \alpha) = \lambda \left(\alpha|\beta_j| + (1-\alpha)\beta_j^2 \right). \tag{5.13}$$

The elastic net intends to combine the strength of both the LASSO and the ridge regression: it explores sparsity of the regression coefficient like the LASSO, and is more stable to compute in high dimensional problems like the ridge regression.

The adaptive LASSO uses the following penalty

$$J_\lambda(\beta_j; \gamma) = \lambda|\beta_j|/|\hat{\beta}_j^{\mathrm{LS}}|^\gamma \tag{5.14}$$

where $\gamma > 0$ is a prespecified parameter. It is worth noting that in the special case when $\gamma = 1$, the adaptive LASSO is the similar to the nonnegative garrote except that it does not impose the condition that β_j has the same sign as $\hat{\beta}_j^{\mathrm{LS}}$. In the special case of orthogonal design, the nonnegative garrote and the adaptive LASSO with $\gamma = 1$ are equivalent.

6 Towards more interpretable models

Thus far, we have considered only general-purpose variable selection techniques. In practice, the predictors are often related. Taking their relationship into account is essential in constructing meaningful models. We give here two specific examples.

6.1 Group variable selection

In many regression problems we are interested in finding important explanatory factors in predicting the response variable, where each explanatory factor may be represented by a group of derived input variables. The most common example is the multi-factor ANOVA problem, in which each factor may have several levels and can be expressed through a group of dummy variables. The goal of ANOVA is often to select important main effects and interactions for accurate prediction, which amounts to the selection of groups of derived input variables. Another example is the additive model with polynomial or nonparametric components. In both situations, each component in the additive model may be expressed as a linear combination of a number of basis functions of the original measured variable. In such cases the selection of important measured variables corresponds to the selection of groups of basis functions. In both of these two examples, variable selection typically amounts to the selection of important factors (groups of variables) rather than individual derived variables, as each factor corresponds to one measured variable and is directly related to the measurement cost.

To illustrate the necessity of group variable selection, we consider the TLI dataset, accessible from the **XTABLE** package of R. It contains math scores and demographic data of 100 randomly selected students participating in the Texas Assessment of Academic Skills (TAAS). One of the goals is to relate the math score of the students (**tlimat**) to their grade (**Grade**), gender (**Sex**), race (**Ethncity**), and whether or not their family is economically disadvantaged (**Disadvg**). Grade ranges from the 3rd grade to 6th grade. Ethnicity can take values from Black, White, Hispanic and Other Races. The other two factors have two levels. For illustration purpose, we apply LASSO to the data. Following common practice, we can introduce dummy variables to represent these factors. For **Grade**, we use the 3rd grade as the baseline whereas for Ethnicity, we use Other Races as the baseline.

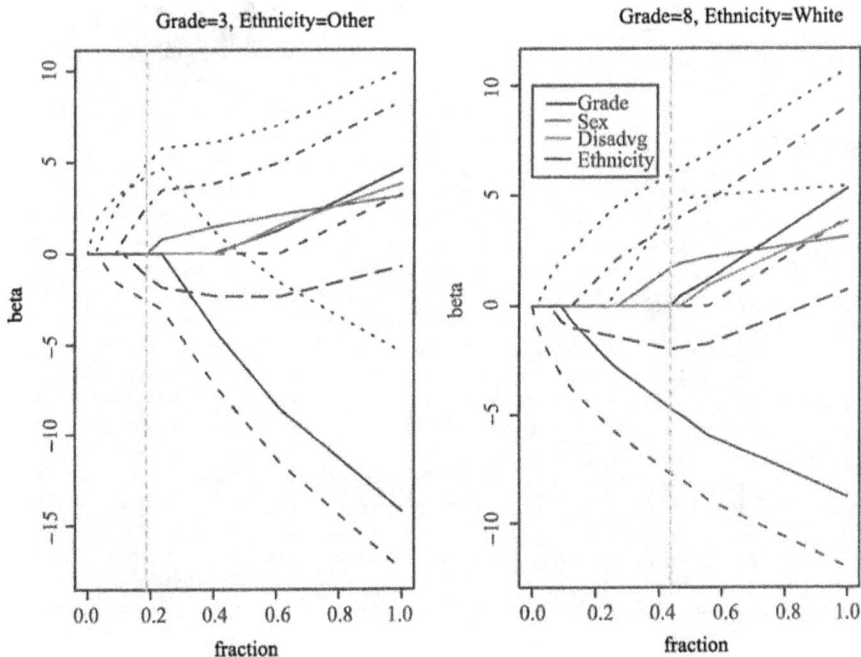

Figure 6.1   LASSO applied to TLI data with different ways of representing the categorical factor. The coefficients corresponding to dummy variables of the same factor is given in the same color.

The left panel of Figure 6.1 gives the solution path of the LASSO. The vertical line corresponds to the tuning parameter selected by ten fold cross validation. The problem, however, is that such a procedure is not invariant to the coding of the

categorical factors. If we instead use the 8th grade and white as the baseline for
Grade and Ethnicity respectively. The same procedure yields the solution path
given in the right panel. Clearly, such ambiguity makes the selected model hard
to interpret.

One way of overcoming this problem is to use the group LASSO (Yuan and
Lin, 2006). Rewrite the regression equation in terms of factors:

$$Y = \sum_{j=1}^{J} X_j' \beta_j + \epsilon, \tag{6.1}$$

where X_j is a p_j dimensional variables corresponding to the jth factor, and β_j is
a coefficient vector of size p_j, $j = 1, ..., J$. The group LASSO estimate is defined

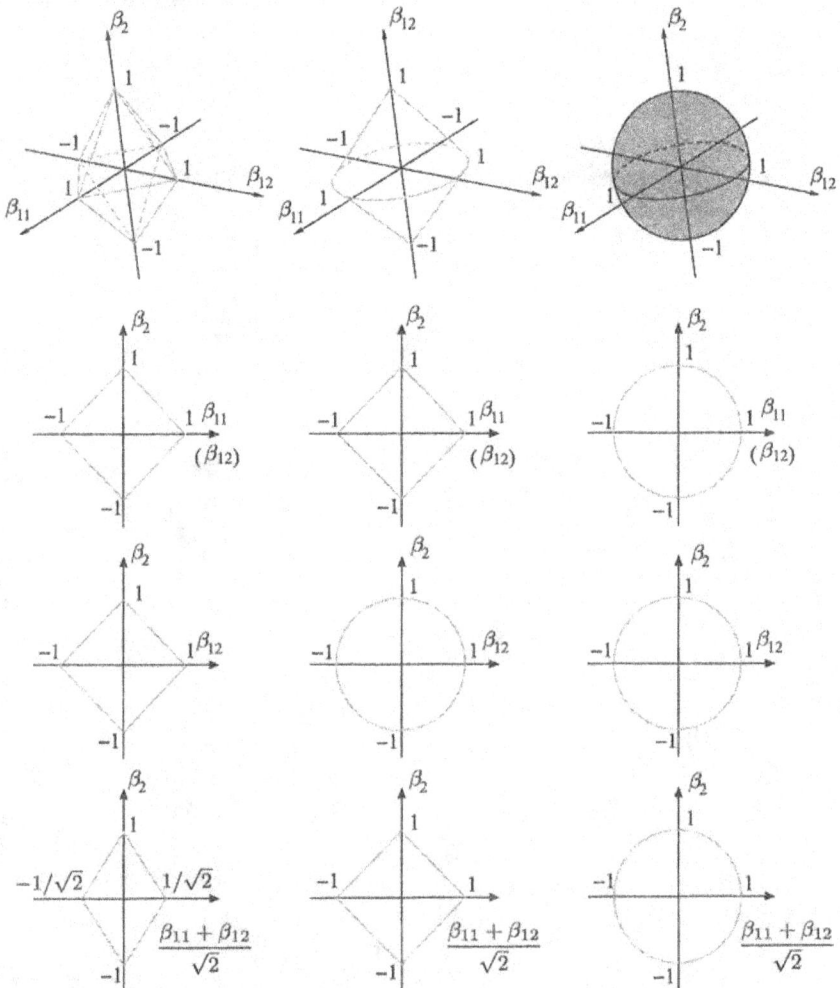

Figure 6.2   The ℓ_1 penalty (left panels), Group LASSO penalty (central panels) and ℓ_2 penalty
(right panels).

as the solution to

$$\left\| \mathbf{Y} - \sum_{j=1}^{J} \mathbf{X}_j \beta_j \right\|^2 + \lambda \sum_{j=1}^{J} ||\beta_j||_{\ell_2}, \qquad (6.2)$$

where $\lambda \geqslant 0$ is a tuning parameter. The penalty function used in the group LASSO is intermediate between the ℓ_1 penalty used in the LASSO and ℓ_2 penalty used in ridge regression. This is illustrated in Figure 6.2. Consider a case in which there are two factors, and the corresponding coefficients are a 2-vector $\beta_1 = (\beta_{11}, \beta_{12})'$ and a scalar β_2. The top panels of Figure 6.2 depict the contour of the penalty functions. The leftmost panel corresponds to the ℓ_1 penalty $|\beta_{11}| + |\beta_{12}| + |\beta_2| = 1$, the central panel corresponds to $||\beta_1|| + |\beta_2| = 1$, and the rightmost panel corresponds to $||(\beta_1', \beta_2)'|| = 1$. The intersections of the contours with planes $\beta_{12} = 0$ (or $\beta_{11} = 0$), $\beta_2 = 0$, and $\beta_{11} = \beta_{12}$, are shown in the next three rows of Figure 6.2. As shown in Figure 6.2, the ℓ_1 penalty treats the three coordinate directions differently from other directions, and this encourages sparsity in individual coefficients. The ℓ_2 penalty treats all directions equally, and does not encourage sparsity. The group LASSO encourages sparsity at the factor level.

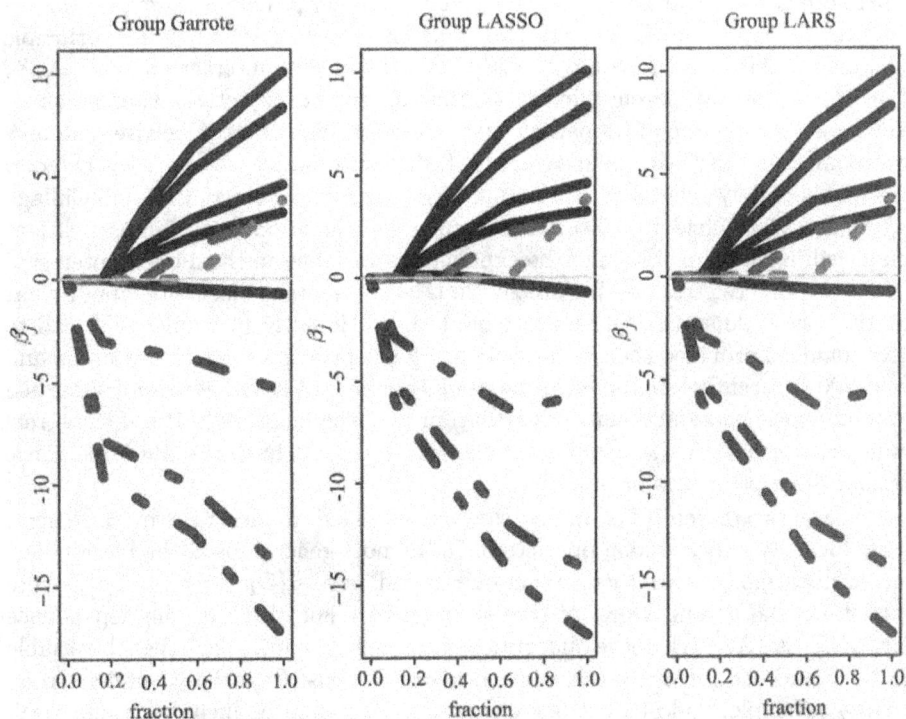

Figure 6.3 Group garrote, group LASSO and group LARS applied to the TLI data. Black lines correspond to dummy variables associated with Grade, red lines to Sex, green lines to Disadvg, and blue lines to Ethnicity.

Yuan and Lin (2006) also considered generalizations of the nonnegative garrote and LARS to handle the group structure. For illustration purpose, Figure 6.3
shows the solution paths of the group garrote, group LASSO and group LARS.
All three methods are invariant to the coding to the categorical factors.

Alternative regularization approaches can also be found in Zhao, Rocha and
Yu (2006) and Zhou and Zhu (2007).

6.2   Enforcing heredity principle

In many practical situations, there is a hierarchical structure among the predictors. Consider, for example, multiple linear regression with both main effects and
two-way interactions where a dependent variable Y and q explanatory variables
$X_1, X_2, ..., X_q$ are related through

$$Y = \beta_1 X_1 + \cdots + \beta_q X_q + \beta_{11} X_1^2 + \beta_{12} X_1 X_2 + \cdots + \beta_{qq} X_q^2 + \epsilon, \qquad (6.3)$$

where $\epsilon \sim \mathcal{N}(0, \sigma^2)$. Commonly used general purpose variable selection techniques
do not distinguish interactions $X_i X_j$ from main effects X_i and can select a model
with an interaction but neither of its main effects, i.e., $\beta_{ij} \neq 0$ and $\beta_i = \beta_j = 0$. It
is therefore useful to invoke the so-called effect heredity principle (Hamada and Wu
1992) in this situation. There are two popular versions of the heredity principle
(Chipman 1996). Under *strong heredity*, for a two-factor interaction effect $X_i X_j$
to be active both its parent effects, X_i and X_j, should be active; whereas under
weak heredity only one of its parent effects needs to be active. Likewise, one may
also require that X_i^2 can be active only if X_i is also active. The strong heredity
principle is closely related to the notion of marginality (Nelder, 1977; McCullagh
and Nelder, 1989; Nelder, 1994) which ensures that the response surface is invariant
under scaling and translation of the explanatory variables in the model. Interested
readers are also referred to McCullagh (2002) for a rigorous discussion about what
criteria a sensible statistical model should obey. Heredity principles will lead to
better models, provided the true model obeys such principles. Li, Sudarsanam and
Frey (2006) recently conducted a meta-analysis of 113 data sets from published
factorial experiments and concluded that an overwhelming majority of these real
studies conform with the heredity principles. This clearly shows the importance
of using these principles in practice.

These two heredity concepts can be extended to describe more general
hierarchical structure among predictors. In its most general form, the hierarchical
relationship among predictors can be represented by sets $\{\mathcal{D}_i : i = 1, \ldots, p\}$, where
$\mathcal{D}_i$ contains the parent effects of the ith predictor. For example, the dependence
set of $X_i X_j$ is $\{X_i, X_j\}$ in the quadratic model (6.3). In order that the ith variable
can be considered for inclusion, all elements of $\mathcal{D}_i$ must be included under strong
heredity principle, and at least one element of $\mathcal{D}_i$ should be included under weak
heredity principle. Other types of heredity principle such as the partial heredity
principle (Nelder, 1998) can also be incorporated in this framework. It is helpful
to think of the dependence structure described by the $\mathcal{D}$'s as a directed graph
where all p variables are the nodes and an edge from i to j is present if and only

if $j \in \mathcal{D}_i$. To elaborate, consider two three-level factors A and B. The hierarchy among all variables can be described by the diagram in Figure 6.4.

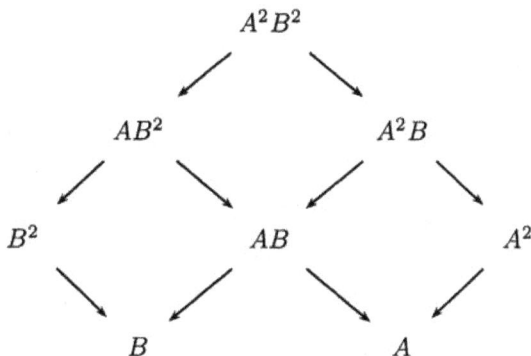

Figure 6.4   Dependence Structure for Two-way Interaction between Three-level Factors.

The dependence sets are then given by

$$\begin{aligned}
\mathcal{D}_{B^2} &= \{B\}, \\
\mathcal{D}_{AB} &= \{A, B\}, \\
\mathcal{D}_{A^2} &= \{A\}, \\
\mathcal{D}_{AB^2} &= \{B^2, AB\}, \\
\mathcal{D}_{A^2B} &= \{AB, A^2\}, \\
\mathcal{D}_{A^2B^2} &= \{AB^2, A^2B\}.
\end{aligned}$$

Both the nonnegative garrote and LARS have been recently generalized to enforce such hierarchical structure among the predictors (Turlach, 2004; Yuan, Joseph and Lin, 2007; Yuan, Joseph and Zou, 2007). In particular, consider the nonnegative garrote. To fix idea, we focus on the strong heredity principles which requires the inclusion of all variables in $\mathcal{D}_j$ if the jth variable is included in the model. Since $\widehat{\beta}_i^{LS} \neq 0$ with probability one, X_i will be selected if and only if scaling factor $d_i > 0$, in which case, d_i behaves more or less like an indicator of the inclusion of X_i in the selected model. Therefore, the strong heredity principles can be enforced by requiring

$$d_i \leqslant d_j, \qquad \forall j \in \mathcal{D}_i \tag{6.4}$$

in computing $d's$ in the original nonnegative garrote procedure. Note that if $d_i > 0$, (6.4) will force the scaling factor for all its parents to be positive and consequently active. Figure 6.5 illustrates the feasible region of the nonnegative garrote with such constraints in contrast with the original nonnegative garrote where no heredity rules are enforced. We consider two effects and their interaction with the corresponding shrinking factors denoted by d_1, d_2 and d_{12} respectively. In both

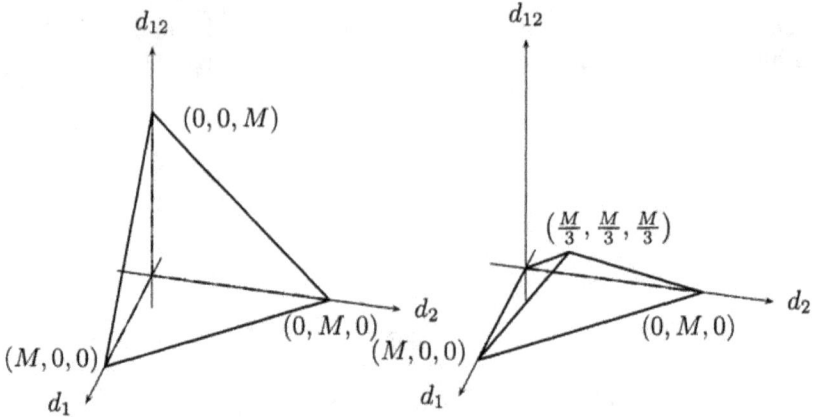

Figure 6.5   Feasible region of the nonnegative garrote with (right) and without (left) strong heredity constraints.

situations, the feasible region is a convex polyhedron in the three dimensional space.

To demonstrate the importance of respecting the heredity principles, we apply the nonnegative garrote, both with or without the additional constraints (6.4) to the prostate data, previously used in Tibshirani (1996). The data consist of the medical records of 97 male patients who were about to receive a radical prostatectomy. The response variable is the level of prostate specific antigen, and there are 8 explanatory variables. The explanatory variables are eight clinical measures: log(cancer volume) (lcavol), log(prostate weight) (lweight), age, log(benign prostatic hyperplasia amount) (lbph), seminal vesicle invasion (svi), log(capsular penetration) (lcp), Gleason score (gleason) and percentage Gleason scores 4 or 5 (pgg45). We consider model (6.3) with main effects, quadratic terms and two way interactions, which gives us a total of 44 predictors. Figure 6.6 gives the solution path for both methods.

Using 10-fold cross validation, the original nonnegative garrote that neglects the effect heredity chooses a six variable model: lcavol, lweight, lbph, gleason^2, lbph:svi and svi:pgg45. Note that this model does not satisfy heredity principle, because gleason^2 and svi:pgg45 are included without any of its parent factors. In contrast, the nonnegative garrote with strong heredity selects a model with seven variables: lcavol, lweight, lbph, svi, gleason, gleason^2 and lbph:svi; and yields smaller estimate of the prediction error.

Alternative approaches for incorporating the strong heredity principle in variable selection have also been introduced by Chipman (1996), Chipman, Hamada and Wu (1997), Efron et al. (2004), Turlach (2004), Zhao, Rocha and Yu (2006), and Choi, Li and Zhu (2007). To enforce the more general heredity principle can be involved. Interested readers are referred to Chipman, Hamada and Wu (1997), Yuan, Joseph and Lin (2007), and Yuan, Joseph and Zou (2007) for further discussions.

Without Heredity   With Heredity

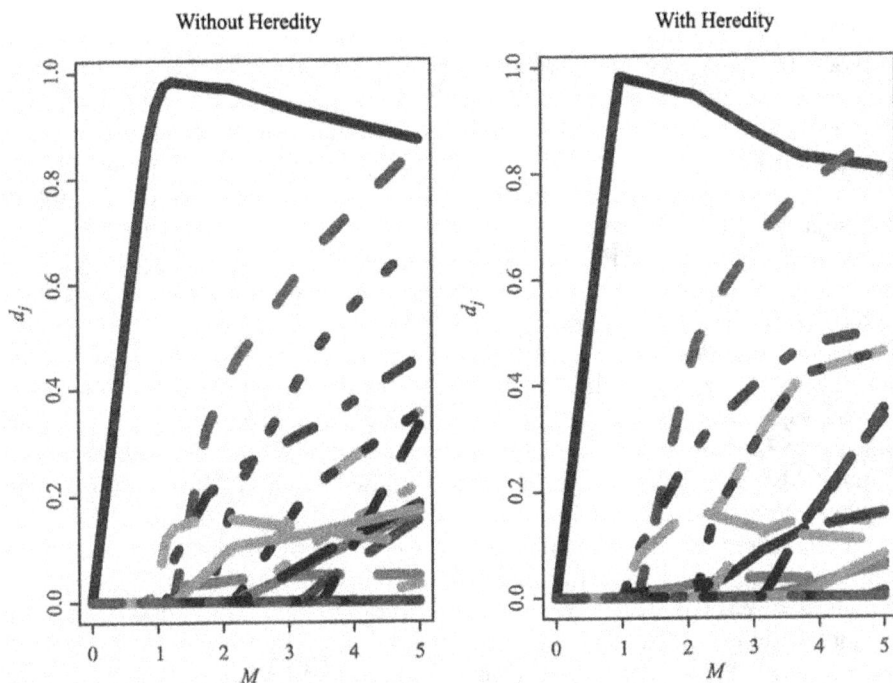

Figure 6.6   Solution Path for Different Versions of the Nonnegative Garrote.

7   Further readings

In this chapter, we provide a selective review of several commonly used variable selection methods, with the focus on highlighting their main ideas. Variable selection, although a classical problem, remains an actively researched area in recent years. What we discuss here is only a corner of an iceberg. There are a large number of interesting topics and literature absent in our discussion as we never intend to give a comprehensive review. We conclude by listing several notable examples.

With the recent advance in science and technology, high dimensional data have become a commonplace. With large datasets computation becomes critical for variable selection. For example, fast computation via the LARS algorithm contributes at least partly to the tremendous success of the LASSO estimator even though it now becomes clear that it may not always be ideal for the purpose of variable selection (Fan and Li, 2001; Zou, 2006). As a result, much attention has been devoted to improved algorithms and efficient computation of various variable selection techniques. See Fu (1999), Osborne, Presnell and Turlach (2000), Zhu, Rosset, Hastie and Tibshirani (2003), Rosset (2004), Park and Hastie (2007), Rosset and Zhu (2007), Yuan and Lin (2007), Zhao and Yu (2007), Friedman, Hastie and Tibshirani (2008), Li and Zhu, (2008), Yuan and Zou (2009), and Zou and Li (2008) among a fast growing literature.

With the emergence of so many variable selection techniques, it becomes important to understand the pros and cons of different approaches. From a theoretical point of view, this can traditionally be measured by the consistency and efficiency. With the number of variables p fixed and n goes to infinity, a variable selection procedure is called consistent if it identifies the true model with probability tending to one; and efficient if the model it selects achieves minimum mean squared error. There is however little agreement between consistency and efficiency. For example, FPE and AIC are known to be efficient whereas BIC is known to be consistent, but not the other way around. More recently, Fan and Li (2001) introduce the concept of oracle property for variable selection in that the relevant variables can be identified with probability tending to one and at the same time their corresponding coefficients estimated as efficient as if the set of variables were given apriori. In particular, the nonnegative garrote, SCAD, and adaptive LASSO are known to enjoy the oracle property, whereas LASSO, elastic net and LARS do not have such property (Fan and Li, 2001; Zhao and Yu, 2006; Zou, 2006; Yuan and Lin, 2007).

Another important issue that deserves perhaps more attention that it has received is post model selection inferences. Routinely, statistical inferences are done given a selected model. This clearly overlooks the potential variation due to model selection. This topic has been touched upon by Pötscher (1991), Shen, Huang and Ye (2004), Leeb and Pötscher (2005), and Leeb and Pötscher (2006) among others.

References

[1] Akaike H. (1969). Fitting autoregressive models for prediction. *Annals of Institute of Mathematical Statistics* **21**, 243-247.

[2] Akaike H. (1973). Information theory and an extension of the maximum likelihood principle. in *2nd International Symposium on Information Theory*, (Petrov, B. and Czàik, F., eds), 261-281, Akademiai Kiadò, Budapest.

[3] Akaike H. (1978). A Bayesian analysis of the minimum AIC procedure. *Annals of Institute of Mathematical Statistics* **30**, 9-14.

[4] Anderson T., Sweeney D. and Williams D. (2003). *Modern Business Statistics*. South-Western College Pub, Americas.

[5] Berger J. and Pericchi L. (1996). The intrinsic Bayes factor for model selection and prediction. *Journal of the America Statistical Association* **91**, 109-122.

[6] Breiman L. (1992). Submodel selection and evaluation in regression: The X-fixed case and little bootstrap. *Journal of the American Statistical Association* **87**, 734-751.

[7] Breiman L. (1995). Better subset regression using the nonnegative garrote. *Technometrics* **37**, 373-384.

[8] Breiman L. (1996). Heuristics of instability and stabilization in model selection. *Annals of Statistics* **24**, 2350-2383.

[9] Brown P. J., Vannucci M. and Fearn T. (1998). Multivariate Bayesian variable selection and prediction. *Journal of the Royal Statistics Society Series B* **60**, 627-642.

[10] Carlin B. and Chib S. (1995). Bayesian model choice via Markov Chain Monte Carlo. *Journal of the Royal Statistics Society Series B* **66**, 473-484.

[11] Chen S. S., Donoho D. L. and Sauders M. A. (1999). Atomic decomposition by basis pursuit. *SIAM J. Scientific Computing* **20**, 33-61.

[12] Chipman H. (1996). Bayesian variable selection with related predictors. *Canadian Journal of Statistics* **24**, 17-36.

[13] Chipman H., Hamada M. and Wu C. F. J. (1997). A Bayesian variable selection approach for analyzing designed experiments with complex aliasing. *Technometrics* **39**, 372-381.

[14] Choi N., Li W. and Zhu J. (2010). Variable selection with the strong heredity constraint and its oracle property. *Journal of the American S'tatistical Association* **105**, 354-364.

[15] Clyde M., Parmigiani G. and Vidakovic B. (1998). Multiple shrinkage and subset selection in wavelets. *Biometrika* **85**, 391-402.

[16] Efron B., Johnstone I., Hastie T. and Tibshirani R. (2004). Least angle regression (with discussions). *The Annals of Statistics* **32**, 407-499.

[17] Fan J. and Li R. (2001). Variable selection via nonconcave penalized likelihood and its oracle properties. *Journal of the American Statistics Association* **96**, 1348-1360.

[18] Foster D. P. and George E. I. (1994). The risk inflation criterion for multiple regression. *Annals of Statistics* **22**, 1947-1975.

[19] Friedman J., Hastie T. and Tibshirani R. (2007). Pathwise coordinate optimization. *Annals of Applied Statistics* **1**, 302-332.

[20] Fu W. J. (1999). Penalized regressions: the bridge versus the lasso. *Journal of Computational and Graphical Statistics* **7**(3), 397-416.

[21] George E. I. and Foster D.P. (2000). Calibration and empirical Bayes variable selection. *Biometrika* **87**, 731-747.

[22] George E. I. and McCulloch R. E. (1993). Variable selection via Gibbs sampling. *Journal of the American Statistics Association* **88**, 881-889.

[23] George E. I. and McCulloch R. E. (1997). Approaches for Bayesian variable selection. *Statistica Sinica* **7**, 339-373.

[24] Green P. (1995). Reversible jump Markov Chain Monte Carlo computation and Bayesian model determination. *Biometrika* **82**, 711-732.

[25] Hamada M. and Wu C. F. J. (1992). Analysis of designed experiments with complex aliasing. *Journal of Quality Technology* **24**, 130-137.

[26] Hastie T., Rosset S., Tibshirani R. and Zhu J. (2005). The entire regularization path for the support vector machine. *Journal of Machine Learning Research* **5**, 1391-1415.

[27] Ishwaran H. and Rao J. (2005). Spike and slab variable selection: frequentist and Bayesian strategies. *Annals of Statistics* **33**, 730-773.

[28] Kass R. and Raftery A. (1995). Bayes factors. *Journal of the American Statistics Association* **90**, 773-795.

[29] Leeb H. (2008). Conditional predictive inference after model selection. To appear in *Annals of Statistics*.

[30] Leeb H. and Pötscher B. (2006). Can one estimate the conditional distribution of post-model-selection estimators? *Annals of Statistics* **34**, 2554-2591.

[31] Leeb H. and Pötscher B. (2005). Model selection and inference: Facts and fiction. *Econometric Theory* **21**, 21-59.

[32] Li X., Sundarsanam N. and Frey D. (2006). Regularities in data from factorial experiments. *Complexity* **11**, 32-45.

[33] Li Y. and Zhu J. (2008). L_1-norm quantile regression. *Journal of Computational and Graphical Statistics* **17**, 163-185.

[34] Linhart H. and Zucchini W. (1986). *Model Selection.* Wiley, New York.

[35] Mallows C. (1973). Some comments on C_p. *Technometrics* **15**, 661-675.

[36] McCullagh P. (2002). What is a statistical model (with discussions). *Annals of Statistics* **30**, 1225-1310.

[37] McCullagh P. and Nelder J. (1989). *Generalized Linear Models*, 2nd ed. Chapman and Hall, London.

[38] McQuarrie A. and Tsai C. (1998). *Regression and Time Series Model Selection.* World Scientific, Singapore.

[39] Miller A. (1990). *Subset Selection in Regression.* Chapman and Hall, London.

[40] Nelder J. (1977). A reformulation of linear models. *Journal of the Royal Statistics Society Series A* **140**, 48-77.

[41] Nelder J. (1994). The statistics of linear models. *Statistics and Computing* **4**, 221-234.

[42] Nelder J. (1998). The selection of terms in response-surface models – how strong is the weak-heredity principle? *The American Statistician* **52**, 315-318.

[43] O'Hagan A. (1995). Fractional Bayes factors for model comparison (with discussion). *Journal of the Royal Statistics Society Series B* **57**, 99-138.

[44] Osborne M., Presnell B. and Turlach B. (2000). A new approach to variable selection in least squares problems. *IMA Journal of Numerical Analysis* **20**, 389-403.

[45] Park M. and Hastie T. (2007). L_1 regularization path algorithm for generalized linear models. *Journal of the Royal Statistical Society, Series B* **69**, 659-677.

[46] Pötscher B. (1991). Effects of model selection on inference. *Econometric Theory* **7**, 163-185.

[47] Rosset S. (2004). Tracking curved regularized optimization solution paths. *Advances in Neural Information Processing Systems 13.* MIT Press, Cambridge.

[48] Rosset S. and Zhu J. (2007). Piecewise linear regularized solution paths. *Annals of Statistics* **35**, 1012-1030.

[49] Schwarz G. (1978). Estimating the dimension of a model. *Annals of Statistics* **6**, 461-464.

[50] Shen X., Huang H. and Ye J. (2004). Inference after model selection. *Journal of the American Statistical Association*, **99**, 751-762.

[51] Shen X. and Ye J. (2002). Adaptive model selection. *Journal of the American Statistical Association*, **97**, 210-221.

[52] Shao J. (1993). Linear model selection by cross-validation. *Journal of the American Statistical Association* **91**, 655-665.

[53] Shao J. (1996). Boostrap model selection. *Journal of the American Statistical*

Association **88**, 486-494.

[54] Shibata R. (1997). Bootstrap estimate of Kullback-Leibler information for model selection. *Statistica Sinica* **7**, 375-394.

[55] Smith M. and Kohn R. (1997). Nonparametric regression using Bayesian variable selection. *Journal of Econometrics* **75**, 317-344.

[56] Spiegelhalter D., Best N., Carlin B. and Linde A. (2002). Bayesian measures of model complexity and fit. *Journal of the Royal Statistical Society B* **64**, 583-639.

[57] Stone M. (1974). Cross-validatory choice and assessment of statistical prediction. *Journal of the Royal Statistics Society Series B* **39**, 111-133.

[58] Tibshirani R. (1996). Regression shrinkage and selection via the lasso. *Journal of the Royal Statistics Society Series B* **58**, 267-288.

[59] Turlach B. (2004). Discussion of "Least angle regression". *The Annals of Statistics* **32**, 481-490.

[60] Wood S., Kohn R., Shively T. and Jiang W. (2002). Model selection in spline nonparametric regression. *Journal of the Royal Statistics Society Series B* **64**, 119-139.

[61] Yuan M., Joseph V. R. and Zou H. (2007). Structured variable selection and estimation. to appear in *Annals of Applied Statistics*.

[62] Yuan M., Joseph V. R. and Lin Y. (2007). An efficient variable selection approach for analyzing designed experiments. *Technometrics* **49**, 430-439.

[63] Yuan M. and Lin Y. (2005). Efficient Empirical Bayes Variable Selection and Estimation in Linear Models. *Journal of the American Statistics Association* **100**, 1215-1225.

[64] Yuan M. and Lin Y. (2006). Model selection and estimation in regression with grouped variables. *Journal of the Royal Statistics Society Series B* **68**, 49-67.

[65] Yuan M. and Lin Y. (2007). On the the nonnegative garrote estimator. *Journal of the Royal Statistics Society Series B* **69**, 143-161.

[66] Yuan M. and Zou H. (2009). Efficient global approximation of generalized nonlinear ℓ_1-regularized solution paths and its applications. *Journal of the American Statistical Association* **104**, 1562-1574.

[67] Zhao P., Rocha G. and Yu B. (2006). Grouped and hierarchical model selection through composite absolute penalties. *Annals of Statistics*.

[68] Zhao P. and Yu B. (2006). On model selection consistency of Lasso. *Journal of Machine Learning Research* **7**, 2481-2514.

[69] Zhao P. and Yu B. (2007). Stagewise lasso. *Journal of Machine Learning Research* **8**, 2701-2726.

[70] Zou H. (2006). The adaptive Lasso and its oracle properties. *Journal of the American Statistical Association* **101**, 1418-1429.

[71] Zou H. and Hastie T. (2005). Regression and variable election via the elastic net. *Journal of the Royal Statistical Society, Series B* **67**, 301-320.

[72] Zou H. and Li R. (2008). One-step sparse estimates in nonconcave penalized likelihood models. *Annals of Statistics* **36**, 1509-1533.

Chapter 5

Bayesian Variable Selection in Regression with Networked Predictors *

Feng Tai,[†]   Wei Pan[‡]   and   Xiaotong Shen [§]

Abstract

This chapter considers Bayesian variable selection in linear regression when the relationships among a possibly large number of predictors are described by a network given *a priori*. A class of motivating examples is to predict some clinical outcomes with high-dimensional gene expression profiles and a gene network, for which it is assumed that the genes neighboring to each other in the network are more likely to participate together in relevant biological processes and thus more likely to be simultaneously included in (or excluded from) the regression model. To account for spatial correlations induced by a predictor network, rather than using an independent (and identical) prior distribution for each predictor's being included in the model as implemented in the standard approach of stochastic search variable selection (SSVS), we propose a Gaussian Markov random field (MRF) and a binary MRF as priors. We evaluate and compare the performance of the new methods against the standard SSVS through both simulated and real data.

Keywords: Binary Markov random field (BMRF), Gaussian Markov random field (GMRF), gene network; high dimension, Markov chain Monte Carlo (MCMC), microarray data; stochastic search variable selection (SSVS).

1   Introduction

Consider linear regression with "large p, small n" data as arising in genomics, in which we predict some clinical outcome using high-dimensional gene expression profiles. In such an application, variable (or gene) selection is crucial for predictive performance and elucidating underlying biological processes. Most existing methods for variable selection are generic, ignoring subject-matter prior knowledge

---

*The authors were partially supported by NIH grants HL65462 and GM081535.

†Division of Biostatistics, School of Public Health, University of Minnesota, Minneapolis, MN 55455, USA.

‡Division of Biostatistics, School of Public Health, University of Minnesota, Minneapolis, MN 55455, USA, Email: weip@biostat.umn.edu

§School of Statistics, University of Minnesota, Minneapolis, MN 55455, USA, Email: xshen@stat.umn.edu

on predictors. For example, a popular Bayesian variable selection method is the Stochastic Search Variable Selection (SSVS) proposed by George and McCulloch (1993, 1997). SSVS introduces a latent binary vector γ to indicate whether a predictor or variable should be included in the model or not, and uses a Bayesian hierarchical model to estimate γ for variable selection. The regression coefficient β_i follows a normal mixture distribution, $\pi(\beta_i|\gamma) = (1-\gamma_i)N(0, v_0) + \gamma_i N(0, v_1)$. Lee *et al.* (2002) applied SSVS to microarray data in the context of classification, using a mixture of a normal and a point mass instead, $\pi(\beta_i|\gamma) = (1-\gamma_i)I_0 + \gamma_i N(0, v_1)$, treating all the genes equally *a priori* by giving independent and identical priors for the probability of a gene being in the final model; i.e. $\pi(\gamma) \equiv 1/2^p$. In Bae and Mallick (2004), instead of using the indicator vector γ for variable selection, they modeled β by assuming $\beta|\Lambda \sim N(0, \Lambda)$, $\Lambda = \text{diag}(\lambda_1, \ldots, \lambda_p)$ and put three different priors on Λ. Variable (gene) selection was based on thresholding the posterior of λ_i to eliminate genes with small variances λ_i.

On the other hand, there has been rapidly accumulating biological knowledge in the form of various gene networks. A gene network can be expressed as an undirected graph with nodes representing genes and edges representing interactions between genes, which provides a natural neighborhood structure for any gene. The importance of incorporating biological knowledge into genomic data analysis has been increasingly recognized. For instance, in a different context of detecting differentially expressed genes, Wei and Li (2007), and Wei and Pan (2008) proposed binary Markov random field (BMRF) and Gaussian MRF (GMRF) models to account for the local dependency of the genes in a network, In the context of linear regression, Li and Li (2008) and Pan et al (2009) proposed network-based penalty functions for variable selection in the framework of penalized regression, in which some smoothness assumption on the regression coefficients is imposed. Here we would like to take a Bayesian approach, which differs from the above penalized regression methods in that we have a less stringent smoothness assumption: we only assume the smoothness of the prior probabilities of the predictors' being selected, rather than of their effect sizes (i.e. regression coefficients). Specifically, we investigate three different spatial priors in the framework of SSVS, targeting applications to regression analysis for high-dimensional microarray data. Instead of treating all the genes independently and identically *a priori*, we assign dependent priors to reflect the relationships among the genes over a gene network. We introduce three different priors to model the potential spatial correlations among the genes based on their network structure. Specifically, we assume the probability of a gene's being informative depends on that of its direct neighbors in the network. In other words, we assume a spatial dependency among γ's as induced by the network.

Markov random field models for binary spatially correlated variables have been widely used in image analysis and spatial statistics for accounting for local dependencies. The basic autologistic model was developed by Besag (1972, 1974) with a broad range of applications, as shown by Heikkinen and Högmander (1994) and Hoeting *et al.* (2000). Weir and Pettitt (2000) proposed a hidden conditional autoregressive Gaussian process to model binary spatially correlated responses. Wei and Pan (2008) used a GMRF to model the prior probabilities of the binary states of some binary variables, and Wei and Pan (2009) compared the performance of the independence, GMRF and BMRF priors in the same context. Smith and Smith (2006) compared three binary Markov random fields, which are popular Bayesian priors for spatial smoothing. Smith and Fahrmeir (2007) extended Bayesian variable selection to a series of spatially linked regressions, incorporating the spatial correlation among the indicators γ by specifying a binary Markov random field prior. It is very similar to, but not exactly the same as, our method. They placed an Ising prior on some binary indicator variables across multiple regressions. A difference between their method and ours is that they had repeated measurements of each covariates from multiple sites, resulting in a matrix of binary indicators $\gamma = (\gamma_1, \ldots, \gamma_N)$ from locations $(1, \ldots, N)$. They modeled spatial correlations across different sites within each covariate. Specifically, for an N-dimension binary vector of covariate j, $\gamma_j = (\gamma_{j1}, \ldots, \gamma_{jN})'$, all elements in γ_j are assumed to be spatially correlated, but for all p covariates, $\gamma_1, \ldots, \gamma_p$ are assumed to be independent, i.e., $p(\gamma) = \prod_{j=1}^{p} \gamma_j$. However, in our method, we only have one "site" ($N = 1$), and consider the spatial correlation between covariates instead of within covariates. All elements of a p-vector $\gamma_N = (\gamma_{1N}, \ldots, \gamma_{pN})$ are spatially correlated based on a given network. During the preparation of this manuscript, we learned the recent work of Li and Zhang (2008), who proposed an Ising model to introduce a spatial prior for γ; Monni and Li (2009) proposed a different network-based prior for γ and considered both linear models for continuous responses and probit models for binary responses. In addition to some differences from theirs in implementations and applications, here we also study a GMRF model and a scaled BMRF (SBMRF) model.

The rest of this chapter is organized as follows. We first review SSVS, then propose our new methods with three Markov random field (MRF) models as priors: GMRF, BMRF and SBMRF. After describing some details on the posterior distributions and sampling schemes, we apply our methods to both simulated and real data, followed by a short discussion.

2 Statistical models

2.1 Review of SSVS

SSVS (George and McCulloch 1993, 1997) starts from the standard linear model

$$f(Y|\beta, \sigma) = N_n(X\beta, \sigma^2 I),$$

where Y is an $n \times 1$ vector of a response variable and $X = (X_1, \ldots, X_p)$ is an $n \times p$ matrix of predictors. The regression coefficient β is a $p \times 1$ unknown vector and σ is an unknown positive scalar.

For variable selection, we define a vector

$$\gamma = (\gamma_1, \ldots, \gamma_p)',$$

where $\gamma_i = 1$ or 0 indicates whether predictor i should be included in or excluded from the model respectively. We model the uncertainty underlying variable selection by a mixture prior $\pi(\beta, \sigma, \gamma) = \pi(\beta|\sigma, \gamma)\pi(\sigma|\gamma)\pi(\gamma)$, which can be conditionally specified as follows,

$$\pi(\beta|\sigma, \gamma) = N_p(0, D_\gamma R_\gamma D_\gamma),$$

where R_γ is a correlation matrix and D_γ is a diagonal matrix with its ith diagonal element denoted by

$$(D_\gamma^2)_{ii} = \begin{cases} v_{0_\gamma} & \text{if } \gamma_i = 0, \\ v_{1_\gamma} & \text{if } \gamma_i = 1. \end{cases}$$

With this prior, each component of β is modeled as coming from a mixture of two normals

$$\pi(\beta_i|\sigma, \gamma) = (1 - \gamma_i)N(0, v_{0_{\gamma(i)}}) + \gamma_i N(0, v_{1_{\gamma(i)}}).$$

The idea of variable selection is that, by setting $v_{0_{\gamma(i)}}$ and $v_{1_{\gamma(i)}}$ "small" and "large" respectively, if the data supports $\gamma_i = 0$ over $\gamma_i = 1$, then β_i should be small enough so that the corresponding predictor X_i plays a negligible role and thus should be excluded from the model. A simple choice for R_γ is $R_\gamma = I$. The residual variance σ^2 is conveniently modeled by an inverse gamma distribution,

$$\pi(\sigma^2|\gamma) = IG(\nu, \lambda).$$

The prior for γ has the form

$$\pi(\gamma) = \prod w_i^{\gamma_i}(1 - w_i)^{1-\gamma_i}.$$

For simplicity, usually $\pi(\gamma) \equiv 1/2^p$ is used to substantially reduce computational cost. We interpret $w_i = P(\gamma_i = 1)$ as the prior probability that β_i is large enough to have X_i included in the model.

Based on data Y, the posterior $\pi(\gamma|Y)$ updates the prior probabilities on each of the 2^p possible values of γ. The γ's with higher posterior probabilities $\pi(\gamma|Y)$ identify the more promising models that are more strongly supported by the data and the prior distribution. MCMC is usually used to explore the posteriors of β, σ and γ.

2.2 Spatial priors for γ

For the standard SSVS, $\pi(\gamma) = \prod w_i^{\gamma_i}(1 - w_i)^{1-\gamma_i}$, which implies the components of γ are *a priori* independent. In other words, the genes are treated independently *a priori*, and are further assumed to have the same prior probabilities to be included in the model by specifying $w_i \equiv w_0$ for all i, where w_0 is a pre-specified constant. In order to account for the dependency among the genes over a gene network, we propose to incorporate biological knowledge of the gene network by specifying a spatial prior for γ over the gene network. A gene network is expressed as an undirected graph with nodes for genes and edges for interactions between genes, which provides a natural neighborhood structure for a Markov Random Field. Here, we consider two different MRF models as priors.

2.2.1 Gaussian Markov Random Field (GMRF)

We define θ_i as a logit transformation of $w_i = Pr(\gamma_i = 1)$

$$\theta_i = \log\left(\frac{w_i}{1 - w_i}\right),$$

and model θ_i by an Intrinsic Gaussian Conditional Autoregression (ICAR) model (Besag and Kooperberg 1995):

$$\theta_i | \theta_{(-i)} \sim N\left(\frac{1}{m_i}\sum_{j \in \delta_i} \theta_j, \frac{T^2}{m_i}\right),$$

where $\theta_{(-i)} = \{\theta_j : j \neq i\}$, δ_i is a set of indices of direct neighbors of gene i, and $m_i = |\delta_i|$ is the size of δ_i as determined by a given gene network. The ICAR model accounts for spatial correlations and smoothness among the prior probabilities of the genes' being included in the model. The same idea can be found in Wei and Pan (2008), but in a different context.

2.2.2 Binary Markov random field (BMRF)

Instead of specifying a full conditional distribution of θ_is as in the ICAR model, a BMRF specifies a full conditional distribution of γ directly,

$$\pi(\gamma_i | \gamma_{(-i)}) \propto \exp(\alpha_0 + \alpha_1 k_i),$$

where $\gamma_{(-i)} = \{\gamma_j : j \neq i\}$, m_{i0} and m_{i1} are the numbers of $\gamma_j = 0$ and $\gamma_j = 1$ for $j \in \delta_i$ respectively, and $k_i = m_{i1} - m_{i0}$. This model is also called an autologistic model. The joint distribution of γ involves a normalizing factor $Z(\alpha)$, which depends on $\alpha = (\alpha_0, \alpha_1)'$ and is analytically intractable. A simple alternative to estimate α is to use a pseudo-likelihood approximation:

$$\mathrm{pl}(\alpha) = \prod_i \pi(\gamma_i | \gamma_{(-i)}).$$

Using the pseudo-likelihood is equivalent to regressing θ_i on k_i,

$$\theta_i = \log\left(\frac{w_i}{1 - w_i}\right) = \alpha_0 + \alpha_1 k_i.$$

Note that α_0 is closely related to the marginal probability $Pr(\gamma_i = 1 | \theta_i)$ for all $k_i = 0$. In practice, we specify α_0 to control the overall number of the genes (or variables) to be selected *a priori*. $\alpha_1 > 0$ is usually assumed, indicating that γ_i has a higher probability to be 1 than 0 if the number of 1's is greater than the number of 0's in its neighborhood. Another alternative is to replace k_i by a scaled $k_i^* = k_i/m_i$, where $m_i = m_{i1} + m_{i0}$ is the neighborhood size for gene i (Wei and Li 2008); we call it the scaled Binary Markov Random Field (SBMRF).

3 Estimation

3.1 Gibbs sampling

We use a Gibbs sampler to simulate posterior distributions. The full conditional posterior distribution for β is a multivariate normal distribution

$$\Pr(\beta|\sigma,\gamma,Y) = N(\Lambda X'Y, \sigma^2\Lambda),$$

where $\Lambda = (X'X + \sigma^2(D_\gamma R_\gamma D_\gamma)^{-1})^{-1}$, and we choose $R_\gamma = I$ for simplicity. σ^2 follows an inverse gamma distribution

$$\Pr(\sigma|\beta,Y) = IG\left(\frac{n}{2}+\nu, \frac{1}{2}\|Y - X\beta\|_2 + \lambda\right).$$

3.1.1 GMRF

For GMRF, we have

$$\Pr(\gamma_i|\beta,\theta) = \text{Ber}\left(\frac{a_i}{a_i+b_i}\right),$$

$$a_i = f(\beta_i|\gamma_i = 1) \cdot \frac{\exp(\theta_i)}{1 + \exp(\theta_i)}, \quad b_i = f(\beta_i|\gamma_i = 0) \cdot \frac{1}{1 + \exp(\theta_i)}. \tag{3.1}$$

The joint distribution of θ given all other parameters under the ICAR specification is

$$\Pr(\theta|\gamma, T^2) \propto \left(\prod_i \frac{\exp(\gamma_i\theta_i)}{1 + \exp(\theta_i)}\right) \exp\left(-\frac{1}{2T^2}\sum_{i\neq j} w_{ij}(\theta_i - \theta_j)^2\right)$$

and using an inverse gamma as the prior of T^2 leads to

$$\Pr(T^2|\theta) = IG\left(\frac{p-1}{2} + 0.5, \frac{1}{2}\sum_{i\neq j} w_{ij}(\theta_i - \theta_j)^2 + 0.005\right).$$

Rather than drawing θ as a vector, it is better to draw it component-wise from

$$\Pr(\theta_i|\gamma, T^2, \theta_{j\neq i}) \propto \left(\frac{\exp(\gamma_i\theta_i)}{1 + \exp(\theta_i)}\right) \exp\left(-\frac{m_i}{2T^2}(\theta_i - \frac{1}{m_i}\sum_{j\in\delta_i}\theta_j)^2\right).$$

Due to the log-concavity of $\Pr(\theta_i|\gamma, T^2, \theta_{j\neq i})$, an adaptive rejection sampling can be directly applied. Under the ICAR specification, the mean of θ_i's is undetermined. Hence, we put a constraint that $\sum_i \theta_i = \theta_0$, where θ_0 is a fixed number to reflect the prior belief of the proportion of the variables to be selected in the model. In practice, we found that sampling T^2 and θs at the same time might cause some convergence problems. Thus, we fixed $T^2 = 0.49$ in the following simulations and real data examples.

3.1.2 BMRF

For BMRF or SBMRF, we have

$$\Pr(\gamma_i|\beta,\theta) = \text{Ber}\left(\frac{a_i}{a_i + b_i}\right),$$

$$a_i = f(\beta_i|\gamma_i = 1) \cdot \frac{\exp(\alpha_0 + \alpha_1 k_i)}{1 + \exp(\alpha_0 + \alpha_1 k_i)}, \quad b_i = f(\beta_i|\gamma_i = 0) \cdot \frac{1}{1 + \exp(\alpha_0 + \alpha_1 k_i)}.$$

$$(3.2)$$

And

$$\Pr(\alpha|\gamma) = \left(\prod_i \frac{\exp(\gamma_i(\alpha_0 + \alpha_1 k_i))}{1 + \exp(\alpha_0 + \alpha_1 k_i)}\right) \pi(\alpha_0)\pi(\alpha_1).$$

To ensure $\alpha_1 > 0$, we use a gamma prior $G(\lambda, \nu)$ for α_1 and have $\pi(\alpha_1) = \alpha_1^{\lambda-1} \exp(-\nu\alpha_1)$. In this way, $Pr(\alpha_1|\gamma)$ is log-concave for $\lambda \geqslant 1$. Thus, an adaptive rejection sampling can be directly applied. In our applications, we used $G(\lambda = 3, \nu = 0.5)$ as the prior, with most of its mass between 0 and 15, which was used by Hoeting *et al.* (2000).

3.2   Computation

To reduce potential bias in parameter estimation, we updated the θ_i and γ_i in a random order. In MCMC sampling, the most costly step is to generate β from a multivariate normal distribution, which requires recomputing the inverse of a large covariance matrix. This step consumes almost the whole computing time due to the high dimensionality of the data. Thus in practice, the computing times are about the same for all four priors, even though the MRF priors have more parameters to estimate. For $p = 1329$ as in a real data example, the time of sampling 100 MCMC samples for all priors differed within 1 second.

3.3   Variable selection and response prediction

Variable selection is based on the marginal frequencies of the variables appearing in the posterior samples, i.e., the posterior mean of γ_is, reflecting the importance of each gene.

We predict a response by $\hat{y}$ based on each MCMC samples:

$$\hat{y} = \frac{1}{B} \sum_t X\hat{\beta}_t,$$

where B is the number of MCMC samples and $\hat{\beta}_t$ is the value of β in the tth MCMC sample. Thus the predictive model is not just only built on those genes with larger $\hat{\gamma}$, but possibly based on other genes. We also tried

$$\hat{y} = \frac{1}{B} \sum_t X\hat{\beta}_t\hat{\gamma}_t,$$

which produced similar results.

4   Results

To evaluate the performance of our proposed network-based SSVS, we conducted both simulations and real data studies with the four SSVS methods: the standard SSVS with an independent prior (SSVS+IND), SSVS with a GMRF prior (SSVS+GMRF), SSVS with a BMRF prior (SSVS+BMRF) and SSVS with a scaled BMRF prior (SSVS+SBMRF).

4.1   Simulations

Simulated data were generated from a linear regression model

$$Y = X\beta + \epsilon.$$

Two simple networks were considered.

1) A simple random network (RanN) consisting of $p = 100$ variables: First, we randomly divided 100 variables into 10 groups, and generated a graph containing 10 subgraphs, each corresponding to one of the 10 groups of variables. Each subgraph was completely connected and there was no edges between any two subgraphs. Then we randomly deleted 300 edges generating a graph having 100 nodes with a total of 271 edges. Next, we randomly added some edges to connect the 10 subgraph together. One of the 10 groups was selected to be informative (with variables numbered from 20 to 34), which contained 15 variables and 50 edges as shown in Figure 4.1. Those informative βs were simulated from $N(0, 2^2)$ and remaining βs were set to 0. Lastly, we simulated X from a multivariate normal distribution, $X \sim MVN(0, \mathbf{I})$.

2) A simple regulatory network (RegN) as used by Li and Li (2008): We had 10 transcription factors (TFs), each of which formed a subnetwork with its 10 regulated genes; there was no connection between any two subnetworks, or between any two regulated genes. The resulting network consisted of 110 genes and 100 edges. We assumed two TFs and their regulated genes were informative with non-zero regression coefficients, while the regression coefficients for the other genes were zero:

$$\beta = (5, \underbrace{\frac{5}{\sqrt{10}}, \ldots, \frac{5}{\sqrt{10}}}_{10}, -3, \underbrace{\frac{-3}{\sqrt{10}}, \ldots, \frac{-3}{\sqrt{10}}}_{10}, 0, \ldots, 0)'.$$

The expression levels of TFs were drawn independently from standard normal, $X_{TF_j} \sim N(0, 1)$, and the expression levels of the genes that TF_j regulated followed $N(0.7X_{TF_j}, 0.51)$.

In both the simulation set-ups, the random error ϵ was iid from $N(0, \sigma^2)$, where $\sigma^2 = \sum \beta_j^2 / r$ with the signal-to-noise ratio (SNR) $r = 2$ or 4. For the random network, we specified $w_i = Pr(\gamma_i = 1) = 15/100 = 0.15$ for the independence (IND) prior, and the constraints $\theta_0 = \text{logit}(0.15)$ for the GMRF model and $\alpha_0 = \text{logit}(0.15)$ for the BMRF and SBMRF models. A similar set-up was used

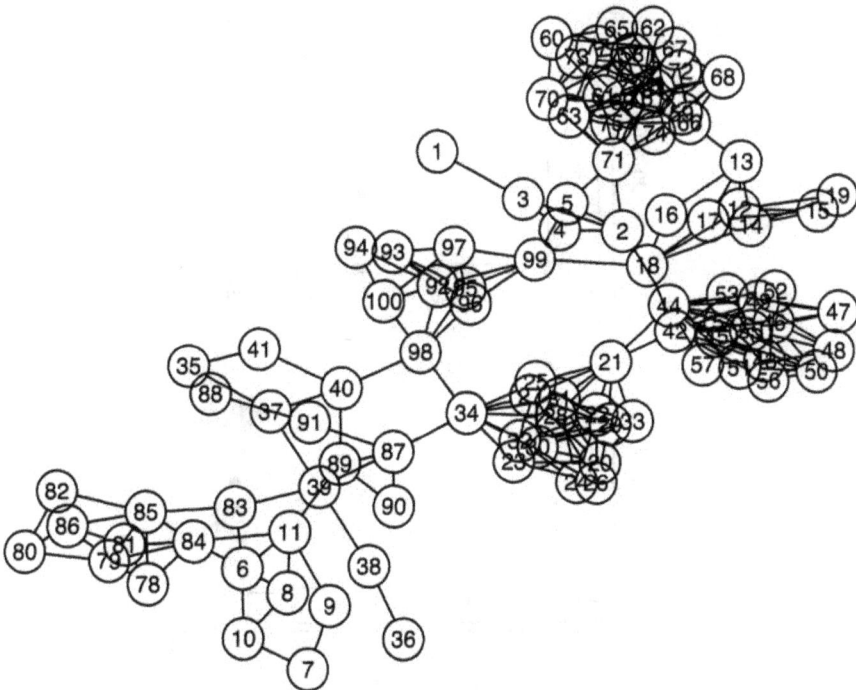

Figure 4.1   A random network used in simulation.

for the regulatory network, except $w_i = 22/110 = 0.2$ and $\theta_0 = \alpha_0 = \text{logit}(0.2)$. For each simulation run, we generated 50 training samples and 100 test samples; the simulation was repeated 100 times. In each run, 10,000 MCMC samples were generated with the first 8000 as the burn-in period. For the GMRF prior, we fixed $T^2 = 0.49$, finding that it worked well in practice. The starting values of θs were randomly generated from $N(\theta_0, 1)$. We randomly picked one simulation sample and applied three different random initial θ's; the results were very stable, indicating convergence. The results shown in Table 4.1. were based on only one initial value of θ. The prediction mean-squared error (PMSE) was calculated for each test data set. In Table 4.1, column *ninfo* shows the number of true informative genes in the top 15 (for RanN) or top 22 (for RegN) most frequently selected genes by each model. SSVS+GMRF had a smaller PMSE than SSVS+IND in all the situations, but selected a smaller proportion of informative genes for the regulatory network. SSVS+SBMRF had a smaller PMSE than SSVS+IND and included more informative genes in all situations. SSVS+BMRF had smaller PMSE than SSVS+IND only for the regulatory network when SNR=4, and included more informative genes except for the random network when SNR=2. If we pool γs from 200 runs (100 each from SNR=2 or 4) for the random network, we found all models performed well in terms of gene selection. Histograms of γ are shown in Figure 4.2. A dot line indicates the cut-off point for distinguishing signal from noise genes. The cut-off points for all models as shown in Figure 4.2 completely

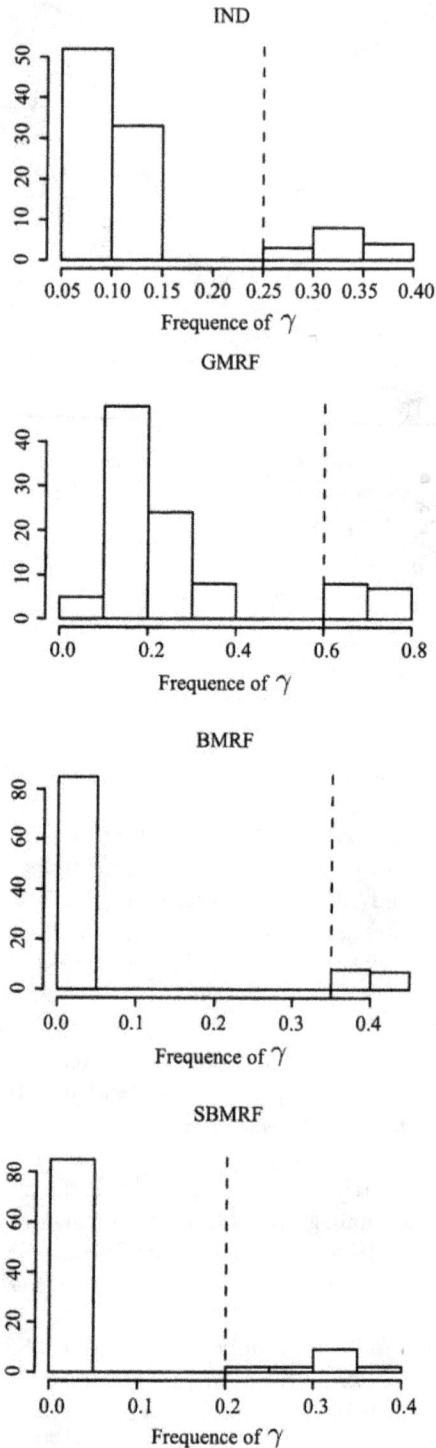

Figure 4.2   Histograms of $\hat{\gamma}_i$ in simulations for the random network.

separated signal and noise genes. In general, the MRF priors better separated signal and noise genes than the independence prior.

Table 4.1   Simulation results

Network	SNR	prior	ninfo	pmse
RanN (15)	2	IND	6.66 (0.15)	62.15 (2.43)
		GMRF	12.96 (0.24)	53.11 (2.07)
		BMRF	4.55 (0.60)	71.19 (3.06)
		SBMRF	9.38 (0.31)	60.11 (2.50)
	4	IND	7.81 (0.13)	34.72 (1.33)
		GMRF	14.63 (0.08)	26.98 (1.07)
		BMRF	9.83 (0.62)	35.31 (2.11)
		SBMRF	11.77 (0.29)	32.78 (1.54)
RegN (22)	2	IND	16.08 (0.18)	55.52 (1.22)
		GMRF	13.36 (0.73)	55.11 (2.62)
		BMRF	21.19 (0.14)	57.99 (1.77)
		SBMRF	21.63 (0.11)	52.66 (1.30)
	4	IND	17.31 (0.15)	33.47 (0.78)
		GMRF	13.27 (0.69)	30.52 (2.10)
		BMRF	21.79 (0.09)	30.23 (1.20)
		SBMRF	21.98 (0.01)	28.86 (0.83)

4.2   Two Real Data Examples

4.2.1 Glioblastoma Data

We applied our proposed methods to a microarray gene expression data set of glioblastoma studied by Horvath *et al.* (2006). Glioblastoma is the most common primary malignant brain tumor of adults and one of the most lethal of all cancers. Patients with this disease have a median survival of 15 months from the time of diagnosis despite surgery, radiation and chemotherapy. Gene expression data from two independent sets of clinical tumor samples ($n = 55$ and $n = 65$) were obtained using Affymetrix HG U133A genechips. The RMA normalization method (Irizarry *et al.*, 2003) was applied to the gene expression data. Here we aimed to build a predictive model for log survival time and to identify biologically important genes. Nine patients who remained alive by the end of study were excluded from analysis, leading to 50 and 61 samples for two data sets respectively. We combined two data sets together and deleted two outliers, whose survival times were extremely short, resulting in a total of 109 subjects. We randomly split the data into two parts with 72 samples in the training and 37 in the test data. The gene network we used was a protein-protein interaction (PPI) network (Chuang *et al.* 2007). We mapped the genes in the microarray data to the PPI network and selected the largest subnetwork, which included 1329 genes. The prior probability for a gene being included in the model, w_i, was set to 0.05 for the independence prior in the standard SSVS, and the constraint θ_0 for the ICAR prior was set to logit(0.05). No intercept ($\alpha_0 = 0$) was fitted for the BMRF and SBMRF priors. We ran a total of 10000 MCMC iterations with a burn-in period of 8000 iterations, and the analysis was based on the last 2000 MCMC samples. The PMSEs for the methods are shown in Table 4.2.

Table 4.2   PMSEs for the glioblastoma data

	IND	GMRF	BMRF	SBMRF
PMSE	0.54	0.55	0.64	0.53

In summary, for this example, the four priors for γ performed pretty similarly in terms of prediction, though SSVS+BMRF performed slightly worse than others with a larger PMSE. For gene selection, as shown in Figure 4.3, $\hat{\gamma}$'s for the independence prior and GMRF were roughly normally distributed around the specified prior at 0.05, and for the BMRF prior it was also normally distributed around 0.02. The BMRF prior seemed to better separate the informative and non-informative genes, however, it also included much more genes. Since our prior was set to reflect the belief of 5% of informative genes in a total of 1329 genes, we plotted the top 66 selected genes for all priors except BMRF (not shown); the network structures looked very similar, though most of the selected genes did not overlap. For this dataset, the similar performance of the methods in terms of PMSE (Table 4.2) and their widely varied genes being selected can be presumably explained by the fact that the genes were barely informative in predicting the survival outcome, as shown by Binder and Schumacher (2008).

4.2.2 NCI-60 Dataset

The NCI-60 cell line data set was generated from a drug discovery project at the National Cancer Institute (NCI). The 60 cell lines from 9 different tissues of origin were exposed to thousands of compounds. Growth inhibitory effects of each compound were measured for each cell line and reported as GI50, the concentration that inhibits growth by 50%. The data set was originally analyzed by Staunton et al (2001) to predict a dichotomized chemosensitivity. Compounds that had a relatively broad and balanced range of effects across the 60 cell lines had been used for analysis. Here, the response variable used was normalized $\log_{10}(GI50)$ values across all cell lines for each compound and there were a total of 232 compounds. Gene expression data were derived using high density Hu6800 Affymetrix microarrays containing 7129 probe sets. The original data were provided as average difference values between perfect match and mismatch scores. The gene expression data used here was pre-processed and contained only 1517 probe sets (Staunton et al., 2001), for which the minimum change in gene expression across all 60 cell lines was greater than 500 average difference units. Data were logged (base 2) and median centered.

The 1517 probe sets corresponded to 1408 unique genes according to their ENTREZ IDs. For probe sets with the same ENTREZ ID, we took the average of their measurements as the expression level for that gene. Mapping to the PPI network, we found that 996 genes formed a connected subnetwork with 7310 edges. The average number of direct neighbors was 14.7, ranging from 1 to 120. The response variable was GI50 for one compound with a relatively high predictive accuracy according to Staunton et al (2001). Data were randomly split into a training set and a test set with sample sizes 40 and 20 respectively. We applied all four methods to the training set for model building and to the test set for

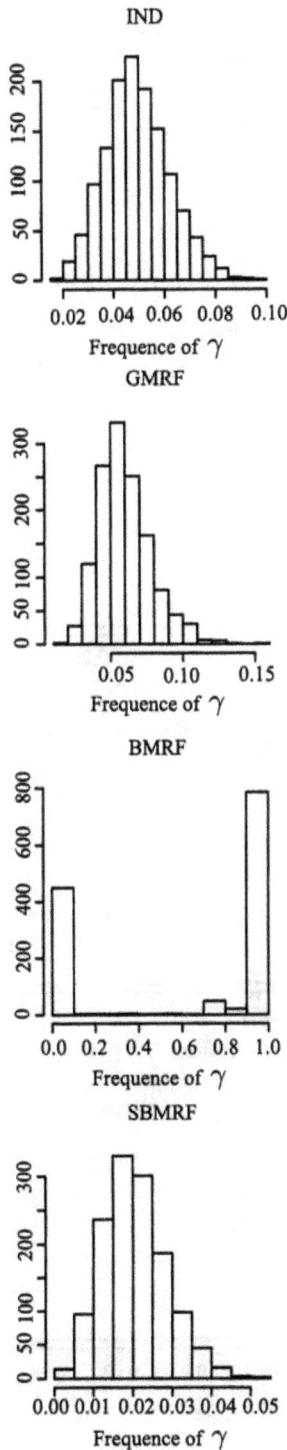

Figure 4.3   Frequencies of the genes being selected for the glioblastoma data.

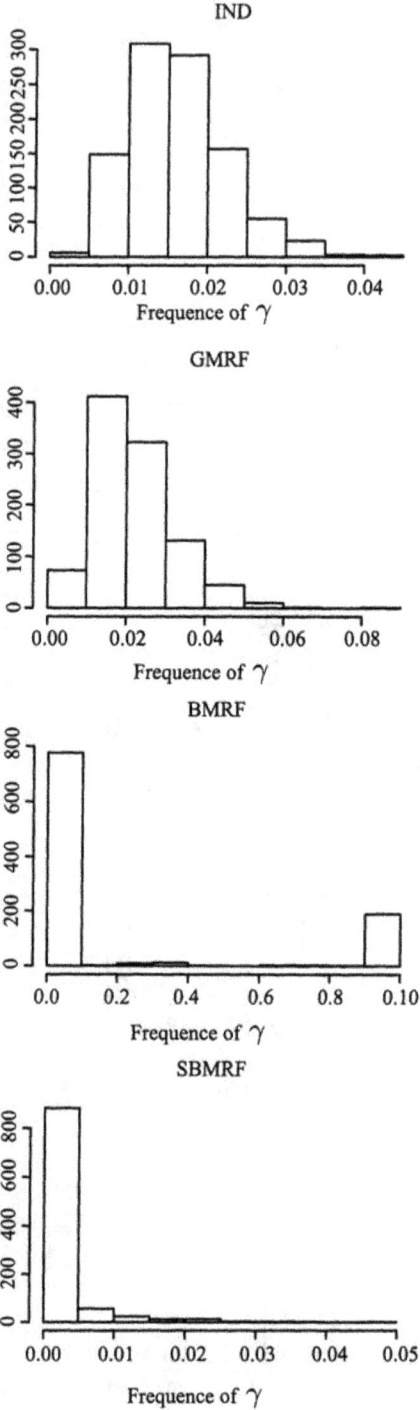

Figure 4.4   Frequencies of the genes being selected for the NCI-60 data.

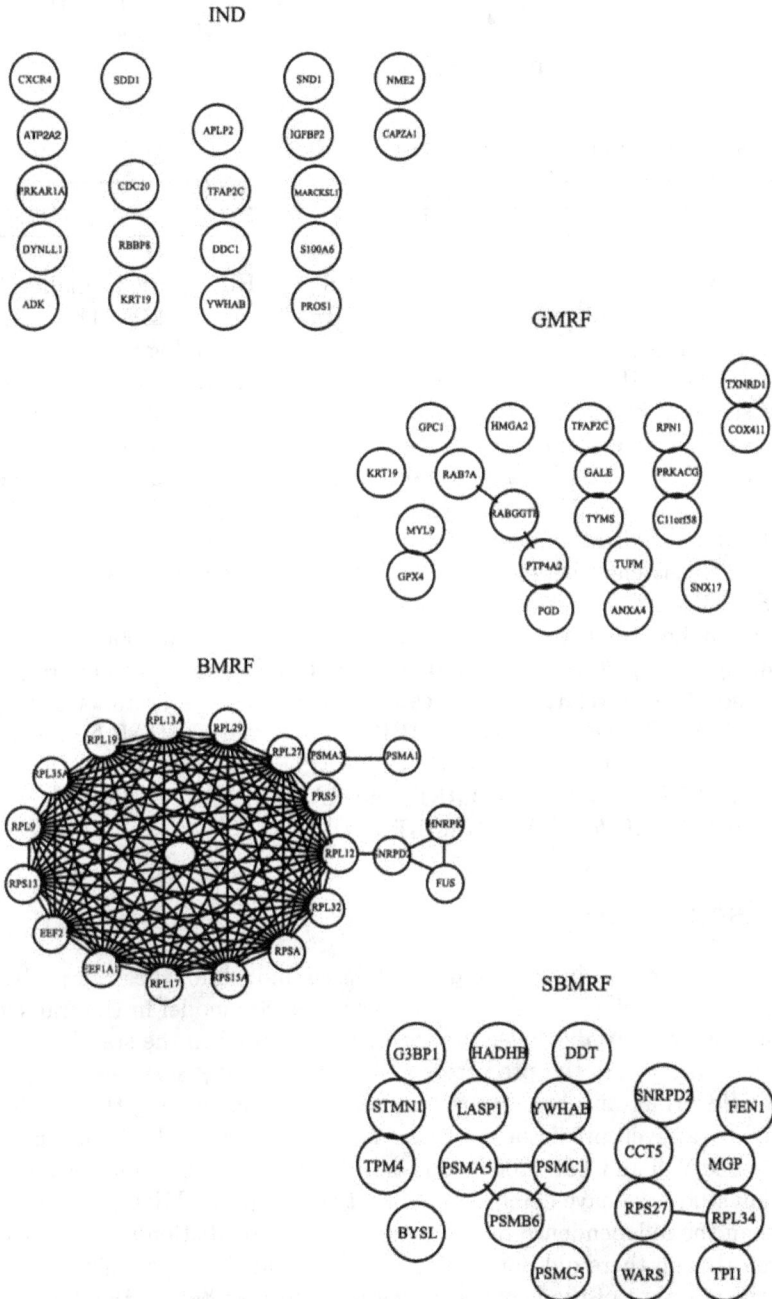

Figure 4.5   Subnetworks of the top 20 selected genes by the methods for the NCI-60 data.

prediction. The results are shown in Table 4.3. For SSVS+GMRF, we set $\theta_0 = $ logit(0.1) and $\alpha_0 = $ logit(0.1) for SSVS+BMRF and SSVS+sBMRF. However, the model size selected by SSVS+BMRF and SSVS+SBMRF was sensitive to α_0, as pointed out by Li and Zhang (2008).

Table 4.3   PMSEs for the NCI-60 data

	IND	GMRF	BMRF	SBMRF
PMSE	0.77	0.56	1.29	0.62

The methods SSVS+GMRF and SSVS+SBMRF yielded smaller PMSEs than SSVS+IND, while SSVS+BMRF had the largest PMSE. The frequencies of the selected genes by the four methods are shown in Figure 4.4. Again the method SSVS+BMRF selected most genes. Figure 4.5 shows the top 20 most frequently selected genes and their associated edges. None of the genes selected by SSVS+IND were connected to each other, while several genes selected by SSVS+GMRF or SSVS+SBMRF were connected. In contrast, most of the genes selected by SSVS+BMRF were highly connected to each other, suggesting that the method SSVS+BMRF seemed to favor the selection of the genes with large degrees, in consistent with its use of k_i, rather than its scaled version k_i^* as used in SSVS+SBMRF.

We searched the Cancer Genes database (Higgins et al 2006) and identified that, among the top 20 genes selected by the four methods, there were respectively 6, 5, 4 and 6 genes related to the cancer gene pathways or functional groups: genes CXCR4, PRKAR1A, PROS1, RBBP8, SOD1 and YWHAB for SSVS+IND; HMGA2, PRKACG, PTP4A2, RPN1 and TYMS for SSVS+GMRF; FUS, RPS13, RPSA and SNRPD2 for SSVS+BMRF; and HADHB, LASP1, PSMC1, SNRPD2, TPM4 and YWHAB for SSVS+SBMRF.

5   Discussion

In this chapter, we have investigated four different models for the prior probabilities of the predictors' being included in a linear regression model in the framework of Stochastic Search Variable Selection (SSVS). Compared to the standard independence prior that treats the predictors as independent *a priori*, the three Markov random field priors aim to capture spatial correlations among the predictors as suggested by a given predictor network. The same idea can be found in Wei and Li (2007) and Wei and Pan (2008), but in a simpler non-regression context. In the simulation study, we have demonstrated that the proposed MRF priors performed better than the independence prior in terms of both prediction and variable selection, even though there did not appear to exist a unanimous winner. For the real data, although some of the new methods still performed better, the difference was smaller.

Although the MRF priors introduce additional parameters into the SSVS model, the increase of computational demand is negligible as compared to the independence prior. Considering the potential gain in prediction and gene selection, but without significant increase in computing time, the MRF priors provide a good means to incorporate network structures to improve statistical efficiency. In particular, it is easier to specify some prior parameter to control the final model size with the GMRF prior, while it is more difficult for the BMRF and SBMRF due to the latter two's dependence on several parameters. We also explored putting a zero point mass on non-informative βs and using conjugate priors for β as mentioned in George (1997), $\beta_\gamma \sim N(0, c\sigma^2(X_\gamma'X_\gamma)^{-1})$. This setup requires $(X_\gamma'X_\gamma)$ to be positive definite, thus can only choose a number of genes no more than the sample size, which may be a shortcoming for the high-dimensional and low-sample-sized setting. Our simulation results (not shown) indicated that it had similar performance in identifying informative genes to the methods presented here, but worse in predictive performance.

Here we have introduced some MRF priors to smooth the prior probabilities of the predictors' being selected over a given predictor network. For the same purpose of incorporating network information into linear regression, Li and Li (2008) and Pan et al (2009) derived network-constrained penalties to induce smoothness in (weighted) regression coefficients β_i's or $|\beta_i|$'s in the framework of penalized regression, in which the smoothness assumption is much stronger than that assumed here. Nevertheless, our methods can be extended to smooth β_i's directly, e.g., by imposing a GMRF prior on β_i's. Alternatively, a penalized approach to smoothing the prior probabilities as done here might be more efficient computationally than the MCMC implementations proposed here. In addition, although we have only applied the proposed methods to linear regression, it is conceptually straightforward to extend them to classification (Mallick et al 2005) and nonlinear regression with generalized linear models (Monni and Li 2009) or the Cox proportional hazards model (Gui and Li 2005). More investigations are necessary.

Acknowledgment
We thanks helpful discussions with Hongzhe Li and Peng Wei.

References

[1] Bae K. and Mallick B. K.(2004). Gene selection using a two-level hierarchical Bayesian model. *Bioinformatics* **20**, 3423-3430.

[2] Besag J.(1972). Nearest-neighbor systems and the auto-logistic model for binary data. *JRSS-B* **34**, 75-83.

[3] Besag J.(1974). Spatial interaction and the statistical analysis of lattice systems. *JRSS-B* **36**, 192-236.

[4] Besag J. and Kooperberg C.(1995). On conditional and intrinsic autoregressions. *Biometrika* **82**, 733-746.

[5] Binder H. and Schumacher M.(2008). Comment on 'Network-constrained regularization and variable selection for analysis of genomic data'. *Bioinformatics* **24**, 2566-2568.

[6] Chuang H. Y. *et al.* (2007). Network-based classification of breast cancer metastasis. *Molecular System Biology* **3**, 140-149.

[7] George E. I. and McCulloch R. E.(1993). Variable selection via gibbs sampling. *Journal of the American Statistical Association* **88**, 881-889.

[8] George E. I. and McCulloch R. E.(1997). Approaches for bayesian variable selection. *Statistica Sinica* **7**, 339-373.

[9] Gilks W. R. and Wild P.(1992). Adaptive rejection sampling for gibbs sampling. *Applied Statiststics* **41**, 337-348.

[10] Gui J. and Li H.(2005). Penalized Cox regression analysis in the high-dimensional and low-sample size settings, with applications to microarray gene expression data. *Bioinformatics* **21**, 3001-3008.

[11] Heikkinen J. and Harriögmande H.(1994). Fully bayesian approach to image restoration with an application in biogeography. *Applied Statistics* **43**, 569-582.

[12] Higgins M. E., Claremont M., Major J. E., Sander C. and Lash A. E.(2006). CancerGenes: a gene selection resource for cancer genome projects. *Nucleic Acids Research* **35** , D721-D726.

[13] Hoeting J. A., Leecaster M. and Bowden D.(2000). An improved model for spatially correlated binary responses. *Journal of Agricultural, Biological and Environmental Statistics* **5**, 102-114.

[14] Lee K. E., Sha N., Dougherty E. R., Vannucci M. and Mallick B. K.(2003). Gene selection: a Bayesian variable selection approach. *Bioinformatics* **19**, 90-97.

[15] Li C. and Li H.(2008). Network-constrained regularization and variable selection for analysis of genomic data. *Bioinformatics* **24**, 1175-1182.

[16] Li F. and Zhang N. R.(2008). Bayesian variable selection in structured high-dimensional covariate spaces with applications in genomics. Manuscript.

[17] Mallick B. K., Ghosh D. and Ghosh M.(2005). Bayesian classification of tumors by using gene expression data. *J. R. Statist. Soc. B* **67**, 219-234.

[18] Monni S. and Li H.(2009). Bayesian variable selection for graph-structured covariates with applications in genomics. Manuscript.

[19] Pan W., Xie B. and Shen X.(2009). Incorporating predictor network in penalized regression with application to microarray data. To appear in *Biometrics*. Available at http://www.biostat.umn.edu./rrs.php as Research Report 2009-001, Division of Biostatistics, University of Minnesota.

[20] Smith D. and Smith M.(2006). Estimation of binary markov random fields using markov chain monte carlo. *Journal of Computational and Graphical Statistics* **15**, 207-227.

[21] Smith M. and Fahrmeir L.(2007). Spatial bayesian variable selection with application to functional magnetic resonance imaging. *Journal of the American Statistical Association* **102**, 417-431.

[22] Wei P. and Pan W.(2008). Incorporating gene networks into statistical tests for genomic data via a spatially correlated mixture model. *Bioinformatics* **24**, 404-411.

[23] Wei P. and Pan W.(2009). Network-based genomic discovery: application and comparison of Markov random field models. To appear in *Applied Statistics*

Available at `http://www.biostat.umn.edu./rrs.php` as Research Report 2009-009, Division of Biostatistics, University of Minnesota.

[24] Wei Z. and Li H.(2007). A markov random field model for network-based analysis of genomic data. *Bioinformatics* **23**, 1537-1544.

Part IV

High-Dimensional Statistics in Genomics

Chapter 6

High-Dimensional Statistics in Genomics*

Hongzhe Li †

Abstract

Large-scale systematic genomic datasets have been generated to inform our biological understanding of both the normal workings of organisms in biology and disrupted pathways that cause human diseases. The integrative analysis of these vast amounts of diverse types of quantitative data, which has become an increasingly important part of genomics and systems biology research, poses many interesting statistical and computational problems, largely driven by the complex interrelationships between these high-dimensional genomic measurements. Partially driven by various types of high-throughput genomic data, high-dimensional statistics have drawn great research attention among statisticians in recent years. Many novel statistical methods have been developed for analysis of high-/ultrahigh-dimensional genomic data, including powerful statistical methods for high-dimensional regression analysis and estimation of Gaussian graphical models. In this chapter, we survey some problems in genomics and systems biology that require development of novel statistical methods, focusing on the problem of identifying active transcription factors and motif finding based on microarray time-course gene expression data, the problem of incorporating prior network information into analysis of genomic data, and the problems related to the analysis of gene expression quantitative traits loci (eQTL) studies. We present some newly developed statistical methods to address these important problems in genomics. We conclude this chapter with directions for future research in high-dimensional statistics in genomics.

Keywords: Motif finding, time-course gene expression data, network and graphs, varying coefficients models, regularization, eQTLs, seemingly unrelated regressions.

1 Introduction

Large-scale systematic genomic datasets have been generated to inform our biological understanding of both the normal workings of organisms in biology and

---

*The author is partly supported by NIH grants R01CA127334 and R01ES009911.
†Department of Biostatistics and Epidemiology, University of Pennsylvania School of Medicine, Philadelpha, PA 19104, USA, E-mail: hongzhe@upenn.edu

disrupted pathways that cause human diseases. The integrative analysis of these vast amounts of diverse types of quantitative data, which has become an increasingly important part of genomics and systems biology research, poses many interesting statistical problems, largely driven by the complex inter relationships between these high-dimensional genomic measurements. Partially driven by high-throughput genomic data, high-dimensional statistics have drawn great research interest among statisticians in the past decade. Many novel statistical methods have been developed for high-/ultrahigh- dimensional data, especially in the area of high-dimensional regression analysis and estimation of Gaussian graphical models, both have many applications in genomics.

1.1   Methods for high-dimensional regression and applications in genomics

The most important development in problem formulation, computational algorithms and theoretical properties is in the area of high-dimensional or ultrahigh-dimensional regression analysis, primarily in the framework of penalized least squares problems via L_1 or other types of regularization. Important penalty functions that can lead to sparse variable selection in regression include Lasso (Tibshirani, 1996) and SCAD (Fan and Li, 2001). In particular, Lasso has the crucial advantage of being a convex problem, which lead to an efficient computational algorithm by coordinate descent (Efron et al., 2004; Friedman et al., 2008; Wu and Lange, 2008) and sparse solutions. Furthermore, Lasso has been shown to have optimal theoretical properties in model selection, sign consistency or sparsistency (Meinshausen and Buhlmann, 2006; Wainwright, 2007; Zhao and Yu, 2006), oracle properties (Meinshausen and Yu, 2009; Zhang and Huang, 2007) and persistence (Greenstein and Ritov, 2006). Zou (2006) proposed a novel adaptive Lasso procedure and presented results on model selection consistency and oracle properties of the parameter estimates. Huang and Xie (2006) established the asymptotic oracle properties of the SCAD-penalized least squares estimators when the number of covariates may increase with the sample sizes. Although major theoretical development has focused on linear regression models, the ideas have also been extended to other regression types such as generalized linear models and Cox proportional hazards models (Gui and Li, 2005; Li and Gui, 2005; Li and Luan, 2006; Meier, van de Geer and Buhlmann, 2008; Zhang and Lu, 2008).

These procedures have been applied to the analysis of gene expression data in order to build predictive models for clinical phenotypes and to identify the genes that are associated with the phenotypes, and to analyze data from genome-wide association (GWA) studies in order to identify the genetic variants and their interactions associated with certain traits or diseases (Wu et al., 2008). Although promising results are obtained, there is still a limitation in terms of the number of variables that these penalized procedures can handle, in the setting of ultrahigh-dimensional data such as the GWA data with 500,000 to 1 million SNPs, especially when the interactions are considered. To deal with this difficulty, Fan and Lv (2008) and Fan, Samworth and Wu (2008) introduced the sure independence screening and iterative sure independence screening procedure to deal with ultrahigh-dimensional data for both linear and generalized linear models. Shi *et al.* (2008) developed Lasso-pattern search algorithm in order to detect interactions among the variables. Bickel *et al.* (2008) developed a hierarchical selection procedure using Lasso that can select relevant interactions among the variables selected as main effects. Another area that needs further research is on inferences after the variables are selected based on these new procedures. One promising approach is based on sample-splitting (Meinshausen, Meier, and Buhlmann, 2008).

There have also been some recent developments in variable selection for semi- or non-parametric regressions in high-dimensional settings. Motivated by applications for the analysis of time-course gene expression data, Wang, Chen and Li (2007) and Wang, Li and Huang (2008) developed a penalized estimation procedure for nonparametric varying-coefficient models using the SCAD penalty. Wei, Huang and Li (2010) developed a similar procedure using Lasso or adaptive Lasso penalties. Meier, van de Geer and Bulhmann (2008), as well as Huang, Horowitz and Wei (2008) and Xue (2009) developed variable selection procedures for nonparametric additive models. The key ideas of these procedures are similar and all involve using the basis function expansions to approximate the nonparametric functions or nonparametric coefficient functions and then using some types of regularization on the sets of the coefficients of the basis functions in order to induce the sparsity of the solutions.

1.2   Methods to account for special structures of the genomic data

New estimating procedures have also been developed in recent years to account for certain structures of the genomic data. These include the group Lasso procedure (Yuan and Lin, 2006; Meier *et al.*, 2008) and other types of group penalty functions (Zhao, Rocha and Yu, 2006; Breheny and Huang, 2008; Wang *et al.*, 2009) when the explanatory variables are grouped or organized in a hierarchical manner. Wei and Li (2007) and Luan and Li (2008) developed a group gradient descent boosting procedure to select the group of genetic features that are related to clinical phenotypes. The elastic net (Enet) procedure (Zou and Hastie, 2005) deals with groups of highly correlated variables and the fused Lasso (Tibshirani *et al.*, 2005) imposes L_1 penalty on the absolute differences of the regression coefficients in order to account for some smoothness of the coefficients. Nardi and Rinaldo (2008) established the asymptotic properties of the group Lasso estimator for linear models. Jia and Yu (2008) provided conditions for model selection consistency of the elastic net when $p >> n$. Zou and Zhang (2009) proposed an adaptive elastic net with a diverging number of parameters and established its oracle property. Among these procedures, the Enet regularization and the fused-Lasso are particularly appropriate for the analysis of genomic data, where the former encourages a grouping effect and the latter often leads to smoothness of the coefficient profiles for ordered covariates. In addition, biological knowledge such as gene ontology can often reasonably divide the genomic data into groups and it is sometimes interesting to identify groups of variables that are related to certain clinical outcomes.

Motivated by a genomic application to account for network information in the analysis of genomic data, Li and Li (2008, 2009) and Li (2009) proposed a network-constrained regularization procedure for fitting linear regression models and for variable selection, where the predictors in the regression model are genomic data measured on the genetic networks, which we call the covariates with a graphical structure. Zhu, Pan and Shen (2009) developed support vector machines with a disease-gene-centric network penalty for high-dimensional microarray data.

1.3   Methods for estimation of Gaussian graphical models

Another important development is in the area of estimating Gaussian graphical models (GGMs) in high-dimensional settings. The key idea behind GGMs is to use partial correlations as a measure of the independence of any two genes. This makes it straightforward to distinguish direct from indirect interactions. Note that partial correlations are related to the inverse of the correlation matrix. Also note that in GGMs missing edges indicate conditional independence. Such models can potentially be used to infer the gene expression network, although no strong evidence has indicated that such models have indeed led to advancement in genomics. For microarray gene expression data, since the number of genes is often much larger than the number of samples, inference of the Gaussian graphical structure can be more challenging. Since the graph structure has one-one correspondence to the off-diagonal elements of the precision or concentration matrix, sparsity condition is often imposed on the off-diagonal elements. Based on this fact, new regularization procedures have been developed to infer Gaussian graphical model structures. Li and Gui (2006) proposed a threshold gradient descent procedure for estimating the sparse concentration matrix, which can be regarded as an approximation of the L_1 penalized estimation. Friedman *et al.* (2008) developed a graphical Lasso algorithm to efficiently implement the L_1 penalized maximum likelihood estimation. The procedure of Li and Gui (2006) is essentially an approximation of the L_1 penalized maximum likelihood estimation. Using the links between the off-diagonal elements of the precision matrix and the marginal regression of one variable on the rest of the variables, Meinshausen and Buhlmann (2006), Rocha, Zhao and Yu (2008), Peng *et al.* (2009) and Fan, Feng and Wu (2009) have developed Lasso-type algorithms to estimate the graphical model structures. Bickel and Levina (2008a, 2008b) proposed methods for estimating the covariance matrix by banding or tapering or hard thresholding the sample covariance matrix and presented theoretical properties of such estimates.

It should be pointed out that it is very difficult to estimate biologically meaningful genetic networks based only on the gene expression data measured from pooled cell samples. One reason is that the ultimate biological unit lies within a single cell and the genetic networks often act at the single-cell level. As the techniques for single-cell measurements mature (Cohen *et al.*, 2008; Tang *et al.*, 2009), methods of building graphical models in high-dimensional settings will be greatly needed.

1.4   Problems and methods for integrative analysis of multiple genomic data

High-dimensional regression analysis and graphical model estimation are often developed for analyzing only one type of genomic data such as gene expression data. In addition, these methods were mainly developed for generic high-dimensional data analysis without utilizing biological knowledge. These two facts may limit the applications of these methods. In modern genomic and systems biology research, large-scale systematic genomic datasets have been generated to inform our biological understanding of both the normal workings of organisms in biology and disrupted pathways that cause human disease. These data include high-throughput sequencing data, microarray data and their descendants, in vivo imaging techniques, microscopy, *etc.* New statistical methods are greatly needed to integrate these different types of genomic data and to effectively utilize our current knowledge on biological pathways/networks and gene functions.

We present in the following sections several statistical problems in genomics and systems biology that require new high-dimensional statistical methods, including the methods for analysis of microarray time course gene expression data (Section 2), the methods for the analysis of genomic data with a graphical structure (Section 3) and methods for analysis of expression quantitative trait loci (eQTLs) studies (Section 4). For each of these problems, we present the biological motivation, statistical formulation and the statistical methods that have recently been developed. We conclude this chapter with directions for future research in high-dimensional statistics in genomics.

2   Identification of active transcription factors using time-course gene expression data

Since many important biological systems or processes are dynamic systems, it is important to study the gene expression patterns over time in a genomic scale in order to capture the dynamic behavior of gene expression. Microarray technologies have made it possible to measure the gene expression levels of essentially all the genes during a given biological process. One important problem with such microarray time-course (MTC) gene expression data is to identify the active transcription factors that affect the gene expression levels over time. One approach of studying gene regulation is to associate gene expression values with oligomer motif abundance by using a simple linear regression for each oligomer of a given length. Figure 2.1 gives an illustration of gene regulation by a transcription factor (TF). Those oligomers with significant coefficients in regression analysis are inferred as potential transcriptional factor binding motifs (TFBMs) (Bussemaker *et al.* 2001; Keles *et al.*, 2002; Gao *et al.*, 2004). Assuming that in response to a given biological condition, the effect of a TFBM is strongest among genes with the most dramatic increase or decrease in mRNA expression. Conlon *et al.* (2003) proposed to use simple linear regression to relate the motif abundance to gene expression by first selecting genes with large changes in expression levels. While these approaches work reasonably well in discovery of regulatory motifs in lower organisms, they often fail to identify mammalian transcriptional factor binding sites (Das *et al.*, 2006). Das *et al.* (2006) proposed to correlate the binding strength of motifs with expression levels using the multivariate adaptive smoothing splines (MARS) of Friedman (2001). These existing regression-based methods model each individual sample or time point separately. To better capture the dynamic relationships in time-course microarray experiments, Wang, Chen and Li (2007) proposed to apply the varying-coefficient (VC) model for the joint modeling of binding motif multivariate expression data measured over time and developed a penalized approach for variable selection for such VC models.

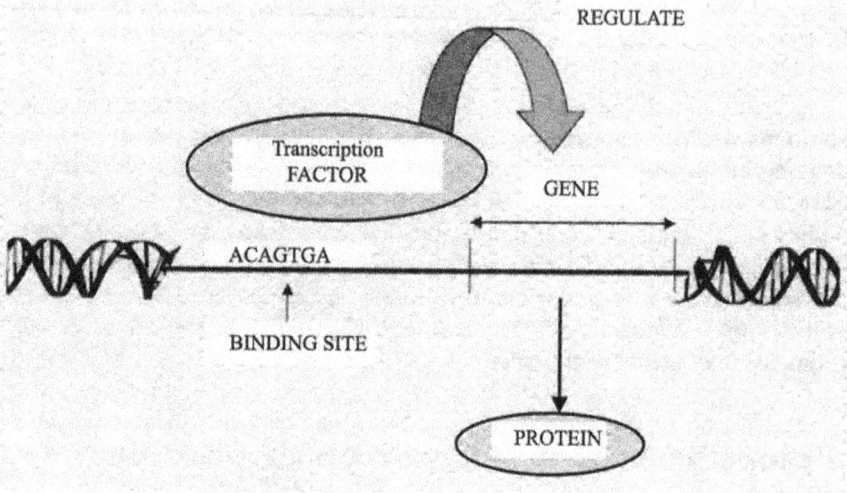

Figure 2.1   Illustration of gene regulation where a transcription factor (TF) binds to the promotor region of a genes at its binding site and regulates the expression of this gene (http://www.cs.uiuc.edu/homes/sinhas/img/DAILYILLINI.jpg).

2.1   Varying coefficient models for transcription factor identification

Wang, Chen and Li (2007) and Wang, Li and Huang (2008) considered the problem of identifying the transcription factors from a large set of candidates (e.g., from the TRANSFAC database) that may explain the variations of gene expression over time in the framework of varying-coefficient models and developed variable selection procedures based on basis function expansion and penalized estimation. In such models, the ith response is a real function $y_i(t), i = 1, \ldots, n, t \in T$ with associated covariate vector $x_i = \{x_{i1}, \ldots, x_{ip}\}$, which is constant in time. Of course, it is only possible to observe the function $y_i(t)$ at a finite number of points, possibly with errors. For the problem of modeling the MTC gene expression data,

$y_i(t)$ is the measure expression data for the ith gene at time point t during a given biological process, and x_{ik} is the binding strength of the kth motif corresponding to the kth TF (Das *et al.*, 2006). The statistical question is to select a set of TFs from a large set of p candidate TFs that can explain partially the variation of gene expression levels over time, where the effects of the TFs on gene expression levels are time-varying.

Let $Y_i(t)$ be the expression level of the ith gene at time t, for $i = 1, \ldots, n$. Wang, Chen and Li (2007) and Wang, Li and Huang (2008) considered the following regression model with functional response,

$$y_i(t) = \mu(t) + \sum_{k=1}^{p} \beta_k(t) x_{ik}(t) + \epsilon_i(t), \qquad (2.1)$$

where $\mu(t)$ is the overall mean effect, $\beta_k(t)$ is the regulation effect associated with the kth transcriptional factor, x_{ik} is the matching score or the binding probability of the kth transcriptional factor on the promoter region of the ith gene, and $\epsilon_i(t)$ is a realization of a zero-mean stochastic process.

Several different ways and data sources can be used to derive the matching score $x_{ik}(t)$ (Wang, Chen and Li, 2007), which is often time-independent. One approach is to derive the score using the position-specific weight matrix (PSWM). Specifically, for each candidate TF k, let P_k be the positive specific weight matrix of length L, b with element $P_{kj}(b)$ being the probability of observing the base b at position j. Then each L-mer l in the promoter sequence of the ith gene is assigned a score S_{ikl} as:

$$S_{ikl} = \sum_{j=1}^{L} \log \frac{P_{kj}(b_{ilj})}{B(b_{ilj})},$$

where b_{ilj} is the nucleotide at position j on the lth sequence for gene i, and $B(b)$ is the probability of observing b in the background sequence. This score always assumes a value between 0 and 1. We then define $x_{ik} = max_l S_{ikl}$, which is the maximum of the matching scores over all the L-mer in the promoter region of the ith gene. The maximum scores can then be converted into the binding probabilities using the method described in Chen *et al.* (2007). Alternatively, we can define the binding probability based on the chromatin immunoprecipitation (ChIP-chip) data. The results produced by a typical ChIP binding experiment for TF k is a set of measures Z_{ik} for the enrichment of each gene i for that TF k. These measures are then standardized, $U_{iK} = (Z_{ik} - \overline{Z_k})/s_{Z_k}$, to have a common mean and standard deviation. For each U_{ik}, a significance test is performed against a null hypothesis of no enrichment, giving a p-value p_{ik} for each gene that is calculated using a standard normal distribution. However, as these p-values cannot be directly interpreted as the probability $x_{ik} = P(\text{TF} k \, j \, \text{binds gene} \, i)$, we adopted the method proposed by Chen *et al.* (2007) to convert p_{ik} into binding probabilities x_{ik} using mixture modeling. Finally, for the motif finding problem (Bussemaker, Li and Siggia, 2001; Colon *et al.*, 2003; Zhang, Wildermuth and Speed, 2008), x_{ik} is the number of the kth word in the promoter sequence of the ith gene.

2.2   Variable selection for varying coefficient models

Suppose that the coefficient $\beta_k(t)$ can be approximated by a basis expansion $\beta_k(t) \approx \sum_{l=1}^{L_k} \gamma_{kl} B_{kl}(t)$, where L_k is the number of basis functions in approximating the function $\beta_k(t)$. Model (2.1) becomes

$$y_i(t_{ij}) \approx \sum_{k=1}^{p} \sum_{l=1}^{L_k} \gamma_{kl} x_{ik}(t_{ij}) B_{kl}(t_{ij}) + \varepsilon_i(t_{ij}). \tag{2.2}$$

The parameters γ_{kl} in the basis expansion can be estimated by minimizing

$$\frac{1}{n} \sum_{i=1}^{n} w_i \sum_{j=1}^{J_i} \left\{ y_i(t_{ij}) - \sum_{k=1}^{p} \sum_{l=1}^{L_k} \gamma_{kl} x_{ik}(t_{ij}) B_{kl}(t_{ij}) \right\}^2, \tag{2.3}$$

where w_i are the weights, taking the value 1 if we treat all observations equally or $1/n_i$ if we treat each subject equally, and J_i is the number of gene expression observations for the ith gene. In typical MTC experiments, all the genes should have the same number of observations, except for some possible missing values. An estimate of $\beta_k(t)$ is obtained by $\hat{\beta}_k(t) = \sum_{l=1}^{L_k} \hat{\gamma}_{kl} B_{kl}(t)$ where $\hat{\gamma}_{kl}$ are minimizers of (2.3). Various basis systems such as Fourier bases, polynomial bases and B-spline bases can be used in the basis expansion.

Now suppose some variables are not relevant in the regression so that the corresponding coefficient functions are zero functions. We introduce a regularization penalty to (2.3) so that these zero coefficient functions will be estimated as identically zero. To this end, it is convenient to rewrite (2.3) using function space notation. Let $\mathcal{G}_k$ denote all functions that have the form $\sum_{l=1}^{L_k} \gamma_{kl} B_{kl}(t)$ for a given basis system $\{B_{kl}(t)\}$. Then (2.3) can be written as $(1/n) \sum_i w_i \sum_j \{y_i(t_{ij}) - \sum_k g_k(t_{ij}) x_{ik}(t_{ij})\}^2$ where $g_k(t) = \sum_{l=1}^{L_k} \gamma_{kl} B_{kl}(t) \in \mathcal{G}_k$. Let $p_\lambda(u), u \geqslant 0$, be a nonnegative penalty function that depends on a penalty parameter λ. It is assumed that $p_\lambda(0) = 0$ and $p_\lambda(u)$ is nondecreasing as a function of u. Let $\|g_k\|$ denote the L_2-norm of the function g_k. We minimize over $g_k \in \mathcal{G}_k$ the penalized criterion

$$\frac{1}{n} \sum_{i=1}^{n} w_i \sum_{j=1}^{J_i} \left\{ y_i(t_{ij}) - \sum_{k=1}^{p} g_k(t_{ij}) x_{ik}(t_{ij}) \right\}^2 + \sum_{k=1}^{p} p_\lambda(\|g_k\|). \tag{2.4}$$

There are two sets of tuning parameters: the L_k's control the smoothness of the coefficient functions and the λ governs variable selection or sparsity of the model. In our implementation of the method, we use B-splines as the basis functions. Thus $L_k = n_k + d + 1$ where n_k is the number of interior knots for g_k and d is the degree of the spline. The interior knots of the splines can be either equally spaced or placed on the sample quartiles of the data so that there are about the same number of observations between any two adjacent knots. We use equally spaced knots for all numerical examples in this chapter. For $g_k(t) = \sum_l \gamma_{kl} B_{kl}(t) \in \mathcal{G}_k$, its squared L_2-norm can be written as a quadratic form in $\gamma_k = (\gamma_{k1}, \ldots, \gamma_{kL_k})^T$. Let $R_k = (r_{ij})_{L_k \times L_k}$ be a matrix with entries $r_{ij} = \int_T B_{ki}(t) B_{kj}(t)\, dt$. Then

$\|g_k\|^2 = \boldsymbol{\gamma}_k^T \boldsymbol{R}_k \boldsymbol{\gamma}_k \triangleq \|\boldsymbol{\gamma}_k\|_k^2$. Set $\boldsymbol{\gamma} = (\boldsymbol{\gamma}_1^T, \ldots, \boldsymbol{\gamma}_p^T)^T$. The penalized weighted least squares criterion (2.4) can be written as

$$
pl(\boldsymbol{\gamma}) = \frac{1}{n} \sum_{i=1}^n w_i \sum_{j=1}^{J_i} \left\{ \left(y_i(t_{ij}) - \sum_{k=1}^p \sum_{l=1}^{L_k} \gamma_{kl} x_{ik}(t_{ij}) B_{kl}(t_{ij}) \right) \right\}^2 + \sum_{k=1}^p p_\lambda(\|\boldsymbol{\gamma}_k\|_k).
$$

(2.5)

There are many ways to specify the penalty function $p_\lambda(\cdot)$. If $p_\lambda(u) = \lambda u$, then the penalty term in (2.4) is similar to that in COSSO. Wang, Chen and Li (2007) and Wang, Li and Huang (2008) used the SCAD penalty function by Fan and Li (2001) and obtained theoretical properties such as consistent variable selection and the oracle property for the proposed method. Wei and Huang (2009) proposed to use the Lasso or the adaptive Lasso penalty function and showed that the group Lasso selects a model of the right order of dimensionality and selects all variables with the norms of the corresponding coefficient functions greater than a certain threshold and is estimation-consistent in the case when the number of variable is of greater order than the sample size. For both approaches, the parameter λ controls the sparsity of the model and the number of basis functions controls the smoothness of the coefficient functions.

Alternatively, if we use a large number of basis functions, one can consider the sparsity-smoothness penalty function proposed by Meier *et al.* (2009) for high-dimensional additive modeling,

$$
p_\lambda(g_k) = \lambda \sqrt{\|g_k\|^2 + \gamma I^2(g_k)},
$$

where

$$
I^2(g_k) = \int (g_k''(t))^2 dt
$$

measures the smoothness of $g_k(t)$. A slightly different penalty function

$$
p_\lambda(g_k) = \lambda \|g_k\| + \gamma I^2(g_k)
$$

was proposed and studied by Chuai and Li (2009).

2.3 Time-varying censored data regression, further discussion and extensions

The time-varying coefficient models can be extended to censored survival data. For example, the effects of gene expression measured at the baseline level on cancer recurrence can be time-varying. Let T and C be the survival time and the censoring time. Also let $\boldsymbol{X}(t) = (X_1(t), \ldots, X_p(t))$ be a possibly time-dependent p-dimensional covariate process on $(0, T)$, which is bounded and predictive. Conditioning on $\boldsymbol{X}$, T and C are assumed to be independent. In practice the triple $(Y_i, \boldsymbol{X}_i(t), \delta_i)$ are observed for ith subject, $i = 1, \ldots, n$, where $Y_i = \min(T_i, C_i)$, $\delta_i = I(T_i \leqslant C_i)$. The target is to estimate the hazard function $h(t|\boldsymbol{x})dt = P(t \leqslant T < t + dt|T > t)$ based on the observations. We can then consider the Cox model with time-varying regression coefficients (Cox, 1972; Anderson and Gill, 1982), we assume

$$h(t|\boldsymbol{x}(t)) = h_0(t) \exp\{\boldsymbol{X}^T(t)\boldsymbol{\beta}(t)\}, \tag{2.6}$$

where $\boldsymbol{\beta}(t) = (\beta_1(t), \ldots, \beta_p(t))^T$ is a vector of p smooth functions of t and $h_0(t)$ is an unspecified baseline hazard function. Wang, Li and Huang (2009) proposed a regularized basis function expansion approach for variable selection for model (2.6) and established both the variable selection consistency and the rate of convergence of the estimates.

The varying-coefficient model (2.1) assumes that gene expression data observed over time are *i.i.d.* across all the genes. In reality, the same TF may regulate the gene expressions of different genes differently. The model is therefore a simplified version of the true model. One interesting extension of model (2.1) is to consider a mixture of varying coefficients models in order to identify the set of genes with similar expression profiles and are also controlled by the same TFs. Another interesting extension is to incorporate the prior gene regulatory network information into identifying the gene transcription modules based on the time-course data. These extensions should provide more flexible and biologically sound statistical models for studying gene regulation by combining gene expression, motif binding and prior regulatory network information.

3 Methods for analysis of genomic data with a graphical structure

Graphs and networks are common ways of depicting information. In biology, many different biological processes are represented by graphs, such as regulatory networks, metabolic pathways and protein-protein interaction networks. This kind of a priori use of graphs is a useful supplement to the standard numerical data such as microarray gene expression data. As an example, Figure 3.1 shows the Kyoto Encyclopedia of Genes and Genomes (KEGG) regulatory network that consists of 30 interconnected regulatory pathways. Li and Li (2008) considered the problem of regression analysis and variable selection when the covariates are linked on a graph, where they introduced a graph-constrained regularization procedure for regression analysis to take into account the neighborhood information of the variables measured on a graph, where a smoothness penalty on the coefficients is defined as a quadratic form of the Laplacian matrix associated with the graph. Li and Li (2009) presented further details and extensions of the methods proposed in Li and Li (2010).

3.1 Variable selection for genomic data with a graphical structure

Consider a weighted graph $G = (V, E, W)$, where $V = \{1, \ldots, p\}$ is the set of vertices that correspond to the p predictors, $E = \{u \sim v\}$ is the set of edges indicating that the predictors u and v are linked on the network and there is an edge between

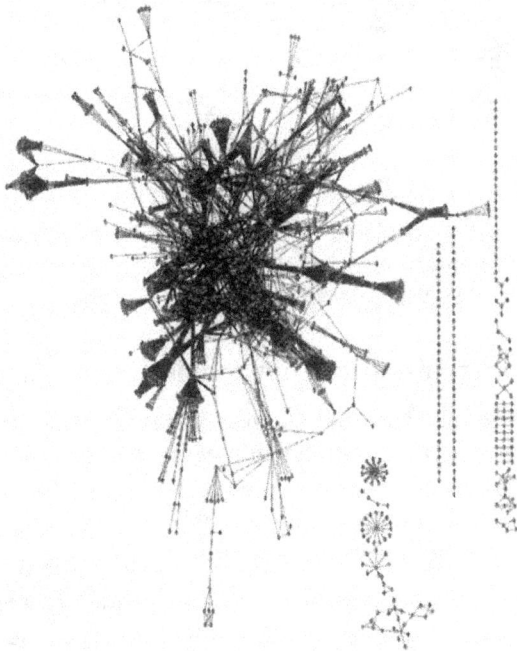

Figure 3.1   Undirected graph of the KEGG regulatory network, consisting of 33 interconnected regulatory pathways. There are a total of 1663 genes (nodes) and 8011 regulatory relationships (edges).

u and v, and W is the set of weights of the edges, where $w(u, v)$ denotes the weight of edge $e = (u \sim v)$. In applications, the weight can be used to measure the uncertainty of the edge between two vertices. For each given sample, we assume that we have numerical measurements on each of the vertices and these measurements are treated as explanatory variables in a regression analysis framework. For the uth node, let x_{iu} be the numerical measurement of the uth vertex on the graph for the ith individual. Further, let $X_u = \{x_{1u}, \ldots, x_{nu}\}^T$ be the measured values at the uth vertex for n i.i.d. samples. Consider the problem of variable selection and estimation where we have design matrix $\boldsymbol{X} = (X_1, X_2, \ldots, X_p) \in R^{n \times p}$ and response vector $\boldsymbol{Y} = (y_1, y_2, \ldots, y_n)^T \in R^n$, and they follow a linear model

$$\boldsymbol{Y} = \mathbf{X}\beta + \varepsilon, \tag{3.1}$$

where $\varepsilon = (\varepsilon_1, \ldots, \varepsilon_n)^T \sim N(0, \sigma^2 I_n)$ and $\beta = (\beta_1, \ldots, \beta_p)^T$. Throughout this Section we assume that the predictors and the response are centered so that

$$\sum_{i=1}^n y_i = 0, \sum_{i=1}^n x_{iu} = 0, \quad \text{and} \quad \frac{1}{n}\sum_{i=1}^n x_{iu}^2 = 1 \quad \text{for } u = 1, \ldots, p.$$

When p is large, we assume that model (3.1) is "sparse", i.e., most of the true regression coefficients β are exactly zero. Without loss of generality we assume the first q elements of vector β are non-zeroes. Denote $\beta_{(1)} = (\beta_1, ..., \beta_q)^T$ and $\beta_{(2)} = (\beta_{q+1}, ..., \beta_p)^T$, then element-wise $\beta_{(1)} \neq 0$ and $\beta_{(2)} = 0$. Now write $X_{(1)}$ and $X_{(2)}$ as the first q and last $p - q$ columns of X, respectively, and let $C = \frac{1}{n} X^T X$, which can then be expressed in the following block-wise form of matrix:

$$C = \begin{pmatrix} C_{11} & C_{12} \\ C_{21} & C_{22} \end{pmatrix}.$$

The goal is to develop a regularization procedure for selecting the true relevant variables. Different from the existing approaches, we particularly account for the fact that the explanatory variables are related on a graph.

3.2 Graph-constrained regularization and variable selection

In order to account for the fact that the p explanatory variables are measured on a graph, we first introduce the Laplacian matrix (Chung, 1997) associated with a graph. Let the degree of the vertex v be $d_v = \sum_{u \sim v} w(u, v)$. We say u is an isolated vertex if $d_u = 0$. Following Chung (1997), we define the Laplacian matrix L for graph G with the uvth element defined by

$$L(u, v) = \begin{cases} 1 - w(u, u)/d_u & \text{if } u = v \text{ and } d_u \neq 0 \\ -w(u, v)/\sqrt{d_u d_v} & \text{if } u \text{ and } v \text{ are adjacent} \\ 0 & \text{otherwise.} \end{cases} \quad (3.2)$$

It is easy to verify that this matrix is semi-positive definite with 0 as the smallest eigenvalue and 2 as the largest eigenvalue when all the weights are 1 (Chung, 1997). In our application, we often set $w(u, u) = 0$ and therefore $L(u, u) = 1$.

The Laplacian matrix has the following interpretations. For a given vector β, the edge derivative of β along the edge $e(u, v)$ at u is defined as

$$\frac{\partial \beta}{\partial e}|_u = \sqrt{w(u, v)} \left(\frac{\beta_u}{\sqrt{d_u}} - \frac{\beta_v}{\sqrt{d_v}} \right),$$

and therefore the local variation of β at u can be measured by

$$\sqrt{\sum \left(\frac{\partial \beta}{\partial e}|_u \right)^2}.$$

The smoothness of vector β with respect to the graph structure can be expressed as

$$\beta^T L \beta = \sum_{u \sim v} \left(\frac{\beta_u}{\sqrt{d_u}} - \frac{\beta_v}{\sqrt{d_v}} \right)^2 w(u, v).$$

For many problems with covariates measured on a graph, we would expect that the neighboring variables are correlated and therefore the regression coefficients would show some smoothness. One way to account for such a dependence of the regression coefficients is to impose a Markov random field (MRF) prior to the collection of β vectors. The MRF decomposes the joint prior distribution of the

β_us into lower-dimensional distributions based on the graph-neighborhood structures. A common MRF model is the Gaussian MRF model that assumes that the joint distribution of β is given by

$$f(\beta) \propto \exp\left\{-\frac{1}{2\sigma^2}\beta^T L\beta\right\},$$

which is an improper density. Based on this Gaussian MRF prior assumption, Li and Li (2008) introduced the following graph-constrained estimation (Grace) of the regression coefficients, denoted by $\hat{\beta}$,

$$\hat{\beta} = \operatorname{argmin}_\beta Q(\beta, \lambda_1, \lambda_2) \tag{3.3}$$

where

$$\begin{aligned}
Q(\beta, \lambda_1, \lambda_2) &= \|\boldsymbol{Y} - \boldsymbol{X}\beta\|_2^2 + \lambda_1|\beta|_1 + \lambda_2\beta^T L\beta \\
&= (\boldsymbol{Y} - \boldsymbol{X}\beta)^T(\boldsymbol{Y} - \boldsymbol{X}\beta) + \lambda_1\sum_u|\beta_u| \\
&\quad + \lambda_2\sum_{u\sim v}\left(\frac{\beta_u}{\sqrt{d_u}} - \frac{\beta_v}{\sqrt{d_v}}\right)^2 w(u, v),
\end{aligned}$$

where L is the Laplacian matrix as defined in (3.2) and the tuning parameters λ_1, λ_2 control the amount of regularization for sparsity and smoothness. For the special case when $\lambda_2 = 0$, the estimate reduces to the Lasso, and when L is the identity matrix, the estimate reduces to the elastic net estimates.

To account for the fact that two variables that are linked on the graph may have different signs of the regression coefficients, we can first perform a standard least square regression when $p < n$ or elastic-net regression when $p \geqslant n$ and denote the estimate as $\tilde{\beta}$. We can then modify the above objective function as

$$\begin{aligned}
Q^*(\lambda_1, \lambda_2, \beta) &= \|\boldsymbol{Y} - \boldsymbol{X}\beta\|_2^2 + \lambda_1\|\beta\|_1 + \lambda_2\beta^T L^*\beta \\
&= \|\boldsymbol{Y} - \boldsymbol{X}\beta\|_2^2 + \lambda_1\sum_{j=1}^p|\beta_j| \\
&\quad + \lambda_2\sum_{u\sim v}\left(\frac{\operatorname{sgn}(\tilde{\beta}_u)\beta_u}{\sqrt{d_u}} - \frac{\operatorname{sgn}(\tilde{\beta}_v)\beta_v}{\sqrt{d_v}}\right)^2 w(u, v),
\end{aligned}$$

where

$$L^*(u, v) = \begin{cases} 1 - w(u, u)/d_u & \text{if } u = v \text{ and } d_u \neq 0 \\ -\operatorname{sign}(\tilde{\beta}_u)\operatorname{sign}(\tilde{\beta}_v)w(u, v)/\sqrt{d_u d_v} & \text{if } u \text{ and } v \text{ are adjacent} \\ 0 & \text{otherwise.} \end{cases}$$

Note that the L^* matrix is still positive semi-definite. We call the β defined by

$$\hat{\beta} = \operatorname{argmin}_\beta Q^*(\beta, \lambda_1, \lambda_2) \tag{3.4}$$

the adaptive Grace (aGrace).

Both optimization problems of (3.3) and (3.4) can be solved efficiently either by reformatting them into a Lasso problem (Li and Li, 2008) or by applying the very efficient coordinate descent algorithm (Friedman *et al.*, 2008). Li (2009), in her PhD dissertation at Penn, derived the estimation error bound of the estimate of β from Grace and gave the conditions for variable selection consistency (see also Li and Li, 2010). Future work should include handling directed graphs because most of the biological networks are represented as directed graphs and establishing the large-sample theory when $p \gg n$. One possibility is to replace the Laplacian matrix L with the Laplacian matrix defined for directed graphs (Chung, 2005). The graph-constrained regularization procedure can be easily extended to GLMs or the Cox models, but the theoretical properties have yet to be developed.

3.3   Further discussion and extensions

Analysis of genomic data with a graphical structure is still in its infancy. One problem that we have to face is the fact that many prior biological networks are not complete or not accurate. Our proposed graph-smoothed penalty functions in the Grace or aGrace procedure heavily depends on the graph structures. Important future research is to assess how robust these procedures are to misspecifications of the network structures. We are currently pursuing the idea of defining some type of data-driven penalty that utilizes both the network structures and also the genomic data observed. Recent work by Xu, Dyer and Owen (2009) indicated that the use of empirical stationary correlations can greatly improve the prediction for semi-supervised learning on graphs. The paper by Li and Li (2008) treats the graph-structured genomic data as covariates in regression analysis. In some applications, we may have multivariate response variables that are linked on a graph. Certainly smoothness conditions can also be improved on the coefficients of the multivariate regression analysis. An interesting approach has been recently reported (Kim, Sohn and Xing, 2009) for this type of problem.

4   Statistical methods in eQTL studies

Another area that requires new statistical methods is in the analysis of expression quantitative traits (eQTL) studies or in the area of genetical genomics (Jansen and Nap, 2001). The key idea of the eQTL studies is that the abundance of a gene transcript is directly modified by polymorphisms in regulatory elements. Consequently, transcript abundance (i.e., gene expression levels) can be considered as a quantitative trait that can be mapped with genetic linkage or association studies (Morley et al., 2004; Cheung *et al.*, 2005). The resulting comprehensive eQTL maps provide an important reference source for categorizing both *cis-* and *trans-* effects of disease-associated SNPs on gene expression, where *cis-*acting eQTLs are typically considered to include SNPs within 100 kb upstream and downstream of the gene that is affected by that eQTL. Figure 4.1 presents an illustration of *cis-*eQTL and *tans-*eQTL. Detailed analyses of the position of mapped *cis-*acting eQTL effects have shown that these are enriched around transcription start sites and within 250 bp upstream of transcription end sites, and they rarely reside more than 20 kb away from the gene. *Cis-*acting variants also seem to occur more often in exonic SNPs. *Trans-*effects are usually weaker than *cis-*effects in humans and in rats, but they are more numerous. It is not known if *trans* effects are mostly mediated through transcription factor variants or through other mechanisms (Cookson *et al.*, 2008).

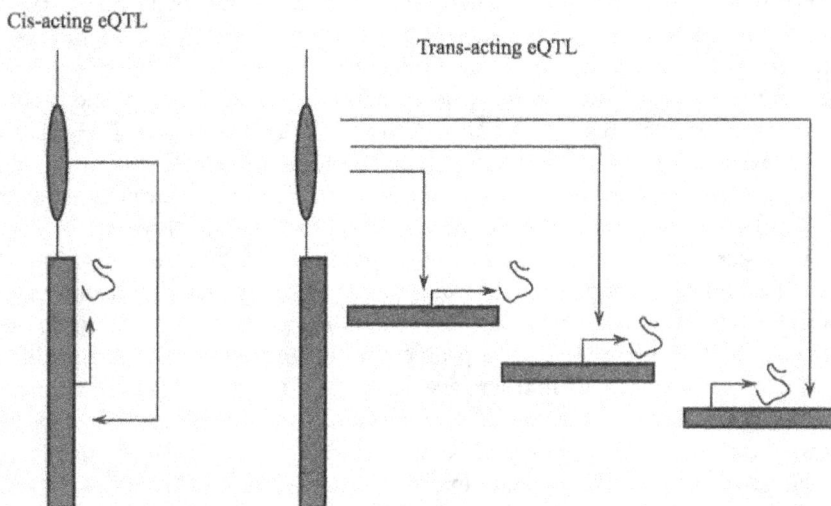

Figure 4.1   Illustration of cis-acting (left) and trans-acting (right) eQTL in genetic analysis of gene expression traits, where eQTLs that are associated with the transcripts of nearby genes are called cis-acting eQTLs and the eQTLs that are associated with transcripts that are a distance away are called tans-acting eQTLs.

Suppose that we have an eQTL study with n independent samples with both genotypes at p SNPs and gene expression data available for T genes. Let X_{ik} be the genotype (coded as 0, 1 and 2) for individual i at the SNP k, and Y_{it} be the expression level for individual i measured for transcript t. So the data can be summarized as (X_{ik}, Y_{it}) for $i = 1, \ldots, n$, $k = 1, \ldots, p$ and $t = 1, \ldots, T$. Further, we let $Y_t = (y_{1t}, \ldots, y_{nt})'$ be the vector of gene expression level for the tth gene, $X_k = (X_{1k}, \ldots, X_{nk})'$ be the vector of of the n genotypes at the kth SNP, and $X = (X_1, \ldots, X_p)$ be the matrix of the genotypes. For a given marker k, we are interested in which transcripts are associated with this marker. The null hypothesis of interest is

H_{kt0} : SNP k is not associated with the transcript t

   vs.

H_{kt} : SNP k is associated with the transcript t.

This constitutes a total of $p \times T$ hypothesis tests. Most of the published papers on the analysis of eQTL data perform this type of single-SNP vs. single-transcript analysis and somehow correct for multiple comparisons using Bonferoni correction, false discovery rate (FDR) controls, and some permutation procedures. The SNPs and the transcripts identified are then summarized as *cis*-eQTLs, *trans*-eQTLs and master *trans*-eQTLs that regulate more genes than other *tans*-eQTLs or eQTL hot spots.

One interesting dataset that has been analyzed by many researchers is the eQTL study of 112 yeast segregants by BY4716 and RM11-1a parental strains (Brem and Kruglyak, 2005). This dataset includes the expression data of 6216 yeast ORFs and 2957 markers. There are 5428 transcripts with fewer than 22 missing values across all 112 segregants. Many markers are in strong linkage disequilibrium (LD). After merging the markers that only differ by one genotype across all 112 samples, we can form 585 marker blocks and select one representative marker for each block. For this dataset, we have $p = 585, T = 5428$ and $n = 112$. For each marker and transcript combination, we can perform a simple two-sample t-test for association between the marker and the transcript expression level. Figure 4.2 shows the plot of the percentiles of such t-statistics for each of the 585 markers. Clearly, some markers are associated with more transcripts than others. Typical next step of eQTL analysis is to select the significant marker-transcript pairs using some type of FDR controlling procedure. In order to derive the optimal FDR control for such a problem, one has to deal with three different types of dependency: (1) the markers are dependent due to LD; (2) the transcripts can be dependent due to sharing of common biological functions; (3) for a given transcript, the test statistics over all the markers are all calculated based on exactly the same set of data on gene expression levels with different partitions of the samples. It is still not clear how to devise such an optimal FDR controlling procedure.

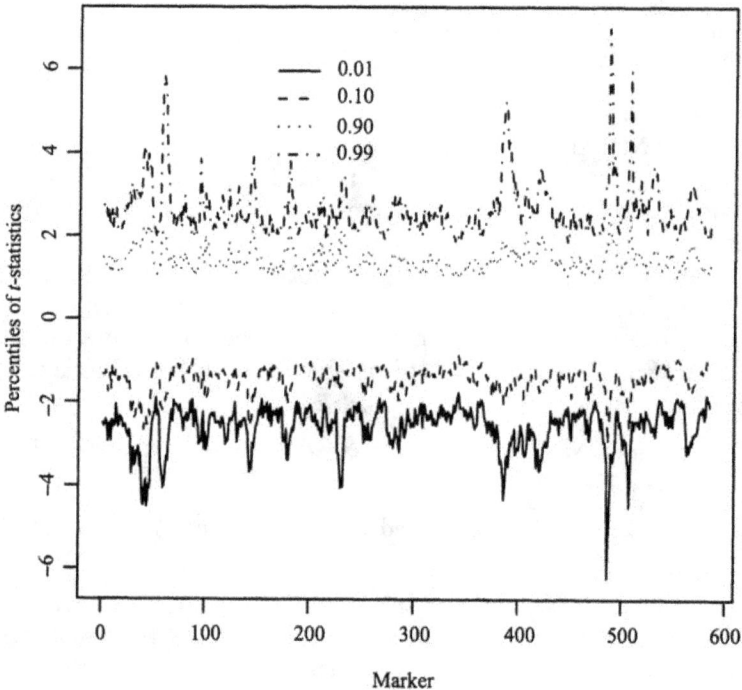

Figure 4.2 Summary of the two-sample t-statistics for 5428 transcripts and 585 markers for the yeast eQTL dataset of Brem and Kruglyak (2005).

However, such single-SNP vs. single-transcript analysis has some limitations: (1) it does not account for the fact that the transcript levels are complex traits and are likely affected by multiple eQTLs; and (2) the transcripts from different genes can be high-correlated (i.e., co-regulated genes). Because of the large number of genes and genetic markers in such analyses, it is extremely challenging to discover how a small number of eQTLs interact with each other to affect mRNA expression levels for a set of (most likely co-regulated) genes. New statistical methods are required to jointly consider all the SNPs and transcripts in the analysis of eQTL data [see Kendziorski and Wang (2006) for a review of statistical methods for eQTL mapping]. In the following, we present two new approaches for analysis of eQTL data.

4.1 Estimation and variable selection for sparse seemingly unrelated regressions (SpSUR)

Zellner (1962) presented the seminal work on seemingly unrelated regression (SUR) systems that comprise a set of regression equations linked by the fact that their disturbances or residual errors are correlated. Zellner showed that estimation regression coefficients based on the generalized least squares can result in a gain in efficiency, especially when the independent variables in different equations are not highly correlated and if the disturbances in different equations are highly correlated. The SUR provides an interesting framework for modeling eQTL data. We consider the regression approach of eQTL analysis since the many transcripts are correlated and different transcripts are often affected by different sets of SNPs. The difficulty is how to deal with the high-dimensionality of the SNP data. Banerjee, Yandell and Yi (2008) proposed Bayesian quantitative trait loci mapping for multiple traits in the framework of SUR models.

Assume that the expression level of gene t can be modeled by the following linear regression model,

$$Y_{it} = \sum_{k=1}^{p} \beta_{tk} x_{ik} + \epsilon_{it},$$

or equivalently we can write this into a matrix/vector form as

$$Y_t = X\beta_t + \epsilon_t, \tag{4.1}$$

where $\beta_t = (\beta_{t1}, \ldots, \beta_{tp})'$ is the vector of the regression coefficients of the SNPs associated with the tth transcript and ϵ_t is a vector of mean zero random error terms. Note that for a given gene transcript t, we expect that there is only a very small number of the SNPs with non zero β_{tk}; therefore, the above model should be sparse. Considering the model (4.1) for each of the T transcripts, $t = 1, \ldots, T$, we have a set of sparse linear regression models, which define the sparse seemingly unrelated regression (SpSUR) models.

In typical eQTL studies, the number of the SNP genotypes is often much larger than the sample size, i.e., $p >> n$. One approach for estimation and variable selection for the SpSUR models is through a combination of Lasso and penalized generalized least squares. Specifically, we can first apply Lasso or other high-dimensional regression methods to the linear regression model (4.1) to select $q_t(\lambda_1)$ variables, where λ_1 is the tuning parameter for Lasso. Denote the Lasso estimate of the non-zero coefficients as $\tilde{\beta}_{1t}$ and the corresponding SNPs with non-zero $\tilde{\beta}_{tk}$ as X_t. For simplicity, assume that there are q variables selected for each transcript. Then we can estimate the disturbance variance-covariance matrix as

$$\Sigma_e(\lambda_1) = \{\sigma_{tt'}\}, \text{ where } \sigma_{tt'} = \frac{\{(Y_t - X_t\tilde{\beta}_{1t})'(Y_t - X\tilde{\beta}_{1t})\}}{n-q}.$$

We now consider the following system of equations

$$Y = X\beta + \epsilon \Leftrightarrow \begin{bmatrix} Y_1 \\ Y_2 \\ \vdots \\ Y_T \end{bmatrix} = \begin{bmatrix} X_1 & 0 & \cdots & 0 \\ 0 & X_2 & 0 & 0 \\ \vdots & \vdots & & \vdots \\ 0 & 0 & \cdots & X_T \end{bmatrix} \begin{bmatrix} \beta_1 \\ \beta_2 \\ \vdots \\ \beta_T \end{bmatrix} + \begin{bmatrix} \epsilon_1 \\ \epsilon_2 \\ \vdots \\ \epsilon_T \end{bmatrix},$$

where note that X_t only includes the SNPs that are selected by the Lasso regression for the tth transcripts. This is the SUR regression setup considered in Zellner (1962). To further select the relevant variables, we can perform the following L_1 regularized generalized least squares

$$Q(\beta; \lambda_1, \lambda_2) = (Y - X\beta)'\Sigma_e(\lambda_1)^{-1}(Y - X\beta_t) + \lambda_2 \sum_{t=1}^{T} |\beta_t|_1,$$

where λ_1 and λ_2 are the two tuning parameters. The powerful coordinate descent algorithm can be used to estimate and select the variables and to estimate the parameters. The theoretical properties of this two-stage regularized estimation of the SpSUR models are yet to be developed.

4.2 Dynamic co-expression analysis

The eQTL studies have so far only focused on identifying the SNPs that affect the mean differential expression, which provide one aspect of differential regulation. However, there are still many important classes of differentially regulated genes that can be studied. Sun, Yuan and Li (2007) proposed to use the liquid association (LA) method (Li, 2002) to analyze eQTL data, where they studied the expression of a pair of genes and treat the variation in their co-expression pattern as a two-dimensional quantitative trait. They applied the LA method to find the gene pairs, whose co-expression patterns, including both signs and strengths, are mediated by genetic variations and mapped these 2D-traits to the corresponding genetic loci. Chen, Xie and Li (2010) proposed an alternative formulation of this problem that can simultaneously consider multiple SNPs.

Suppose there are n *i.i.d.* samples and let X, Y be the expression levels of the gene pair under study and $\mathbf{Z} = (\mathbf{z}_1, \dots, \mathbf{z}_p)$ be the set of candidate mediating variables, which can be the SNPs in eQTL studies. Let $(x_i, y_i)_{i=1,\dots,N}$ denote the normalized expression level of X and Y in the ith sample. $(\mathbf{z}_i)_{i=1,\dots,N}$ is a vector denoting the normalized SNP genotype scores of the SNP set in the ith sample. We assume the expression level of genes X,Y conditioned on $\mathbf{Z}$ has a bivariate normal distribution with the correlation determined by linear combination of the elements of $\mathbf{Z}$, that is,

$$\begin{pmatrix} X \\ Y \end{pmatrix} \mid (\mathbf{Z} = \mathbf{z}) \sim N(\mathbf{0}, \Sigma(\mathbf{z})), \tag{4.2}$$

where

$$\Sigma(\mathbf{z}) = \begin{pmatrix} 1 & \rho(\mathbf{z}) \\ \rho(\mathbf{z}) & 1 \end{pmatrix}$$

is the covariance matrix and

$$\rho(\mathbf{z}) = \frac{1 - e^{\boldsymbol{\beta}^T \mathbf{z} + c}}{1 + e^{\boldsymbol{\beta}^T \mathbf{z} + c}},$$

models the correlation between X and Y as a function of $\mathbf{Z}$, where $\boldsymbol{\beta}$ is the coefficient vector and c is the intercept. Note that in this model (4.2), we assume that both X and Y are standardized to have mean zero and variance 1.

When the dimension of the mediating set Z is large, our goal is to select these genes or SNPs that can mediate the correlations between two genes X and Y. For a given candidate mediating SNP set Z, we expect that most of the SNPs in $\mathbf{Z}$ are irrelevant. We consider the following general penalized likelihood formulation,

$$O_1(\boldsymbol{\beta}) = \sum_{i=1}^{n} \left(\frac{x_i^2 + y_i^2 - 2\rho(\boldsymbol{\beta}, \mathbf{z}_i) x_i y_i}{(1 - \rho^2(\boldsymbol{\beta}, \mathbf{z}_i))} + \log(1 - \rho^2(\boldsymbol{\beta}, \mathbf{z}_i)) \right) + \sum_{i=1}^{p} p_\lambda(\boldsymbol{\beta}_i)$$

where $p_\lambda(\boldsymbol{\beta}_i)$ is a penalty function and λ is a parameter controlling the degree of sparsity. Some popular penalty functions include the L_1 or Lasso penalty or or adaptive Lasso L_1 penalty proposed by Zou (2006). Alternatively, we can also consider the SCAD penalty proposed by Fan and Li (2001). Chen, Xie and Li (2010) have studied this approach and have established the variable selection consistency and the oracle property of the estimates when $p < n$.

5   Discussion and future direction

Genomic and systems biology research has generated many types of high-dimensional genomic data. The ultimate goal of systems biology research is to elucidate the inter-relationships among tens of thousands of genes based on various types of genomic data. The problems are intrinsically statistical and require new statistical methods to fully understand these data, especially in the area of high dimensional statistics. In this chapter, we presented several problems and statistical solutions for integrative analysis of multiple types of genomic data. In order to account for the sparsity of the data, methods of regularization are employed in the solutions. There are, however, many such datasets being generated and the development of new methods is therefore required. As an example, the US National Institutes of Health (NIH) launched "The Cancer Genome Atlas" (TCGA) project, which is a comprehensive and coordinated effort to accelerate our understanding of the molecular basis of cancer through the application of genome analysis technologies, including large-scale genome sequencing by systematically exploring the entire spectrum of genomic changes involved in human cancer. Another example is the NIH Genotype-Tissue Expression (GTEx) project, which aims to provide to the scientific community a resource with which to study human gene expression and regulation and its relationship to genetic variation. This project will collect and analyze multiple human tissues from donors who are also densely genotyped, to assess genetic variation within their genomes. By analyzing global RNA expression within individual tissues and treating the expression levels of genes as quantitative traits, variations in gene expression that are highly correlated with genetic variation can be identified as expression quantitative trait loci, or eQTLs. As we reviewed, statistical methods for the analysis of eQTLs are still in their infancy and greatly needed.

A new generation of deep sequencing technologies, including Applied Biosystems' SOLiD, Helicos BioSciences' HeliScope, Illumina's Solexa, and Roche's 454 Life Sciences sequencing systems, has delivered on promises of sequencing DNA at unprecedented speed, thereby enabling impressive scientific achievements and novel biological applications (Schuster, 2008). These high-throughput sequencing technologies have already been applied for studying genome-wide transcription levels (mRNA-Seq) (Mortazavi *et al.*, 2008; Nagalakshmi *et al.*, 2008; Wang *et al.*, 2008), transcription factor binding sites (ChIP-Seq) (Johnson *et al.*, 2007; Robertson *et al.*, 2007; Schmid *et al.*, 2007), chromatin structure, DNA methylation status (Barski *et al.*, 2007; Cokus *et al.*, 2008) and metagenomics (Eckberg *et al.*, 2005). While sequencing-based technologies provide high-resolution measurements of various biological quantities, these new biotechnologies also raise novel statistical and computational challenges, in areas such as image analysis, base-calling and read-mapping in initial analysis, peak finding and differential comparisons in comparative experiments, and mixture modeling in metagenomics. Another exciting technology is in the area of single-cell dynamic proteomics using a combination of CD tagging technology and quantitative fluorescent microscopy (Cohen *et al.*, 2008), which can yield quantitative traces of protein amounts and localization over time for each individual cell for over 1000 proteins. Such data provide great opportunities to understand the dynamics of biological systems. Novel statistical methods are being developed to meet these challenges (Ji *et al.*, 2008; Ellis and Wong, 2008; Jiang and Wong, 2009). We expect to see more such methods to be developed in the near future, especially in the integrative analysis of multiple high-resolution genomic data at single cell levels in the context of biological knowledge.

References

[1] Anderson P. and Gill R. (1983). Cox's regression models for counting processes: a large sample study. *Annals of Statistics* **10**, 1100-1120.

[2] Bannerjee S., Yandell B. S. and Yi N. (2008). Bayesian quantitative trait loci mapping for multiple traits. *Genetics* **179**, 2275-89.

[3] Barski A., Cuddapah S., Cui K., Roh T. Y., Schones D. E., Wang Z., Wei G., Chepelev I. and Zhao K. (2007). High-resolution profiling of histone methylations in the human genome. *Cell* **129**, 823-837.

[4] Bickel P. and Levina L. (2008a). Covariance regularization by thresholding. *Annals of Statistics* **34**, 2577-2604.

[5] Bickel P. and Levina L. (2008b). Regularized estimation of large covariance matrices. *Annals of Statistics* **36**, 199-227.

[6] Bickel P. L., Ritov Y. and Tsybakov A. B. (2008). Hierarchical selection of variables in sparse high-dimensional regression. *Technical report, Statistics*, UC Berkeley.

[7] Breheny P. and Huang J. (2008). Penalized methods for bi-level variable selection. *Technical report*, University of Iowa.

[8] Brem R. B. and Kruglyak L. (2005). The landscape of genetic complexity across 5,700 gene expression traits in yeast. *Proceedings of National Academy of Scieces, U S A.* **102**(5), 1572-1577

[9] Bussemaker H. J., Li H. and Siggia E. D. (2001). Regulatory element detection using correlation with expression. *Nature Genetics* **27**, 167-171.

[10] Chen G., Jensen S., and Stockert C. (2007). Clustering of genes into regulons using integrated moeling(COGRIM). *Genome Biology* **8**, R4.

[11] Chen J., Xie J. and Li H. (2010). A penalized likelihood approach for dynamic co-expression analysis. *Biometrics*, in press.

[12] Cheung V. G., Spielman R. S., Ewens K. G., Weber T. M., Morley M. and Burdick J. T. (2005). Mapping determinants of human gene expression by regional and whole genome association. *Nature* **437**, 1365-1369.

[13] Chuai S. and Li H. (2009). Variable selection for models with time-dependent coefficients in high-dimensional settings with applications to genomics data. *Manuscript.*

[14] Chung F. (2005). Laplacians and the Cheeger inequality for directed graphs. *Annals of Combinatorics* **9**, 1-19.

[15] Chung F. (1997). *Spectral Graph Theory*, Vol. 92 of CBMS Regional Conferences Series. American Mathematical Society, Providence.

[16] Cohen A. A., Geva-Zatorsky N., Eden E., Frenkel-Morgenstern M., Issaeva I., Sigal A., Milo R., Cohen-Saidon C., Liron Y., Kam Z., Cohen L., Danon T., Perzov N. and Alon U. (2008). Dynamic proteomics of individual cancer cells in response to a drug. *Science* **322**(5907), 1511-1516.

[17] Cokus S. J., Feng S., Zhang X., Chen Z., Merriman B., Haudenschil C. D., Pradhan S., Nelson S. F., Pellegrini M., Jocobsen S. E. (2008). Shotgun bisulphite sequencing of the Arabidopsis genome reveals DNA methylation patterning. *Nature* **452**, 215-219.

[18] Conlon E. M., Liu X. S., Lieb J. D. and Liu J. S. (2003). Integrating regulatory motif discovery and genome-wide expression analysis. *Proceedings of National Academy of Sciences* **100**, 3339-3344.

[19] Cookson W., Liang L., Abecasis G., Moffatt M. and Lathrop M. (2009). Mapping complex disease traits with global gene expression. *Nature Review*

Genetics **10**, 184-194.

[20] Cox D. R. (1972). Regression models and life-tables (with discussion). *Journal of the Royal Statistical Society, Series B* **34**, 187-220.

[21] Das D., Nahle Z. and Zhang M. Q. (2006). Adaptively inferring human transcriptional subnetworks. *Molecular Systems Biology*, msb410067-E1.

[22] Eckerbeg P. B., Bik E. M., Bernstein C. N., Purdom E., Dethlefsen L., Sargenet M., Gill S. R., Nelson K. E. and Relman D. A. (2005). Diversity of the human intestinal microbial flora. *Science* **308**, 1635-1638.

[23] Efron B., Hastie T., Johnstone I. and Tibshirani R. (2004). Least angle regression. *Annals of Statistics* **32**, 407-499.

[24] Ellis B. and Wong W. H. (2008). Learning causal Bayesian network structures from experimental data. *Journal of the American Statistical Association* **103**, 778-789.

[25] Fan J., Feng Y. and Wu Y. (2009). Network exploration via the adaptive lasso and scad penalties. *Annals of Applied Statistics* **3**, 521-541.

[26] Fan J. and Li R. (2001). Variable selection via nonconcave penalized likelihood and its oracle properties. *Journal of American Statistical Association* **96**, 1348-1360.

[27] Fan J. and Lv J. (2008). Sure independence screening for ultra-high dimensional feature space (with discussion). *Journal of Royal Statistical Society B* **70**, 849-911.

[28] Fan J., Samworth R. and Wu Y. (2008). Ultrahigh dimensional variable selection: beyond the lienar model. *Manuscript*.

[29] Friedman J. H. (1991). Multivariate Adaptive Regression Splines (with discussion). *Annals of Statistics* **19**, 1.

[30] Friedman J. H., Hastie T, Tibshirani R. (2008). Sparse inverse covariance estimation with the graphical lasso. *Biostatistics* **9**(3), 432-441.

[31] Friedman J., Hastie T., Hoefling H. and Tibshirani R. (2007). Pathwise coordinate optimization. *Annals of Applied Statistics* **1**, 302-332.

[32] Gao F., Foat B.C. and Bussemaker H. J. (2004). Defining transcriptional networks through integrative modeling of mRNA expression and transcription factor binding data. *BMC Bioinformatics* **5**, 31.

[33] Greenstein E. and Ritov Y. (2004). Persistence in high dimensional linear predictor-selection and the virtue of over-parametrization. *Bernoulli* **10**, 971-988.

[34] Gui J. and Li H. (2005). Penalized Cox regression analysis in the high-dimensional and low-sample size settings, with applications to microarray gene expression data. *Bioinformatics* **21**, 3001-3008.

[35] Huang J., Horowitz J. L. and Wei F. (2008). Variable selection in nonparametric additive models. *Technical Report*, University of Iowa.

[36] Huang J. and Xie H. (2007). Asymptotic oracle properties of SCAD-penalized least squares estimators. *IMS Lecture Notes - Monograph Series, Asymptotics: Particles, Processes and Inverse Problems* **55**, 149-166.

[37] Jansen R. C. and Nap J. P. (2006). Genetical genomics: the added value from segregation. *Trends in Genetics* **17**, 388-391.

[38] Ji H., Jiang H., Ma W., Johnson D. S., Myers R. M. and Wong W. H. (2008).

An integrated software system for analyzing ChIP-chip and ChIP-seq data. *Nature Biotechnology* **26**, 1293-1300.

[39] Jia J. and Yu B. (2008). On model selection consistency of elastic net when $p >> n$. *Technical Report 756, Statistics*, UC Berkeley.

[40] Jiang H. and Wong W. H. (2009). Statistical inferences for isoform expression in RNA-Seq. *Bioinformatics* **25**(8), 1026-1032.

[41] Johnson D. S., Mortazavi A. et al. (2007). Genome-wide mapping of in vivo protein-DNA interactions. *Science* **316**, 1497-1502.

[42] Kanehisa M. and Goto S. (2002). KEGG: Kyoto encyclopedia of genes and genomes. *Nucleic Acids Research* **28**, 27-30.

[43] Keles S., Van Der Laan M. and Eisen M. B. (2002). Identification of regulatory elements using a feature selection method. *Bioinformatics* **18**, 1167-1175.

[44] Kendziorski C. and Wang P. (2006). A review of statistical methods for expression quantitative trait loci mapping. *Mammalian Genome* **17**(6), 509-517.

[45] Kim S., Sohn K. and Xing E. P. (2009). A multivariate regression approach to association analysis of a quantitative trait network. *Bioinformatics* **25**(12), i204-i212.

[46] Li C. (2009). Statistical methods for analysis of graph-constrained genomic data. *PhD Dissertation*, University of Pennsylvania.

[47] Li C. and Li H. (2008). Network-constrained regularization and variable selection for analysis of genomic data. *Bioinformatics* **24**, 1175-1182.

[48] Li C. and Li H. (2010). Variable selection and regression analysis for graph-structured covariates with an application to genomics. *Annals of Applied Statistics*, in press.

[49] Li H., and Gui J. (2006). Gradient directed regularization for sparse Gaussian concentration graphs, with applications to inference of genetic networks. *Biostatistics* **7**, 302-317.

[50] Li H. and Luan Y. (2005). Boosting proportional hazards models using smoothing spline, with applications to high-dimensional microarray data. *Bioinformatics* **21**, 2403-2409.

[51] Li K. C. (2002). Genome-wide coexpression dynamics: Theory and application. *Proceedings of National Academy of Sciences*, **99**(26), 16875-16880.

[52] Luan Y. and Li H. (2008). Group additive regression models for analysis of genomic data. *Biostatistics* **9**, 100-113.

[53] Meier L., van de Geer S., and Buhlmann P. (2008). High-dimensional additive modeling. To appear in the *Annals of Statistics*, arXiv:0806.4115v3

[54] Meier L., van de Geer S. and Buhlmann P. (2008). The group Lasso for logistic regression. *Journal of the Royal Statistical Society, Series B* **70**, 53-71.

[55] Meinshausen N. and Buhlmann B. (2006). High dimensional graphs and variable selection with the lasso. *Annals of Statistics* **34**, 1436-1462.

[56] Meinshausen N., Meier L. and Buhlmann P. (2008). P-values for high-dimensional regression. Preprint, arXiv:0811.2177v2.

[57] Meinshausen N. and Yu B. (2009). Lasso-type recovery of sparse representations for high-dimensional data. *Annals of Statistics* **37**, 246-270.

[58] Morley M., Molony C. M., Weber T., Devlin J. L., Ewens K. G., Spielman R. S., and Cheung V. G. (2004). Genetic analysis of genome-wide variation

in human gene expression. *Nature* **430**, 743-747.

[59] Mortazavi A., Williams B. A., McCue K., Schaeffer L., and Wold B. (2008). Mapping and quantifying mammalian transcriptomes by RNA-Seq. *Nature Methods* **5**, 621-628.

[60] Nagalakshmi U., Wang Z., Waern K., Shou C., Raha D., et al. (2008). The transcriptional landscape of the yeast genome defined by RNA sequencing. *Science* **320**, 1344-1349.

[61] Nardi Y. and Rinaldo A. (2008). On the asymptotic properties of the group lasso estimator for linear models. *Electronic Journal of Statistics* **2**, 605-633.

[62] Peng J., Wang P., Zhou N. F. and Zhu J. (2009). Partial correlation estimation by joint sparse regression model. *Journal of the American Statistical Association* **104** (486), 735-746

[63] Robertson G., et al. (2007). Genome-wide profiles of STAT1 DNA association using chromatin immunoprecipitation and massively parallel sequencing. *Nature Methods* **4**, 651-657.

[64] Rocha G., Zhao P. and Yu B. (2008). A path following algorithm for Sparse Pseudo-Likelihood Inverse Covariance Estimation (SPLICE). *Technical Report 759*, Statistics Department, UC Berkeley.

[65] Schmid, et al. (2007). ChIP-Seq data reveal nucleosome architecture of human promoters. *Cell*, **131**, 831-832.

[66] Schuster S. C. (2008). Next-generation sequencing transforms today's biology *Nature Methods* **5**, 16-18.

[67] Shi W., Wahba G., Wright S., Lee K., Klein R. and Klein B. (2008). LASSO-Patternsearch algorithm with applications to ophthalmology and genomic data. *Statistics and Its Interface(SII)* **1**, 137-153.

[68] Sun W., Yuan S. and Li K. C. (2008). Trait-trait dynamic interaction: 2D-trait eQTL mapping for genetic variation study. *MBC Genomics* **9**, 242.

[69] Tang F., Barbacioru C., Wang Y., Nordman E., Lee C., Xu N., Wang X., Bodeau J., Tuch B. B., Siddiqui A., Lao K. and Surani, A. (2009). mRNA-Seq whole-transcriptome analysis of a single cell. *Nature Methods* **6**, 377-382.

[70] Tibshirani R. J. (1996). Regression shrinkage and selection via the lasso. *Journal of Royal Statistical Society B* **58**, 267-288.

[71] Tibshirani R., Saunders M., Rosset S., Zhu J. and Knight K. (2005). Sparsity and smoothness via the fused lasso. *Journal of Royal Statistical Society SER B* **67**, 91-108.

[72] Wainwright M. (2006). *Sharp thresholds for high-dimensional and noisy recovery of sparsity*. Available at arXiv.math.ST/0605740.

[73] Wang E. T., Sandberg R., Luo S., Khrebtukova I., Zhang L., Mayr C., Kingsmore S. F., Schroth G. P. and Burge C. B. (2008). Alternative isoform regulation in human tissue transcriptomes. *Nature* **456**, 470-476.

[74] Wang L., Chen G., Li H. (2007). Group SCAD regression analysis for microarray time course gene expression data. *Bioinformatics* **23**, 1486-1494.

[75] Wang, L., Li, H. and Huang, J. (2008). Variable selection in nonparametric varying-coefficient models for analysis of repeated measurements. *Journal of the American Statistical Association* **103**, 1556-1569.

[76] Wang L., Li H. and Huang J. Z. (2009). Variable selection for the Cox model

with time-varying regression coefficients. *Manuscript.*

[77] Wang S., Nan B., Zhou N. and Zhu J. (2009). Hierarchically penalized Cox regression for censored data with grouped variables and its oracle property. *Biometrika* **96**, 307-322.

[78] Wei F., Huang J. and Li H. (2010). Varying-coefficient model in high-dimensional linear regression. *Statistica*, in press.

[79] Wei Z. and Li H. (2007). Nonparametric pathways-based regression models for analysis of genomic data. *Biostatistics* **8**(2), 265-284.

[80] Wu T. T. and Lange K. (2008). Coordinate descent algorithms for lasso penalized regression. *Annals of Applied Statistics* **2**, 224-244.

[81] Wu T. T., Chen Y. F., Hastie T., Sobel E. and Lange K. (2009). Genome-wide association analysis by Lasso penalized logistic regression. *Bioinformatics* **25**, 714-721.

[82] Xu Y., Dyer J.S. and Owen A.B. (2009). Empirical stationary correlations for semi-supervised learning on graphs. *Technical report*, Stanford University.

[83] Xue L. (2009). Consistent variable selection in additive models. *Statistica Sinica* **19**, 1281-1296.

[84] Yuan M. and Lin Y. (2006). Model selection and estimation in regression with grouped variables. *Journal of Royal Statistical Society B* **68**, 49-67.

[85] Zellner A. (1962). An efficient method of estimating seemingly unrelated regression equations and tests for aggregation bias. *Journal of the American Statistical Association* **57**, 348-368.

[86] Zhang C. and Huang J. (2006). The sparsity and bias of the Lasso selection in high-dimensional linear regression. *Annals of Statistics* **36**, 1567-1594.

[87] Zhang H. H and Lu W. (2007). Adaptive LASSO for Cox's proportional hazards model. *Biometrika* **94**, 691-703.

[88] Zhang N., Wildermuth Wildermuth M. C., and Speed T. P. (2008). Transcription factor binding site prediction with multivariate gene expression data. *Annals of Applied Statistics* **2**, 332-365.

[89] Zhao P., Rocha G., and Yu B. (2009). Grouped and hierarchical model selection through composite absolute penalties. *Annals of Statistics* **37**, 3468-3497.

[90] Zhao P. and Yu B. (2006). On model selection consistency of Lasso. *Journal of Machine Learning Research* **7**, 2541-2567.

[91] Zhu, Pan W. and Shen X. (2009). Support vector machines with disease-gene-centric network penalty for high dimensional microarray data. *Statistics and its Inferences*, in press.

[92] Zou H. (2006). The adaptive lasso and its oracle properties. *Journal of the American Statistical Association* **101**, 1418-1429.

[93] Zou H. and Hastie T. (2005). Regularization and variable selection via the elastic net. *Journal of Royal Statistical Society SER B* **67**, 301-320.

[94] Zou H. and Zhang H. H. (2009). On the adaptive elasticnet with a diverging number of parameters. *Annals of Statistics* **37**, 1733-1751.

Chapter 7

An Overview on Joint Modeling of Censored Survival Time and Longitudinal Data

Runze Li [*]   and   Jian-Jian Ren [†]

Abstract

To facilitate statistical analysis studies encountered in many scientific research fields, recently statisticians started paying attention to jointly modeling survival data and longitudinal data. This is a considerably difficult problem due to censoring (especially complicated types of censoring) on the survival time, and dimensionality issue and measurement errors on the observed longitudinal covariates. By now, various modeling and estimation methods have been studied by different authors. In this chapter, we present a selective review of recent developments on this topic, and provide discussions on related further research problems.

Keywords: Accelerated life model, Cox's model; double censoring, failure time, intensive longitudinal data, interval censoring; likelihood, maximum likelihood estimator; right censoring.

1 Introduction

There has been a vast literature on longitudinal data analysis, viz., Diggle, Liang and Zeger (1994), Lin and Ying (2001), van der Laan and Robins (2003), Fitzamurice, Laird and Ware (2004), Fan and Li (2004), Zeng and Cai (2005), Fan, Huang and Li (2007), among others. In Fan and Li (2006), a selective review is provided on recent developments of nonparametric and semiparametric techniques for longitudinal data. As part of longitudinal data analysis, the interrelationship between time-to-event (*survival time*) variable and longitudinal covariates is often the primary research interest in medical and epidemiological studies. Here, the *longitudinal covariates* refer to covariate variables that are recorded at multiple,

---

[*]The author's research was partially supported by NSF grant DMS-0348869 and NIDA grant P50 DA10075; Mailing Address: Department of Statistics, The Pennsylvania State University, University Park, PA 16802-2111, USA, E-mail: rli@stat.psu.edu

[†]The author's research was partially supported by NSF grants DMS-0604488 and DMS-0905772, and NIDA grant DA-10075; Mailing Address: Department of Mathematics, University of Central Florida, Orlando, FL, 32816, USA, E-mail: jren@mail.ucf.edu

possibly subject-dependent time points during the study. Due to the challenges encountered in some important clinical trials on AIDS and cancer research, recently the statisticians started paying attention to jointly modeling survival data and longitudinal data; see DeGruttola and Tu (1994), Tsiatis, DeGruttola and Wulfsohn (1995), Wulfsohn and Tsiatis (1997), Bycott and Taylor (1998), Henderson, Diggle and Dobson (2000), Tsiatis and Davidian (2001), Brown and Ibrahim (2003), van der Laan and Robins (2003), Zeng and Cai (2005), Tseng, Hsieh and Wang (2005), Hsieh, Tseng and Wang (2006), Elashoff, Li and Li (2007), Song and Wang (2008), Ding and Wang (2008), among others. A review paper on this topic has been written by Tsiatis and Davidian (2004).

The joint modeling procedures on survival and longitudinal data have broad applications in many scientific research fields, such as medical research, epidemiology, social and behavioral sciences, etc., but it is a considerably difficult problem. The challenge is threefold. First, survival data are subject to right censoring in practical situations due to death or drop-outs, while there have been examples where the survival data are subject to complicated types of censoring, such as double censoring and interval censoring. Second, the observed longitudinal covariate data are unbalanced, especially in the case of intensive longitudinal data, which causes dimensionality problems and difficulties in development of modelings and estimation procedures. Third, the longitudinal covariates are subject to measurement errors in many practical situations. Note that different from *conventional* longitudinal covariates, the *intensive longitudinal covariates* and the multi-phase intensive longitudinal covariates refer to situations where considerably more repeated measurements are taken within a relative short period of time. The intent of collecting these intensive or multi phase intensive longitudinal covariates data is to conduct micro analytic assessment of the patterns of change, and the emergence of a rapidly expanding literature on collection of intensive longitudinal data can be found in social and personality psychology (Csikszentmihalyi and Larson, 1987; de Vries, 1992), in the substance abuse and health science domain (Shiffman *et al.* 1994), in child development studies (Cole *et al.*, 2008), etc.; see a recent book on intensive longitudinal data by Walls and Schafer (2006).

So far, survival models such as proportional hazards model and accelerated life model with longitudinal covariate processes have been considered in the literature of joint modeling survival time and longitudinal covariates. The estimation methods which have been considered include Bayesian, likelihood, semiparametric and estimating equation procedures, which all rely on certain parametric modeling or specification of the likelihood for the joint model parameters. Most existing works deal with right censored survival data, and theoretical properties of the existing procedures are available under certain special parametric assumptions. The issues of measurement errors and random effects have been studied via various methods. However, there have not been any consideration on issues of model checking, and there have not been any special considerations or procedures developed for intensive longitudinal covariates.

This chapter is organized as follows. Section 2 describes the data settings on joint modeling survival and longitudinal data, and reviews various types censored data in survival analysis with some examples. Section 2 also discusses intensive longitudinal data problems, measurement errors, and dimensionality issues for joint modeling problem. Section 3 reviews existing works for right censored survival data with longitudinal covariates, where more emphases are given to more recent works after Tsiatis and Davidian (2004). Section 4 reviews existing works for interval censored survival data with longitudinal covariates. Finally, Section 5 discusses issues and further research problems on joint modeling survival time and longitudinal data.

2 Survival data with longitudinal covariates

2.1 Data setting and various types of censoring

Throughout this article, we use the following notations for joint modeling survival time and longitudinal covariates:

$$(T_i, \boldsymbol{Z}_i, \boldsymbol{\mathcal{X}}_i), \quad i = 1, 2, \dots, n \tag{2.1}$$

where T_i is the survival time of the ith individual, $\boldsymbol{Z}_i$ is a vector of q explanatory variables observed on covariate variables $\boldsymbol{Z} = (Z_1, \dots, Z_q)^{\mathrm{T}}$ which do not change over the course of the studies, and the observed longitudinal data on covariate process $\boldsymbol{X}(t) = (X_1(t), \dots, X_p(t))^{\mathrm{T}}$ are

$$\boldsymbol{\mathcal{X}}_i = (\boldsymbol{X}_i(t_{i1}), \dots, \boldsymbol{X}_i(t_{i,J_i})), \tag{2.2}$$

with $\boldsymbol{X}_i(t_{ij})$ as the explanatory vector observed on covariate process $\boldsymbol{X}(t)$ at time t_{ij} on individual $i = 1, \dots, n$. In the case that T_i is subject to censoring, such as right censoring, double censoring and interval censoring, the actually observed data in practice on data (2.1) are

$$(V_i, \delta_i, \boldsymbol{Z}_i, \boldsymbol{\mathcal{X}}_i), \quad i = 1, 2, \dots, n, \tag{2.3}$$

where (V_i, δ_i) represents the observed censored data on T_i. In the case that there are measurement errors on observed longitudinal covariates $\boldsymbol{X}_i(t_{ij})$, the actually observed data in practice on data (2.3) are

$$(V_i, \delta_i, \boldsymbol{Z}_i, \boldsymbol{\mathcal{W}}_i), \quad i = 1, 2, \dots, n, \tag{2.4}$$

where $\boldsymbol{\mathcal{W}}_i$ is actually observed in practice on $\boldsymbol{\mathcal{X}}_i$, denoted by

$$\boldsymbol{\mathcal{W}}_i = (\boldsymbol{W}_i(t_{i1}), \dots, \boldsymbol{W}_i(t_{i,J_i})). \tag{2.5}$$

The interest of the study is the effects of covariates $(\boldsymbol{Z}, \boldsymbol{X}(t))$ on the survival time T using available data (2.3) or (2.4). As follows, we specify the meaning of right censoring, double censoring and interval censoring, respectively, while longitudinal covariates, measurement errors, data examples and related issues are reviewed later in Sections 2.2 and 2.3.

Right censored data: Under right censoring, observed (V_i, δ_i) in (2.3) or (2.4) are given by

$$V_i = \begin{cases} T_i, & \text{if } T_i \leqslant C_i, & \delta_i = 1 \\ C_i, & \text{if } T_i > C_i, & \delta_i = 0 \end{cases} \tag{2.6}$$

where C_i is the *right censoring variable* and is independent of $(T_i, \boldsymbol{Z}_i, \boldsymbol{X}_i)$. In practical situations, right censoring C_i occurs usually due to drop-outs or the end time of the study or any other possible factors; see examples in Cox and Oakes (1984), Kalbfleisch and Prentice (2002), among others. In statistical literature, this type of censoring has been extensively studied in the past few decades for the case without any covariates $(\boldsymbol{Z}_i, \boldsymbol{X}_i)$ in (2.3), or without longitudinal covariates $\boldsymbol{X}_i$ in (2.3); see Kaplan and Meier (1958), Cox (1972), Breslow (1974), Efron (1977), Tsiatis (1981), Gill (1983), Fleming and Harrington (1991), Stute and Wang (1993), Andersen, Borgan, Gill and Keiding (1993), Murphy and van der Vaart (2000), Ren and Zhou (2007), Ren (2009), among others.

Doubly censored data: Under double censoring, observed (V_i, δ_i) in (2.3) or (2.4) are given by

$$V_i = \begin{cases} T_i, & \text{if } D_i < T_i \leqslant C_i, & \delta_i = 1 \\ C_i, & \text{if } T_i > C_i, & \delta_i = 2 \\ D_i, & \text{if } T_i \leqslant D_i, & \delta_i = 3 \end{cases} \qquad (2.7)$$

where C_i and D_i are *right* and *left censoring variables*, respectively, and they are independent of $(T_i, \boldsymbol{Z}_i, \boldsymbol{X}_i)$ with $P\{C_i < D_i\} = 1$. This type of censoring has been studied for the case without covariates by Turnbull (1974), Chang and Yang (1987), Chang (1990), Gu and Zhang (1993), Ren (1995), Mykland and Ren (1996), Wellner and Zhan (1997), Chen and Zhou (2003), among others. The case with time-independent covariates $\boldsymbol{Z}_i$ was studied by Zhang and Li (1996), Ren and Gu (1997), Ren (2003b), among others. It should be noted that in statistical literature, Turnbull (1974) first studied doubly censored child development data from Leiderman *et al.* (1973), which was a doubly censored survival data set without any covariates, while one recent example of such case was encountered in a study of primary breast cancer (Peer *et al.*, 1993; Ren and Peer, 2000). By now, it is well-known that compared with right censored survival data, the problem of Turnbull's doubly censored survival data is far more complicated and challenging; see techniques used in the works mentioned above. And the problem of doubly censored child development data with longitudinal covariates has not been studied in the literature.

Interval censored data: Under interval censoring, observed (V_i, δ_i) in (2.3) or (2.4) are given by

$$V_i = C_i, \ \delta_i = I\{T_i \leqslant C_i\}, \qquad (2.8)$$

where C_i is independent of $(T_i, \boldsymbol{Z}_i, \boldsymbol{X}_i)$. This types of censoring is also called *current status data* or *interval censored Case 1 data* , and has been studied for the case without covariates by Finkelstein and Wolfe (1985), Groeneboom and Wellner (1992), Geskus and Groeneboom (1999), Ren (2003a), Banerjee and Wellner (2005), Ren (2008a, 2008b), among others. The case with time-independent covariates $\boldsymbol{Z}_i$ was studied by Huang (1996), van der Laan and Robins (2003), among others. So far, there is only a little bit works done on the problem of interval censored data with longitudinal covariates, which is reviewed in Section 4 of this article. In the literature, there have been interval censored data examples given, such as those from cancer research, and some discussions on this topic are given in Section 2.3.

2.2 Longitudinal data and measurement errors

This subsection discusses issues related to the longitudinal covariates and measurement errors in the data setting given in (2.3) or (2.4). More detailed descriptions on this topic can be found in books by Diggle, Liang and Zeger (1994), van der Laan and Robins (2003), Fitzamurice, Laird and Ware (2004), and Walls and Schafer (2006).

Longitudinal data: In the health science and in social and behavioral science, longitudinal studies play an important role in enhancing our understanding of the development of disease or the development of certain *event* of interest. The distinguishing feature of longitudinal studies is that the participants or subjects under the study have measurements taken repeatedly throughout the duration of the study, denoted by covariate process $X(t)$ in our data setting (2.3), thereby permitting the direct assessment of changes in the response variable over time. By obtaining measurements $X_i(t_{ij})$ of the same individual repeatedly through time at time points t_{ij}, longitudinal data provide information to address fundamental questions concerning the assessment of within-subject changes in the response variable of interest. There is much natural heterogeneity among individuals in terms of how disease or *event* of interest develop and progress. This heterogeneity is due to genetic, environmental, social and behavioral factors. A longitudinal study design permits the discovery of individual characteristics that can explain these within-subject differences in changes that affect health outcomes or response variable of interest over time. In summary, the fundamental objective of a longitudinal analysis is the assessment of within-subject changes in the response and the explanation of systematic differences among individuals in their changes.

Joint modeling survival time T_i and longitudinal data $X_i(t_{ij})$ in our data setting (2.3) provides tools for statistical analysis to characterize the relationship between longitudinal covariates and survival time, i.e., the relationship between covariate process and time to progression or death. For instance, in studies of prostate cancer, characterizing the association between features of the longitudinal prostate specific antigen process and cancer recurrence may be of interest. Obviously, the joint modeling procedures on survival and longitudinal data have broad applications in many scientific research fields, such as medical research, epidemiology, social and behavioral sciences, etc.

In practice, the implementation of a joint modeling procedure is complicated by the fact that the covariate process $X_i(t)$ is available only intermittently at time point t_{ij} for each subject and is possible subject to measurement error. In particular, traditional sources of longitudinal data, such as panel surveys, clinical trials, studies of human growth or disease development, etc., usually have only a few repeated measurements for each individual during the course of study. This makes it more difficult to obtain good estimate for the covariate process $X(t)$ based on limited available longitudinal data. Recently, there has been a rapidly growing literature on collecting *intensive longitudinal data*, which provide far more information for the estimation of covariate process $X(t)$ than traditional longitudinal data, and is briefly reviewed below.

Intensive longitudinal data: Different from *conventional* longitudinal covariates, the intensive longitudinal covariates and the multi-phase intensive longitudinal covariates refer to situations where considerably more repeated measurements are taken within a relative short period of time. As the capacity of modern computers to store and process information continues to increase, and as new technologies expand the possibilities for data capture in diverse areas of science, longitudinal databases with a much higher intensity and volume of data are becoming commonplace. In recent years, collecting intensive longitudinal data has been rapidly developing in behavioral science which relies on handled computers, beepers, web interfaces, and other tools for data collection. The intent of collecting these intensive or multi-phase intensive longitudinal covariates data is to conduct micro-analytic assessment of the patterns of change, and the emergence of a rapidly expanding literature on collection of intensive longitudinal data can be found in social and personality psychology (Csikszentmihalyi and Larson, 1987; de Vries, 1992), in the substance abuse and health science domain (Shiffman *et al.* 1994), in child development studies (Cole *et al.*, 2008), etc.; see a recent book on intensive longitudinal data by Walls and Schafer (2006). As follows, we briefly describe some common methods that have been or are being used for collecting intensive longitudinal data.

One source of intensive longitudinal data comes from human participants' diary studies. Individuals are asked to complete by paper and pencil, usually at the end of the day, a log of their experiences or actions. The primary reason for obtaining daily reports is to reduce bias and variance in the measurement process by having participants respond while the experiences are still relatively fresh in their minds. Recent technological developments have made the collection of diary data more convenient for respondents and researchers alike. Subjects are now given small electronic devices (e.g., hand-hold computers) which prompt them at various times throughout the day to ask questions and record responds. Among psychologists, the frequent recording of thoughts, feelings, or actions by electronic means has been referred to as the experience-sampling method (Csikszentmihalyi and Larson, 1987) and ecological momentary assessment (Stone and Shiffman, 2002). Other techniques used for collecting intensive longitudinal data involve automatic sensing of physical behaviors (e.g., number of steps taken) or bodily states (e.g., blood pressure or glucose levels) by using ambulatory devices that passively monitor participants in their natural environments outside of a laboratory. Frequently, ambulatory telemetric measurement devices and technology-enabled questionnaires have been employed in these studies. And devices measuring actions and states are also used in laboratories and clinics. Moreover, audio and video recordings of individuals as they interact with their physical environment or with one another, which are subsequently reviewed and coded by researchers, can also generate high volumes of intensive longitudinal data.

Dimensionality issues for joint modeling: The observed longitudinal covariate data $X_i(t_{ij})$ in (2.3) are unbalanced, especially in the case of intensive longitudinal data which can be of very large volume of data as described above. From mathematical point of view, due to the curse of dimensionality one needs to specify a lower-dimensional working model on observed longitudinal covariates $\mathcal{X}_i$ in (2.3) for joint modeling procedures, because the dimension of $\mathcal{X}_i$ generally varies from subject to subject; in the case of intensive longitudinal covariates such a variation of the dimension can be rather great. The existing working models used for joint modeling survival and longitudinal data are mainly proportional hazards model and accelerated life model (Cox and Oakes, 1984), which will be reviewed in Section 3 and Section 4, respectively.

Measurement errors: In longitudinal data, one source of variability is random measurement error. For some health outcomes, for example, height and weight, variation due to measurement error can be almost negligible (or can be made negligible with the use of more sophisticated measurement instruments). However, for many other outcomes, the variability due to measurement error can be quite substantial; see Fitzamurice, Laird and Ware (2004) for more discussions on this. It should be noted though that measurement error is a ubiquitous component of almost all studies, longitudinal or other types of statistical analysis.

For the main focus of this article, we note that in practice, the longitudinal covariates $X_i(t_{ij})$ in (2.3) may be subject to measurement errors; that is $X_i(t_{ij})$ is a latent variable and what actually observe is its surrogate variable $W_i(t_{ij})$ in (2.5) which satisfies the following assumption:

$$W_i(t_{ij}) = X_i(t_{ij}) + e_i(t_{ij}), \quad i = 1, \ldots, n; j = 1, \ldots, J_i; \qquad (2.9)$$

where $e_i(t_{ij})$ is a zero-mean random error vector at time t_{ij} for the ith individual. Statistical procedures for measurement error data have been well developed for independent observations; see Fuller (1987), Liang, Härdle and Carroll (1999), Carroll *et al.* (2006), Liang, Wang and Carroll (2007), Liang *et al.* (2008), among others. However, in most longitudinal studies, there are no replicates of $W_i(t_{ij})$ for a given time point t_{ij}, thus these methods are not directly applicable for the problem of joint modeling survival and longitudinal data. There have been various methods to handle this problem, and they are reviewed in Section 3.

2.3   Data examples

This subsection gives a few real data examples encountered in practice which are of the data setting given in (2.3) or (2.4).

Example 1. (*AIDS Data*). In the HIV clinical trials considered by Tsiatis, DeGruttola and Wulfsohn (1995), T_i is time to progression to AIDS or death of a patient, which is subject to right censoring (2.6). For each patient, time-independent covariates Z_i include treatment assignment, demographic information, etc., while the longitudinal covariates X_i include measurements of immunologic and virologic status such as CD4 count and viral RNA copy number which are taken at subsequent clinic visits. This is an example of *right censored survival data with longitudinal covariates*.

Example 2. (*Cancer Data*). In the carcinogenicity experiment data considered by Hoel and Walburg (1972) and Finkelstein and Wolfe (1985), T_i is the tumor onset of a mouse, which is subject to interval censoring (2.8) because the mouse was examined at sacrifice or death time C_i for evidence of a malignancy. For each mouse, time-independent covariates Z_i include doses of the suspected carcinogen, etc., while the longitudinal covariates $\boldsymbol{\mathcal{X}}_i$ include measurements such as weight, etc., (see discussions in van der Laan and Robins, 2003; pages 249-250) which are taken on the mouse over time. This is an example of ***interval censored survival data with longitudinal covariates***.

3   Joint modeling with right censored data

So far, the existing models used for joint modeling survival and longitudinal data are mainly Cox's proportional hazards model and accelerated life model (Cox and Oakes, 1984). A review paper on this topic using the proportional hazards model was written by Tsiatis and Davidian (2004). In this section, Section 3.1 summarizes some overview key points and some selected works reviewed by Tsiatis and Davidian (2004), then Section 3.2 and Section 3.3 review some more recent works on joint modeling right censored survival data (2.6) with longitudinal covariates (2.3) or (2.4) via the proportional hazards model and the accelerated life model, respectively.

3.1   Overview

Currently, most existing methods for joint modeling survival time and longitudinal covariates are mainly on the line of Tsiatis and Davidian (2004). In these works, the observed longitudinal covariate $\boldsymbol{\mathcal{X}}_i$ in (2.1) is modeled as a time-dependent covariate process $\{\boldsymbol{X}_i(t); t \geqslant 0\}$ which satisfies (2.2). Denoting $\boldsymbol{X}_i^H(t) = \{\boldsymbol{X}_i(s); 0 \leqslant s < t\}$ as the history of the longitudinal process up to time t, the following Cox proportional hazards model is assumed:

$$\lambda(t; \boldsymbol{Z}_i, \boldsymbol{X}_i^H(t)) = \lambda_0(t) \exp\left(\boldsymbol{Z}_i^{\mathrm{T}} \boldsymbol{\beta}_Z + \boldsymbol{X}_i(t)^{\mathrm{T}} \boldsymbol{\beta}_X\right), \tag{3.1}$$

where $\boldsymbol{\beta}_Z \in \mathbb{R}^q, \boldsymbol{\beta}_X \in \mathbb{R}^p$ are regression parameters, $\lambda_0(t)$ is an arbitrary unspecified baseline hazard function, and $\lambda(t; \boldsymbol{Z}_i, \boldsymbol{X}_i^H(t))$ is the conditional hazard function of T given $\boldsymbol{Z} = \boldsymbol{Z}_i, \boldsymbol{X}^H(t) = \boldsymbol{X}_i^H(t)$ in the following sense:

$$\lambda(t; \boldsymbol{Z}_i, \boldsymbol{X}_i^H(t)) = \lim_{\Delta \to 0} \Delta^{-1} P\{t \leqslant T_i < t + \Delta \,|\, T_i \geqslant t, \boldsymbol{Z}_i, \boldsymbol{X}_i^H(t)\}. \tag{3.2}$$

It should be noticed that model (3.1) is different from the usual Cox's proportional hazards model with time-dependent covariates. In (3.1), $\boldsymbol{X}_i(t)$ is only observed at time t_{ij}, while in the Cox's model with time-dependent covariates, $\boldsymbol{X}_i(t)$ is assumed to be observed fully at any time t; see Kalbfleisch and Prentice (2002), Tsiatis and Davidian (2004). Thus, the usual partial likelihood formula cannot be applied in joint model (3.1) for right censored survival data (2.6) with longitudinal covariates given by (2.3) or (2.4).

Since the covariate process $\boldsymbol{X}_i(t)$ is only observable at time points $t_{i1} < \cdots < t_{i,J_i} \leqslant V_i$, various parametric or semiparametric process assumptions on $\boldsymbol{X}_i(t)$ are considered by various authors in order to estimate the covariate process $\boldsymbol{X}_i(t)$. As follows, we review some common methods used for modeling the longitudinal covariate process $\boldsymbol{X}_i(t)$ given in (2.2) and for modeling the measurement error $\boldsymbol{W}_i(t)$

given in (2.5), and we review some main approaches used for estimation, such as approximation approach, likelihood approach, Bayesian approach and estimating equations approach.

Modeling longitudinal covariate process: Generally, a standard approach for modeling the longitudinal covariate process is to characterize $X_i(t)$ through a subject-specific random effect vector $\boldsymbol{\alpha}_i$ by the following:

$$X_i(t) = \boldsymbol{\nu}(t)^{\mathrm{T}}\boldsymbol{\alpha}_i \tag{3.3}$$

where $\boldsymbol{\nu}(t)$ is a vector of functions of time t, and (3.3) includes polynomial functions as a special case. An example of (3.3) is $\boldsymbol{\nu}(t) = (1, t)^{\mathrm{T}}$, which gives the following simple linear model:

$$X_i(t) = \alpha_{0i} + \alpha_{1i}t, \quad \boldsymbol{\alpha}_i = (\alpha_{0i}, \alpha_{1i})^{\mathrm{T}}, \tag{3.4}$$

and was used by DeGruttola and Tu (1994), Tsiatis, DeGruttola and Wulfsohn (1995), Wulfsohn and Tsiatis (1997), Dafni and Tsiatis (1998), Bycott and Taylor (1998), among others.

Alternative to model (3.3), Taylor, Cumberland and Sy (1994), Henderson, Diggle and Dobson (2000), Wang and Taylor (2001), among others, considered the following model:

$$X_i(t) = \boldsymbol{\nu}(t)^{\mathrm{T}}\boldsymbol{\alpha}_i + U_i(t) \tag{3.5}$$

where $U_i(t)$ is a zero-mean stochastic process, usually assumed to be independent of random effect variable $\boldsymbol{\alpha}_i$ and time-independent covariate $\boldsymbol{Z}_i$. In particular, Henderson, Diggle and Dobson (2000) considered $U_i(t)$ to be a stationary Gaussian process. Note that in contrast to (3.3), model (3.5) allows the trend to vary with time t and induces a within-subject autocorrelation structure which may reflect fluctuations in the process about a smooth trend.

The usual assumptions made in models (3.3) and (3.5) are that conditioning on $\boldsymbol{Z}_i$, the random effect $\boldsymbol{\alpha}_i$ is normally distributed which represents within-subject variations. Depending on the situation, the mean and covariance matrix of $\boldsymbol{\alpha}_i$ may depend on components of $\boldsymbol{Z}_i$. A common specification is that $\boldsymbol{\alpha}_i$ is independent of $\boldsymbol{Z}_i$ so that the same normal model holds for all subjects under the study.

Modeling measurement errors: When the longitudinal covariate process $X_i(t)$ under model (3.3) or (3.5) is subject to measurement error, the actually observed longitudinal covariates are $W_i(t)$ given by (2.9), where the usual assumptions made are that the random error variable is normally distributed, i.e.,

$$e_i(t_{ij}) \sim \mathrm{N}(0, \sigma^2), \tag{3.6}$$

and that $e_i(t_{ij})$ are independent of $\boldsymbol{\alpha}_i$ (and $U_i(t)$ for all $t \geqslant 0$ if present). Under model (3.5), the $e_i(t_{ij})$ represent deviations due to measurement error and 'local' variation which, on a sufficiently short time scale, may be assumed to be independent across j. Under model (3.3), the $e_i(t_{ij})$'s represent measurement error and variation due to both local and longer-term within-subject autocorrelation process. As noted by Tsiatis and Davidian (2004), model (3.3) along with (2.9) and (3.6) specifies a standard linear mixed effect model.

Parameter estimation: There are several estimation procedures for regression parameters in the proportional hazards model (3.1). We summarize those commonly used as follows.

(a) *Approximation approach*:

To estimate the regression parameters in the proportional hazards model (3.1), there have been some works in the literature that approximate the longitudinal covariate process $X_i(t)$ in (2.4), then substitute them into the usual partial likelihood function. For instance, Tsiatis, DeGruttola and Wulfsohn (1995) considered models (3.1) and (3.3) with data (2.4), and developed an approximation to the hazard for T_i given the actually *observed* covariate history $W_i^H(t)$ for the estimation by maximizing the usual partial likelihood with $X_i(t)$ for each failure time replaced by an estimate of $E\{X_i(t) \mid W_i^H(t), T_i \geqslant t\}$. In particular, for event time t at which the value of $X_i(t)$ is needed in the usual partial likelihood function, they used the empirical Bayes estimator for $X_i(t)$ based on a standard fit of the mixed effects model defined by (3.3) and (2.9) with assumption (3.6) for all subjects still in the risk set, i.e., satisfying $T_i \geqslant t$. This approach thus requires fitting as many linear mixed effects models as there are event times in the data set, which is an advantage over those more computationally intensive methods; see discussions in Tsiatis and Davidian (2004). Other works using similar approximation approach can be found in Self and Pawitan (1992), Bycott and Taylor (1998), Dafni and Tsiatis (1998), among others.

(b) *Likelihood approach*:

Another method for the estimation problem considered in the literature by various authors is the likelihood approach, which is based on specification of a likelihood function for the parameters in models (3.1), (3.3) and (2.9) with assumption (3.6) and a model assumption on the survival time T_i. The usual form of a likelihood function is given by

$$\prod_{i=1}^{n} \int \left(\lambda_0(V_i) \exp\{\boldsymbol{\beta}_Z^T \boldsymbol{Z}_i + X_i(V_i)\beta_X\} \right)^{\delta_i}$$

$$\times \exp\left(- \int_0^{V_i} \lambda_0(t) \exp\{\boldsymbol{\beta}_Z^T \boldsymbol{Z}_i + X_i(t)\beta_X\} \, dt \right)$$

$$\times \frac{1}{(2\pi\sigma^2)^{J_i/2}} \exp\left(- \sum_{j=1}^{J_i} \frac{[W_i(t_{ij}) - X_i(t_{ij})]^2}{2\sigma^2} \right) f(\boldsymbol{\alpha}_i \mid \boldsymbol{Z}_i, \boldsymbol{\eta}) \, d\boldsymbol{\alpha}_i, \quad (3.7)$$

where $f(\boldsymbol{\alpha}_i \mid \boldsymbol{Z}_i, \boldsymbol{\eta})$ is the conditional p.d.f. of $\boldsymbol{\alpha}_i$ given $\boldsymbol{Z}_i$, and is usually assumed to be multivariate normal with parameters $\boldsymbol{\eta}$. The rationale of such a formulation of the likelihood function for joint modeling survival and longitudinal covariates was given in Tsiatis and Davidian (2004) via a discretized version of the problem in a spirit similar to arguments of Kalbfleisch and Prentice (2002).

DeGruttola and Tu (1994) considered a longitudinal model of the form (3.3) along with a lognormal model for T_i and developed an EM algorithm to maximize the resulting loglikelihood. Wulfsohn and Tsiatis (1997) derived an EM algorithm based on a loglikelihood for models (3.1), (3.3) and (2.9) with assumption (3.6). Henderson, Diggle and Dobson (2000) discussed likelihood inference for models (3.1) and (3.5). Song, Davidian and Tsiatis (2002) relaxed the model assumptions in Wulfsohn and Tsiatis (1997) from the normality of the random effects α_i to a smooth density. A theoretical explanation on the robustness against departure from the normal random effects assumption in likelihood formulation (3.7) was provided in Hsieh, Tseng and Wang (2006), which is reviewed in Section 3.2, part (c). For more works on the likelihood approach and semiparametric approach (Tsiatis and Davidian, 2001), see the review and the discussions presented in Tsiatis and Davidian (2004).

(c) *Bayesian approach:*

The Bayesian approach has also been considered in the literature for joint modeling survival time and longitudinal covariate data. Faucett and Thomas (1996) applied the Bayesian method to models (3.1) and (3.3), and developed the implementation via Markov chain Monte Carlo (MCMC) method. Xu and Zeger (2001) extended this approach to allow model (3.5). Wang and Taylor (2001) incorporated model (3.5) into a Bayesian framework using MCMC to fit a joint model to an HIV data set. Brown and Ibrahim (2003) considered a semiparametric Bayesian joint model for models (3.1) and (3.3) which makes no parametric assumptions on the random effects. For more works on the Bayesian approach under the setting of joint modeling, see the review and the discussions presented in Tsiatis and Davidian (2004).

(d) *Estimating equations approach:*

As mentioned above, both Bayesian and likelihood procedures rely on parametric assumptions on the joint model parameters. To minimize the reliance on these parametric modeling assumptions, Tsiatis and Davidian (2001) developed a set of unbiased estimating equations for the parameters involved in models (3.1) and (3.3), which yield consistent estimators with no assumptions on the random effects α_i. The idea used in their derivation is the conditional score of Stefanski and Carroll (1987), which suggests unbiased estimating equations for regression parameters in the proportional hazards model (3.1) via treating the random effects α_i as 'nuisance parameters' and conditioning on an appropriate 'sufficient statistic'. For details on the rationale and the derivation of these estimating equations, see Tsiatis and Davidian (2001), and see Tsiatis and Davidian (2004) for additional discussions.

3.2 Proportional hazards model

This subsection reviews some more recent works on joint modeling right censored survival data (2.6) with longitudinal covariates (2.3) or (2.4) via Cox's proportional hazards model (3.1).

(a) Ding and Wang (2008):

As reviewed in Section 3.1, a parametric longitudinal model such as (3.3) is assumed to facilitate the likelihood approach for the estimation problem in joint modeling procedures, and the existing methods primarily rely on the parametric mixed effects models for the longitudinal process $X_i(t)$, where the unobserved random effects were treated as missing data and imputed by the EM-algorithm. Ding and Wang (2008) pointed out that the related computations in practice can be quite intensive as multi-dimensional integrations are involved in the E-steps and the computational cost increases drastically as the number of random effects increases. Also, the EM-algorithm becomes unstable when the parameters increase, and it does not converge sometimes.

In their paper, Ding and Wang (2008) considered the models (3.1) and (2.9) with data (2.4), where they assumed:

$$\mathrm{E}\{e_i(t_{ij})\} = 0 \quad \text{and} \quad \mathrm{Var}\{e_i(t_{ij})\} = \sigma_e^2 \tag{3.8}$$

for all $i = 1, \ldots, n; j = 1, \ldots, J_i$. For modeling of the covariate process $X_i(t)$, They proposed the following *nonparametric multiplicative random effect model*:

$$X_i(t) = \alpha_i \mu(t), \quad \mathrm{E}\{\alpha_i\} = 1, \quad \mathrm{Var}\{\alpha_i\} = \sigma_\alpha^2, \tag{3.9}$$

where α_i is the random effect variable, and in contrast to model (3.3) used in other previous works mentioned in the previous subsection, the mean function $\mu(t) = \mathrm{E}\{X_i(t)\}$ here is assumed to be completely unknown. Note that model (3.9) contains only one random effect variable α_i for the longitudinal covariate process, but it is flexible to capture the shapes of both the population mean function $\mu(t)$ and individual longitudinal profiles. Here, the random effect α_i describes the variation of each subject about the population mean function $\mu(t)$. Also, note that model (3.9) does not require any stationary assumptions.

Ding and Wang (2008) noted that at first glance, the proposed multiplicative random effect model (3.9) appears restrictive, but its applicability is actually much broader. This is elucidated by their pointing out that longitudinal data are closely related to *functional data*, and that the expository paper by Rice (2004) illuminated this by viewing longitudinal data as scattered realizations of functional data, which are considered as random functions on an interval. They gave a functional principal component representation for the longitudinal covariate process $X_i(t)$, and connected it with several real data examples, such as the primary biliary cirrhosis sequential data in Murtaugh *et al.* (1994), the AIDS data in Yao, Müller and Wang (2005), the hip cycle data in Rice and Wu (2001), and the data from the time course of $^{14}C-folate$ in plasma of healthy adults in Yao *et al.* (2003). Ding and Wang (2008) pointed out that all these data sets exhibit a resemblance of the mean function $\mu(t)$ and the first eigenfunction in the functional principal component representation, and the longitudinal trajectories from different subjects have different amplitude but similar overall shape. These suggest the appropriateness of the proposed multiplicative random effect model (3.9) for these data examples.

To obtain maximum likelihood estimators for those parameters $\beta_Z, \beta_X, \lambda_0(\cdot)$, $\sigma_e^2, \sigma_\alpha^2$ in models (3.1), (2.9), (3.8)-(3.9) with data (2.4), Ding and Wang (2008) obtained the following likelihood function:

$$
\prod_{i=1}^{n} \int \Big(\lambda_0(V_i) \exp\{Z_i\beta_Z + X_i(V_i)\beta_X\} \Big)^{\delta_i}
$$

$$
\times \exp\left(- \int_0^{V_i} \lambda_0(t) \exp\{Z_i\beta_Z + X_i(t)\beta_X\} \, dt \right)
$$

$$
\times \frac{1}{(2\pi\sigma_e^2)^{J_i/2}} \exp\left(- \sum_{j=1}^{J_i} \frac{[W_i(t_{ij}) - X_i(t_{ij})]^2}{2\sigma_e^2} \right)
$$

$$
\times \frac{1}{(2\pi\sigma_\alpha^2)^{1/2}} \exp\left(- \frac{(\alpha_i - 1)^2}{2\sigma_\alpha^2} \right) d\alpha_i, \tag{3.10}
$$

which involves the unobserved random effects α_i. The authors gave an EM-algorithm to maximize the likelihood in (3.10) by calculating the expectation of the functions of the missing random effects and then maximizing the complete likelihood which is the quantity in the integrand of (3.10):

$$
\prod_{i=1}^{n} \Big(\lambda_0(V_i) \exp\{Z_i\beta_Z + X_i(V_i)\beta_X\} \Big)^{\delta_i}
$$

$$
\times \exp\left(- \int_0^{V_i} \lambda_0(t) \exp\{Z_i\beta_Z + X_i(t)\beta_X\} \, dt \right)
$$

$$
\times \frac{1}{(2\pi\sigma_e^2)^{J_i/2}} \exp\left(- \sum_{j=1}^{J_i} \frac{[W_i(t_{ij}) - X_i(t_{ij})]^2}{2\sigma_e^2} \right)
$$

$$
\times \frac{1}{(2\pi\sigma_\alpha^2)^{1/2}} \exp\left(- \frac{(\alpha_i - 1)^2}{2\sigma_\alpha^2} \right). \tag{3.11}
$$

A complication for this type of likelihood approach is that the maximum likelihood estimates actually do not exist if $\lambda_0(t)$ is completely unrestricted. For treatment on this issue, see Ding and Wang (2008). In the M-step of their proposed EM-algorithm, the authors used B-spline to approximate the unknown mean function $\mu(t)$, and selected the number of knots and degrees based on the Akaike Information Criterion (AIC).

The simulation studies presented in Ding and Wang (2008) showed that their proposed EM-algorithm converges fairly quickly, and the computational cost is cut down drastically due to the use of a single random effect in their proposed nonparametric multiplicative random effect model (3.9).

(b) Zeng and Cai (2005):

As reviewed in Section 3.1, the likelihood approach has been used by many researchers to develop joint modeling procedures for survival time and longitudinal covariates. However, so far there is few work on the theoretical justification of the asymptotic properties for the maximum likelihood estimators. The paper by Zeng and Cai (2005) is probably the only one that formally investigated these issues under *their* model assumptions. Although their model assumptions are not exactly on the line of the settings described and outlined in Section 2 and in Sections 3.1-3.2, the results in Zeng and Cai (2005) are nonetheless of interests and are summarized as follows.

In their paper, Zeng and Cai (2005) studied the joint modeling problem mainly under the following model assumptions:

(A.1) The unobserved subject-specific random effect vector $\boldsymbol{\alpha}$ of dimension q is normally distributed with mean zero and covariance $\boldsymbol{\Sigma}_\alpha$.

(A.2) Let T denote the survival time, T_0 the end time of the study, $X(t)$ the longitudinal covariate process, and $Y(t)$ the longitudinal response process. And denote the longitudinal history prior to time t for these longitudinal processes by $X^H(t) = \{X(s); 0 \leqslant s < t\}$ and $Y^H(t) = \{Y(s); 0 \leqslant s < t\}$, respectively. For any $t \in [0, T_0]$, the covariate process $X(t)$ is fully observed, and conditional on $\boldsymbol{\alpha}, X^H(t), Y^H(t)$ and $T \geqslant t$, the distribution of $X(t)$ depends only on $X^H(t)$.

(A.3) Conditional on $\boldsymbol{\alpha}, X^H(t), Y^H(t), X(t)$ and $T \geqslant t$, the hazard function of T at time t satisfies the following proportional hazards model:

$$\lambda(t; \boldsymbol{\alpha}, X^H(t), Y^H(t)) = \lambda_0(t) \exp\{(\tilde{\boldsymbol{A}}(t) \circ \boldsymbol{\gamma})^{\mathrm{T}}\boldsymbol{\alpha} + \boldsymbol{A}(t)^{\mathrm{T}}\boldsymbol{\beta}\}, \qquad (3.12)$$

where $\boldsymbol{A}(t)$ and $\tilde{\boldsymbol{A}}(t)$ are sub-process of $X(t)$, notation $\circ$ represents the component-wise product of two vectors with the same dimension, and $\boldsymbol{\beta}, \boldsymbol{\gamma}$ are unknown parameters.

(A.4) Conditional on $\boldsymbol{\alpha}, X^H(t), Y^H(t), X(t)$ and $T \geqslant t$, the response longitudinal process $Y(t)$ satisfies:

$$Y(t) = \tilde{\boldsymbol{B}}(t)^{\mathrm{T}}\boldsymbol{\alpha} + \boldsymbol{B}(t)^{\mathrm{T}}\boldsymbol{\eta} + e(t) \qquad (3.13)$$

where $\boldsymbol{B}(t)$ and $\tilde{\boldsymbol{B}}(t)$ are sub-process of $X(t)$, $e(t)$ is a white noise process with variance σ_e^2, and $\boldsymbol{\eta}$ is unknown regression parameter.

(A.5) Survival time T is subject to right censoring as in (2.6).

(A.6) Conditional on $\boldsymbol{\alpha}, X^H(t), Y^H(t), X(t)$ and T, the hazard function of the right censoring variable C (see (2.6)) at time t depends only on $X^H(t)$ and $X(t)$ for any $t < T$.

Zeng and Cai (2005) pointed out that if one is interested in studying the simultaneous effects of the covariate process $X(t)$ on response process $Y(t)$ and survival time T, after adjusting for other covariates at the entry and subject-specific effects, the sub-processes $\boldsymbol{A}(t), \tilde{\boldsymbol{A}}(t), \boldsymbol{B}(t)$ and $\tilde{\boldsymbol{B}}(t)$ can be chosen to include treatment variable and/or covariates measured at the entry of the study.

Note that one of the major differences between the above listed model assumptions (A.1)-(A.6) by Zeng and Cai (2005) and the settings described and outlined in Section 2 and in Sections 3.1-3.2 is that although the proportional hazards model is considered in (3.12) for studying the relation between the survival time and the covariate process $X(t)$, above assumption (A.2) requires $X(t)$ to be fully observed. In contrast, the covariate process $X(t)$ in model (3.1) is only observed at time points t_{ij}. Also, the focus of the joint modeling in Zeng and

Cai (2005) is the effects of the covariate process $X(t)$ simultaneously on response process $Y(t)$ and survival time T, while the focus of modeling procedures and methods described in Section 2 and Sections 3.1-3.2 is the effects of the longitudinal covariate process $X(t)$ on the survival time T. Below, we summarize the main theoretical results in Zeng and Cai (2005).

For each subject $i = 1, \ldots, n$, denote the observed data by

$$
\begin{aligned}
&(V_i, \delta_i, \boldsymbol{A}_i(t), \tilde{\boldsymbol{A}}_i(t)), && \boldsymbol{Y}_i = (Y_i(t_{i1}), \ldots, Y_i(t_{i,J_i}))^{\mathrm{T}}, \\
&\boldsymbol{B}_i = (\boldsymbol{B}_i(t_{i1}), \ldots, \boldsymbol{B}_i(t_{i,J_i})), && \tilde{\boldsymbol{B}}_i = (\tilde{\boldsymbol{B}}_i(t_{i1}), \ldots, \tilde{\boldsymbol{B}}_i(t_{i,J_i})).
\end{aligned}
\tag{3.14}
$$

And denote the parameter for the model (A.1)-(A.6) by

$$
\boldsymbol{\theta} = (\boldsymbol{\beta}, \boldsymbol{\gamma}, \boldsymbol{\eta}, \boldsymbol{\Sigma}_\alpha, \sigma_e^2) \quad \text{and} \quad \Lambda_0(t) = \int_0^t \lambda_0(s)\, ds, \tag{3.15}
$$

where the dimension of $\boldsymbol{\theta}$ is p when writing $\boldsymbol{\theta}$ as a vector for all involved parameters. Then, Zeng and Cai (2005) gave the following likelihood which is proportional to the observed likelihood function for $(\boldsymbol{\theta}, \Lambda_0)$:

$$
\begin{aligned}
\prod_{i=1}^{n} \int &\left\{ (2\pi\sigma_e^2)^{-J_i/2} \exp\left(-(\boldsymbol{Y}_i - \boldsymbol{B}_i^{\mathrm{T}}\boldsymbol{\eta} - \tilde{\boldsymbol{B}}_i^{\mathrm{T}}\boldsymbol{\alpha})^{\mathrm{T}}(\boldsymbol{Y}_i - \boldsymbol{B}_i^{\mathrm{T}}\boldsymbol{\eta} - \tilde{\boldsymbol{B}}_i^{\mathrm{T}}\boldsymbol{\alpha})/(2\sigma_e^2) \right) \right. \\
&\times \lambda(V_i)^{\delta_i} \exp\left(\delta_i[\,(\tilde{\boldsymbol{A}}_i(V_i) \circ \boldsymbol{\gamma})^{\mathrm{T}}\boldsymbol{\alpha} + \boldsymbol{A}_i(V_i)^{\mathrm{T}}\boldsymbol{\beta}\,] \right. \\
&\left. - \int_0^{V_i} \exp[\,(\tilde{\boldsymbol{A}}_i(s) \circ \boldsymbol{\gamma})^{\mathrm{T}}\boldsymbol{\alpha} + \boldsymbol{A}_i(s)^{\mathrm{T}}\boldsymbol{\beta}\,]\, d\Lambda(s) \right) \bigg\} \\
&\times (2\pi)^{-q/2} |\boldsymbol{\Sigma}_\alpha|^{-1/2} \exp\left(-\boldsymbol{\alpha}^{\mathrm{T}}\boldsymbol{\Sigma}_\alpha^{-1}\boldsymbol{\alpha}/2 \right) d\boldsymbol{\alpha}.
\end{aligned}
\tag{3.16}
$$

A modified function of (3.16) was obtained from replacing $\lambda(V_i)$ by $\Lambda\{V_i\}$, which is the jump size of $\Lambda(\cdot)$, and the logarithm of this modified function is denoted by $L(\boldsymbol{\theta}, \Lambda)$. Then, the maximum likelihood estimator (MLE) $(\hat{\boldsymbol{\theta}}, \hat{\Lambda})$ for $(\boldsymbol{\theta}, \Lambda_0)$ is the solution that maximizes $L(\boldsymbol{\theta}, \Lambda)$, and can be computed by the EM-algorithm. Define the profile log-likelihood function of $\boldsymbol{\theta}$ by

$$
\ell(\boldsymbol{\theta}) = \max_{\Lambda} L(\boldsymbol{\theta}, \Lambda). \tag{3.17}
$$

Zeng and Cai (2005) established the following theorems on the asymptotic properties of the MLE $(\hat{\boldsymbol{\theta}}, \hat{\Lambda})$ under some additional technical assumptions which can be found in their paper, and are omitted here for brevity of this review.

Theorem 3.1. *Under model assumptions (A.1)-(A.6) and some additional technical assumptions, we have*

$$
\|\hat{\boldsymbol{\theta}} - \boldsymbol{\theta}\| + \sup_{t \in [0, T_0]} |\hat{\Lambda}(t) - \Lambda_0(t)| \xrightarrow{a.s.} 0, \quad \text{as } n \to \infty.
$$

Theorem 3.2. *Under model assumptions (A.1)-(A.6) and some additional technical assumptions, $\sqrt{n}(\hat{\Lambda} - \Lambda_0)$ weakly converges to a Gaussian process on $[0, T_0]$,* and $\sqrt{n}(\hat{\boldsymbol{\theta}} - \boldsymbol{\theta})$ *converges in distribution to a multivariate normal distribution with mean zero and a variance that attains the semiparametric efficiency bound for* $\boldsymbol{\theta}$ *as* $n \to \infty$.

Theorem 3.3. *Under model assumptions* (A.1)-(A.6) *and some additional technical assumptions,* $2[\ell(\hat{\boldsymbol{\theta}}) - \ell(\boldsymbol{\theta})]$ *converges in distribution to a chi-squared distribution with p degrees of freedom as* $n \to \infty$, *and moreover,*

$$-\frac{\ell(\hat{\boldsymbol{\theta}} + h_n e) - 2\ell(\hat{\boldsymbol{\theta}}) + +\ell(\hat{\boldsymbol{\theta}} - h_n e)}{nh_n^2} \xrightarrow{P} e^{\mathrm{T}} \boldsymbol{I} e, \quad as\ n \to \infty$$

where $h_n = O_p(n^{-1/2})$, e *is any vector in* $\mathbb{R}^p$ *with unit norm, and* $\boldsymbol{I}$ *is the efficient information matrix for* $\boldsymbol{\theta}$.

(c) Other works:

In Hsieh, Tseng and Wang (2006), models (3.1), (3.3) and (2.9) with assumption (3.6) and observed data (2.4) were considered for joint modeling survival time and longitudinal covariates on the line of the likelihood approach by Wulfsohn and Tsiatis (1997) aforementioned. Hsieh, Tseng and Wang (2006) pointed out that in the joint modeling framework, there have been no distributional or asymptotic theory that is available to date, and even the standard errors, defined as the standard deviation of the parametric estimators or the maximum likelihood estimators (MLE), are difficult to obtain. They proposed the use of bootstrap procedures to overcome the difficulties of reliable estimates for the standard errors for the MLEs. More interestingly, they provided a theoretical explanation on the robustness against departure from the normal assumption in (3.7) on the random effects $\boldsymbol{\alpha}$ in (3.3). We outline their arguments on the robustness as follows.

Foe simplicity, consider the case without the time-independent covariates $\boldsymbol{Z}_i$ in model (3.1) and the observed data (2.4). To recover the unobserved longitudinal covariate process $X_i(t)$, assume a parametric random effects model, say, $\boldsymbol{\alpha}_i$ in (3.3) has a known density function $g(\cdot\,;\gamma)$ with γ as the unknown parameter. Thus, the parameter that specifies the joint model is $\theta = (\beta, \lambda_0, \gamma, \sigma^2)$, where the regression parameter $\beta = \beta_X$ in (3.1) and the baseline hazards function $\lambda(\cdot)$ is nonparametric. Let $f(\boldsymbol{W}_i; \gamma, \sigma^2)$ and $f(\boldsymbol{W}_i \,|\, \boldsymbol{\alpha}_i; \sigma^2)$ be the marginal and conditional density of $\boldsymbol{W}_i$, respectively, and let $f(V_i, \delta_i \,|\, \boldsymbol{\alpha}_i; \theta)$ be the conditional p.d.f. according to (3.1) and (3.3). Under the assumption of an 'uninformative' time schedule described in Tsiatis and Davidian (2004), the contribution of the ith individual to the joint likelihood is given by

$$L_i(\theta) = f(\boldsymbol{W}_i; \gamma, \sigma^2) E_i^* \{ f(V_i, \delta_i \,|\, \boldsymbol{\alpha}_i; \theta) \}, \tag{3.18}$$

where E_i^* denotes conditional expectation with respect to the posterior density of $\boldsymbol{\alpha}_i$ given $\boldsymbol{W}_i$, which is $g(\boldsymbol{\alpha}_i \,|\, \boldsymbol{W}_i; \gamma, \sigma^2) = f(\boldsymbol{W}_i \,|\, \boldsymbol{\alpha}_i; \sigma^2) g(\boldsymbol{\alpha}_i; \gamma) / f(\boldsymbol{W}_i; \gamma, \sigma^2)$. There are two points to be noticed here. First, $E_i^* \{ f(V_i, \delta_i \,|\, \boldsymbol{\alpha}_i; \theta) \}$ carries information on the longitudinal data. If it is ignored, the marginal statistical inference based on $f(\boldsymbol{W}_i; \gamma, \sigma^2)$ alone would be inefficient and even biased (due to informative dropouts). This sheds light on how a joint modeling approach eliminates the bias incurred by a marginal approach and also why it is more efficient. Second, the random effect structure $g(\cdot; \gamma)$ and $\boldsymbol{W}_i$ are relevant to the information for survival parameters β and $\lambda_0(\cdot)$ only through the posterior density $g(\boldsymbol{\alpha}_i \mid \boldsymbol{W}_i; \gamma, \sigma^2)$, which can be approximated well by a normal density through a Laplace approximation technique (Tierney and Kadane, 1986) when reasonably large numbers of longitudinal measurements are available per subject. This explains why the joint modeling procedure in Wulfsohn and Tsiatis (1997) is robust against departure from the normal prior assumption on $\boldsymbol{\alpha}_i$.

In **Elashoff, Li and Li (2007)**, a joint model for longitudinal covariates and survival data in the presence of multiple failure types was considered. This is in contrast to those works previously reviewed which are all concerned with a single failure time or survival time. The joint model by Elashoff, Li and Li (2007) provides a flexible approach to handle possible nonignorable missing data in the longitudinal measurements due to dropout. It is an extension of previous joint models with a single failure type, which offers a possible way to model informatively censored events as a competing risk. More specifically, their model consists of a linear mixed effects submodel for the longitudinal outcome (similar to Zeng and Cai (2005)) and a proportional cause-specific hazards frailty submodel (Prentice *et al.*, 1978) for the competing risks survival data, which are linked together by some latent random effects. The authors developed an EM-algorithm to obtain the maximum likelihood estimators for the parameters involved in their proposed model, and they also proposed to use a profile likelihood method to estimate the standard errors of these maximum likelihood estimators.

In **Song and Wang (2008)**, the joint modeling of survival and longitudinal data was studied via two regression models. The primary one is the proportional hazards model with time-varying coefficient:

$$\lambda(t; \boldsymbol{X}_i^H(t)) = \lambda_0(t) \exp\left(\boldsymbol{X}_i(t)^{\mathrm{T}} \boldsymbol{\beta}(t)\right), \tag{3.19}$$

which is essentially the same as model (3.1), except that the regression parameter β is allowed to change over time in (3.19). The second model is for longitudinal data which are assumed to follow a random effects model, and is actually the same as model (3.3). The authors point out that based on the trajectory of a subject's longitudinal data, some covariates in the survival model are functions of the unobserved random effects, and that estimated random effects are generally different from the unobserved random effects which leads to covariate measurement error. They proposed two asymptotically equivalent semiparametric estimators for joint modeling of survival and longitudinal data under the framework of above time-varying coefficient proportional hazards model (3.19), and established the consistency and asymptotic normality of their proposed estimators.

3.3   Accelerated life model

So far, the literature on joint modeling of survival and longitudinal data has mainly concentrated on the use of the Cox proportional hazards model (3.1) to characterize the relationship between the survival time and the longitudinal covariates.

However, there exist situations in practice where the proportionality assumption in (3.1) does not hold; for instance, see the fecundity data example given Tseng, Hsieh and Wang (2005). For such situations, the accelerated life model (Cox and Oakes, 1984), or called accelerated failure time model , is an attractive alternative. But as for the traditional accelerated life model with time-dependent covariates, handling the accelerated failure time structure in the joint modeling setting is much harder than for the proportional hazards model. In fact, even the issue of maximum likelihood estimation under the accelerated life model with time-dependent covariates has not been resolved. Thus, in the literature there has been few work that considers using the accelerated life model for joint modeling the survival time and the longitudinal covariate data. Actually, the only work on this topic that we are aware of is the paper by Tseng, Hsieh and Wang (2005). As follows, we review the main results of this paper.

In Tseng, Hsieh and Wang (2005), the authors considered the joint modeling's data setting described in (2.4) and (2.9) with assumption (3.6) and with right censoring data (2.6), and they considered model (3.3) with the following parametric assumption on the random effects α_i:

$$\alpha_i \sim \mathrm{N}_p(\mu, \Sigma), \tag{3.20}$$

where p is the dimension of the multivariate normal distribution, μ is the mean and Σ is the covariance matrix. The survival model linking the survival time T and the longitudinal covariate process $X(t)$ that the authors considered is based on the form of the *accelerated life model* with time-dependent covariate by Cox and Oakes (1984) given by:

$$S(t \mid X^H(t)) = S_0(\psi\{X^H(t); \beta\}) \tag{3.21}$$

where $S_0(t)$ is a baseline survival function, $S(t \mid X^H(t))$ is the survival function for an individual with the history of the covariate process which is denoted by $X^H(t)$ as in (3.6), and $\psi\{X^H(t); \beta\}$ is given by

$$\psi\{X^H(t); \beta\} = \int_0^t \exp\left(\beta X(s)\right) ds. \tag{3.22}$$

Note that when $\beta = 0$, we have in (3.21):

$$S_0(\psi\{X^H(t); 0\}) = S_0(t). \tag{3.23}$$

For an absolutely continuous $S_0(t)$, the hazard rate function for an individual with covariate history $X^H(t)$ corresponding to the accelerated life model (3.21) can be expressed as the following:

$$\lambda(t \mid X^H(t)) = \lambda_0(\psi\{X^H(t); \beta\}) \exp\left(\beta X(t)\right), \tag{3.24}$$

where $\lambda_0(t)$ is the baseline hazard function for $S_0(t)$, and $\exp(\beta X(t))$ is the derivative of $\psi\{X^H(t); \beta\}$ with respect to t. Furthermore, under the model assumption

(3.3) on the longitudinal covariate process, above accelerated life model (3.24) for the ith individual is written by

$$\lambda(t \mid X_i^H(t)) = \lambda(t \mid \beta, \boldsymbol{\alpha}_i) = \lambda_0(\psi(t; \beta, \boldsymbol{\alpha}_i)) \exp\left(\beta \boldsymbol{\nu}(t)^{\mathrm{T}} \boldsymbol{\alpha}_i\right), \tag{3.25}$$

where

$$\psi(t; \beta, \boldsymbol{\alpha}_i) = \int_0^t \exp\left(\beta X(s)\right) ds = \int_0^t \exp\left(\beta \boldsymbol{\nu}(t)^{\mathrm{T}} \boldsymbol{\alpha}_i\right) ds. \tag{3.26}$$

The goal is to construct an estimator for the regression parameter β in the accelerated life model (3.25) without assuming a parametric baseline hazard function $\lambda_0(t)$, and to obtain estimators for other parameters involved in model assumptions for the longitudinal covariate process.

For each subject $i = 1, \ldots, n$, denote the observed data by

$$(V_i, \delta_i, \boldsymbol{W}_i, \boldsymbol{t}_i) \tag{3.27}$$

where $(V_i, \delta_i, \boldsymbol{W}_i)$ is given in (2.4)-(2.5) with $\boldsymbol{W}_i = (W_{i1}), \ldots, W_{i,J_i})$ and $W_{ij} = W_i(t_{ij})$, and $\boldsymbol{t}_i = (t_{i1}, \ldots, t_{i,J_i})$ from (2.2). Assume that the non-informative censoring and measurement schedule time points t_{ij} are independent of the future covariate history and random effects $\boldsymbol{\alpha}_i$. Tseng, Hsieh and Wang (2005) obtained the following observed likelihood under the joint modeling assumptions (2.9), (3.3) and (3.25) along with (3.6) and (3.20):

$$L(\theta) \equiv L(\beta, \mu, \Sigma, \sigma^2, \lambda_0) \tag{3.28}$$

$$= \prod_{i=1}^n \int \left\{ \prod_{j=1}^{J_i} f(W_{ij} \mid \boldsymbol{\alpha}_i, \boldsymbol{t}_i, \sigma^2) \right\} f(V_i, \delta_i \mid \boldsymbol{\alpha}_i, \boldsymbol{t}_i, \beta, \lambda_0) f(\boldsymbol{\alpha}_i \mid \mu, \Sigma) \, d\boldsymbol{\alpha}_i,$$

where $f(W_{ij} \mid \boldsymbol{\alpha}_i, \boldsymbol{t}_i, \sigma^2)$ and $f(\boldsymbol{\alpha}_i \mid \mu, \Sigma)$ are the p.d.f.'s of $N(\boldsymbol{\nu}(t)^{\mathrm{T}} \boldsymbol{\alpha}, \sigma^2)$ and $N_p(\mu, \Sigma)$, respectively; and from censoring and model assumptions (2.6) and (3.25)-(3.26), function $f(V_i, \delta_i \mid \boldsymbol{\alpha}_i, \boldsymbol{t}_i, \beta, \lambda_0)$ is given by

$$f(V_i, \delta_i \mid \boldsymbol{\alpha}_i, \boldsymbol{t}_i, \beta, \lambda_0) = \left(\lambda_0(\psi(t; \beta, \boldsymbol{\alpha}_i)) \exp\left(\beta \boldsymbol{\nu}(t)^{\mathrm{T}} \boldsymbol{\alpha}_i\right)\right)^{\delta_i} \times$$

$$\exp\left\{ \int_0^{\psi(V_i; \beta, \boldsymbol{\alpha}_i)} \lambda_0(t) \, dt \right\}. \tag{3.29}$$

Under the assumption that $\lambda_0(t)$ is constant between two consecutive failure times so that $\lambda_0(t)$ is a step function, the authors provided an EM-algorithm to maximize the likelihood function (3.28) to obtain the maximum likelihood estimator $\hat{\theta}$ for θ denoted by (3.28). Since the calculation of standard error for the estimator $\hat{\beta}$ for β turns out to be a difficult issue because of the missing information about the random effects, they proposed the use of a bootstrap procedure to estimate the standard errors of $\hat{\beta}$.

4   Joint modeling with interval censored data

So far, the literature on joint modeling of survival and longitudinal data has been mainly on *right* censored survival data (2.6) with longitudinal covariates, which has been selectively reviewed in Section 3. The problem of joint modeling *interval* censored survival data, or current status data (2.8), with longitudinal covariates is obviously important; see Example 2 given in Section 2.3. But the modeling and estimation for such problem are quite challenging, and in the literature there has been few work on this topic. In fact, many issues regarding interval censored data (2.8) without any covariates are not yet well understood thus far. To our best knowledge, part of the general methods developed in van der Laan and Robins (2003) is applicable to joint modeling interval censored survival data (2.8) and longitudinal data, as well as to interval censored Case 2 data (Groeneboom and Wellner, 1992). We provide a brief review on these as follows.

In van der Laan and Robins (2003), the cancer data example described in Example 2 of Section 2.3 was studied on pages 248-250. The Cox proportional hazards model was considered in their studies, and via the general methods developed in their book, they proposed a locally efficient estimator for the regression parameter under their model. They pointed out that their estimator is consistent and asymptotically normal under the *idealized* experiment described on page 249 of van der Laan and Robins (2003).

Moreover, the general methods developed in van der Laan and Robins (2003) was also applicable to joint modeling interval censored Case 2 data (Groeneboom and Wellner, 1992) and longitudinal data. In Section 6.7 of van der Laan and Robins (2003), the following data structure was considered:

$$Y_i = (\tilde{X}_i^H(\mathcal{T}_0), U_{i1}, \cdots, U_{i,J_i}), \quad i = 1, \ldots, n, \tag{4.1}$$

for

$$U_{ij} = (t_{ij}, I\{T_i \leqslant t_{ij}\}, X_i(t_{ij})), \quad j = 1, \ldots, J_i, \tag{4.2}$$

where T_i is the survival time of the ith individual, t_{ij} are the *random* monitoring time, $X_i(t)$ is the longitudinal covariate process which is only observed at time points t_{ij}, and for a fixed time point $\mathcal{T}_0$, $\tilde{X}_i^H(t) = \{\tilde{X}_i(s); s \leqslant t\}$ is the history up to time t of a covariate process $\tilde{X}_i(t)$ that is fully observed on time interval $[0, \mathcal{T}_0]$. Note that since the event time T_i is of interest, it is often the case in practice that not all $(t_{ij}, I\{T_i \leqslant t_{ij}\})$'s are informative about T_i. From all $(t_{ij}, I\{T_i \leqslant t_{ij}\})$, we can obtain interval $[a_i, b_i]$ satisfying $a_i \leqslant T_i \leqslant b_i$, which are defined *interval censored Case 2 data* by Groeneboom and Wellner (1992). Thus, data (4.1)-(4.2) can be expressed by

$$([a_i, b_i], \tilde{X}_i^H(\mathcal{T}_0), X_i(t_{i1}), \ldots, X_i(t_{i,J_i})) \tag{4.3}$$

for $i = 1, \cdots, n$, which is of the data setting for interval censored Case 2 data with longitudinal covariates. Van der Laan and Robins (2003) discussed some possible models to be used for data (4.3), including the proportional hazards model. Also, they pointed out that the general methods developed in Section 6.1 of their book can be applied to construct estimators for these models.

5   Further studies

Obviously, there is much work to be done for the problem of joint modeling censored survival time and longitudinal data. For instance, less reliance on the parametric assumptions for the modeling would be desirable, and model diagnosis and model checking procedures should be developed to assess the validity of the existing model assumptions. Also, there are theoretical challenges regarding efficiency and the asymptotic distributions of the parametric estimators in the joint modeling framework; in fact, so far no distributional or asymptotic theory is available to date, and even the standard errors, defined as the standard deviations of the parametric estimators, are difficult to obtain. As mentioned earlier, the model assumptions for the work by Zeng and Cai (2005) do not fall in the line with the main stream works, such as those reviewed in Tsiatis and Davidian (2004).

Hsieh, Tseng and Wang (2006) pointed out that currently theoretical gaps remain to validate the asymptotic properties of the estimates in Wulfsohn and Tsiatis (1997) and the validity of the bootstrap standard error (SE) estimates. The theory of profile likelihood is well developed for parametric setting (Patefield, 1977), but not immediately applicable in semiparametric settings (Murphy and van der Vaart, 2000; Fan and Wong, 2000). The complexity caused by profiling nonparametric parameters with only implicit structure creates additional difficulties in theoretical and computational developments for joint modeling survival time and longitudinal data. Further efforts will be required in future work to resolve these issues and to provide reliable precision estimates for statistical inference under the joint modeling setting.

Also, Tseng, Hsieh and Wang (2005) pointed out that many important issues on the joint modeling survival and longitudinal data via the accelerated life model remain to be resolved. For instance, the asymptotic theory of the estimators is not yet available. In fact, this is the true even for the simpler case of a proportional hazards model.

Clearly, further research is necessary on complicated types of censored survival data, such as doubly censored data (2.7) and interval censored data (2.8). Finally, we note that as reviewed in Section 2.2, recently the methods on collecting intensive longitudinal data have been developed rapidly, it would be of interest to study how intensive longitudinal covariates in the joint modeling setting can help enhance the performance of joint modeling procedures on censored survival time with longitudinal covariates.

References

[1] Andersen P. K., Borgan O., Gill R. D. and Keiding N. (1993). *Statistical Models Based on Counting Processes*. Springer-Verlag, New York.

[2] Banerjee M. and Wellner J. A. (2005). Confidence intervals for current status data. *Scandinavia Journal of Statistics* **32**, 405-424.

[3] Breslow N. E. (1974). Covariance analysis of censored survival data. *Biometrics* **30**, 89-99.

[4] Brown E. R. and Ibrahim J. G. (2003). A Bayesian semiparametric joint

hierarchical model for longitudinal and survival data. *Biometrics* **59**, 221-228.

[5] Bycott P. and Taylor J. (1998). A comparison of smoothing techniques for CD4 data measured with error in a time-dependent Cox proportional hazards model. *Statist. Medicine* **17**, 2061-2077.

[6] Chang M. N. (1990). Weak convergence of a self-consistent estimator of the survival function with doubly censored data. *The Annals of Statistics* **18**, 391-404.

[7] Chang M. N. and Yang G. L. (1987). Strong consistency of a nonparametric estimator of the survival function with doubly censored data. *The Annals of Statistics* **15**, 1536-1547.

[8] Chen K. and Zhou M. (2003). Non-parametric hypothesis testing and confidence intervals with doubly censored data. *Lifetime Data Analysis* **9**, 71-91.

[9] Carroll R. J., Ruppert D., Stefanski L. A. and Crainiceanu C. M. (2006). *Measurement Error in Nonlinear Models*. Second Edition. Chapman and Hall, New York.

[10] Cole P. M., Hall S. E., Tan P. Z., Zhang Y., Crnic K. A., Blair C.B. and Li R. (2008). The development of emotion regulation in early childhood. (Submitted)

[11] Cole P. M., Luby J. and Sullivan M. W. (2008). Emotions and the development of depression: Bridging the gap. *Child Development Perspectives*. (In press)

[12] Cox D. R. (1972). Regression models and life-tables (with Discussion). *Journal of the Royal Statistical Society, Series B* **34**, 187-220.

[13] Cox D. R. and Oakes D. (1984). *Analysis of Survival Data*. Chapman & Hall, London.

[14] Csikszentmihalyi M. and Larson R. W. (1987). Validity and reliability of the experience sampling method. *Journal of Nervous and Mental Disease* **175**, 526-536.

[15] Dafni U. G. and Tsiatis A. A. (1998). Evaluating surrogate markers of clinical outcome measured with error. *Biometrics* **54**, 1445-1462.

[16] DeGruttola V. and Tu X. M. (1994). Modeling progression of CD-4 lymphocyte count and its relationship to survival time. *Biometrics* **50**, 1003-1014.

[17] DeVries M. W. (1992). *The Experience of Psychopathology: Investigating Mental Disorders in Their Natural Settings*. Edited by M. W. de Vries. Cambridge University Press.

[18] Diggle P. J., Liang K. Y., and Zeger S. L. (1994). *Analysis of Longitudinal Data*. Clarendon Press, Oxford.

[19] Ding J. and Wang J. L. (2008). Modeling longitudinal data with nonparametric multiplicative random effects jointly with survival data. *Biometrics* **64**, 546-556.

[20] Efron B. (1977). The efficiency of Cox's likelihood function for censored data. *Journal of American Statistical Association* **72**, 557-565.

[21] Elashoff R. M., Li G. and Li N. (2007). A joint model for longitudinal measurements and survival data in the presence of multiple failure types. *Biometrics* **63**, 1-10.

[22] Fan J. Huang T. and Li R. (2007). Analysis of longitudinal data with semi-

parametric estimation of covariance function. *Journal of American Statistical Association* **102**, 632-641.

[23] Fan J. and Li R. (2004). New estimation and model selection procedures for semiparametric modeling in longitudinal data analysis. *Journal of American Statistical Association* **99**, 710-723.

[24] Fan J. and Li R. (2006). An overview on nonparametric and semiparametric techniques for longitudinal data. *Frontiers of Statistics* (Fan, J. and Koul, H., eds.), 277-304. Imperial College Press.

[25] Fan J. and Wong W. H. (2000). Comment on 'On profile likelihood' by Murphy and van der Vaart. *Journal of American Statistical Association* **95**, 468-471.

[26] Faucett C. J. and Thomas D. C. (1996). Simultaneously modeling censored survival data and repeatedly measured covariates: a Gibbs sampling approach. *Statist. Medicine* **15**, 1663-1685.

[27] Finkelstein D. M. and Wolf R. A. (1985). A semiparametric model for regression analysis of interval-censored failure time data. *Biometrics* **41**, 933-945.

[28] Fitzamurice G. M., Laird N. M. and Ware J. H. (2004). *Applied Longitudinal Analysis*. John Wiley and Sons, New York.

[29] Fleming T. R. and Harrington D. P. (1991). *Counting Processes and Survival Analysis*. John Wiley and Sons, New York.

[30] Fuller W. A. (1987). *Measurement Error Models*. John Wiley and Sons, New York.

[31] Geskus R. and Groeneboom P. (1999). Asymptotically optimal estimation of smooth functionals for interval censoring, Case 2. *The Annals of Statistics* **27**, 627-674.

[32] Gill R. D. (1983). Large sample behavior of the product-limit estimator on the whole line. *The Annals of Statistics* **11**, 49-58.

[33] Groeneboom P. and Wellner J. A. (1992). *Information Bounds and Nonparametric Maximum likelihood Estimation*. Birkhäuser.

[34] Gu M. G. and Zhang C. H. (1993). Asymptotic properties of self-consistent estimators based on doubly censored data. *The Annals of Statistics* **21**, 611-624.

[35] Härdle W., Liang H. and Gao J. (2000). *Partially Linear Models*. Springer-Verlag, New York.

[36] Henderson R., Diggle P. and Dobson A. (2000). Joint modeling longitudinal measurements and event time data. *Biostatistics* **4**, 465-480.

[37] Hoel D. G. and Walburg H. E. (1972). Statistical analysis of survival experiments. *Journal of the National Cancer Institute* **49**, 361-372.

[38] Hsieh F., Tseng Y. K. and Wang J. L. (2006). Joint modeling of survival and longitudinal data: Likelihood approach revisited. *Biometrics* **62**, 1037-1043.

[39] Huang J. (1996). Efficient estimation for the proportional hazards model with interval censoring. *The Annals of Statistics* **24**, 540-568.

[40] Kalbfleisch J. D. and Prentice R. L. (2002). *The Statistical Analysis of Failure Time Data*. Second Edition. John Wiley and Sons, New York.

[41] Kaplan E. L. and Meier P. (1958). Nonparametric estimation from incomplete observations. *Journal of the American Statistical Association*, **53**, 457-481.

[42] Leiderman P. H., Babu D., Kagia J., Kraemer H. C. and Leiderman C. F.

(1973). African infant precocity and some social influences during the first year. *Nature* **242**, 247-249

[43] Liang H., Härdle W. and Carroll R. (1999). Estimation in semiparametric partial linear errors-in-variables model. *The Annals of Statistics* **27**, 1519-1536.

[44] Liang H., Thurston S., Ruppert D., Apanasovich T. and Hauser R. (2008). Additive partial linear models with measurement errors. *Biometrika* **95**, 667-678.

[45] Liang H., Wang S. J., and Carroll R. (2007). Partially linear models with missing response variables and error-prone covariates. *Biometrika* **94**, 185-198.

[46] Lin D. Y. and Ying Z. (2001). Semiparametric and nonparametric regression analysis of longitudinal data (with discussion). *Journal of the American Statistical Association* **96**, 103-113.

[47] Mykland P. and Ren J. (1996). Algorithms for computing self-consistent and maximum likelihood estimators with doubly censored data. *The Annals of Statistics* **24**, 1740-1764.

[48] Murphy S. A. and van der Vaart A. W. (2000). On profile likelihood. *Journal of American Statistical Association* **95**, 449-465.

[49] Murtaugh P. A., Dickson E. R., Van Dam G. M., Malincho M., Grambsch P. M., Langworthy A. L. and Gips C. H. (1994). Primary biliary cirrhosis: Prediction of short-term survival based on repeated patient visits. *Hepatology* **20**, 126-134.

[50] Patefield W. M. (1977). On the maximum likelihood function. *Sankhya, Series B* **39**, 92-96.

[51] Peer P. G., Van Dijck J. A., Hendriks J. H., Holland R. and Verbeek A. L. (1993). Age-dependent growth rate of primary breast cancer. *Cancer* **71**, 3547-3551.

[52] Prentice R. L., Kalbfleisch J. D., Peterson A. V., Flournoy N., Farewell V. T. and Breslow N. E. (1978). The analysis of failure times in the presence of competing risks. *Biometrics* **34**, 541-554.

[53] Ren J. (1995). Generalized Cramer-von Mises tests of goodness of fit with doubly censored data. *The Annals of Institute of Statistical Mathematics* **47**, 525-549.

[54] Ren J. (2001). Weighted empirical likelihood ratio confidence intervals for the mean with censored data. *The Annals of Institute of Statistical Mathematics*, **53**, 498-516.

[55] Ren J. (2003a). Goodness of fit tests with interval censored data. *Scandinavia Journal of Statistics* **30**, 211-226.

[56] Ren J. (2003b). Regression M-estimators with non-i.i.d. doubly censored data. *The Annals of Statistics* **31**, 1186-1219.

[57] Ren J. (2008a). Weighted empirical likelihood in some two-sample semiparametric models with various types of censored data. *The Annals of Statistics* **36**, 147-166.

[58] Ren J. (2008b). Smoothed weighted empirical likelihood ratio confidence intervals for quantiles. *Bernoulli* **14**, 725-748.

[59] Ren J. (2009). Estimation bias of the MLE for the Cox model. (In submission)

[60] Ren J. and Gu M. G. (1997). Regression M-estimators with doubly censored data. *The Annals of Statistics* **25**, 2638-2664.

[61] Ren J. and Peer P. G. M. (2000). A study on effectiveness of screening mammograms. *International Journal of Epidemiology* **29**, 803-806.

[62] Ren J. and Zhou M. (2007). Full likelihood inferences in the Cox model. (In submission)

[63] Rice J. (2004). Functional and longitudinal data analysis: Perspectives on smoothing. *Statistics Sinica* **14**, 631-647.

[64] Rice J. and Wu C. (2001). Nonparametric mixed effects models for unequally sampled noisy curves. *Biometrics* **57**, 253-259.

[65] Self S. and Pawitan Y. (1992). Modeling a marker of disease progression and onset of disease. *AIDS Epidemiology: Methodological Issues* (Edited by N. P. Jewell, K. Dietz and V. T. Farewell). Birkhäuser, Boston.

[66] Shiffman S., Fischer L. A., Paty J. A., Gnys M., Hickcox M. and Kassel J. D. (1994). Drinking and smoking: A field study of their association. *Annals of Behavioral Medicine* **16**, 203-209.

[67] Shiffman S., Gnys M., Richards T. J., Paty J. A. and Hickcox M. (1996). Temptations to smoke after quitting: A comparison of lapsers and maintainers. *Health Psychology* **15**, 455-461.

[68] Song X., Davidian M. and Tsiatis A. A. (2002). A semiparametric likelihood approach to joint modeling of longitudinal and time-to-event data. *Biometrics* **58**, 742-753.

[69] Song X. and Wang C. Y. (2008). Semiparametric approaches for joint modeling of longitudinal and survival data with time-varying coefficients. *Biometrics* **64**, 557-566.

[70] Stefanski L. A. and Carroll, R.J. (1987). Conditional scores and optimal scores in generalized linear measurement error models. *Biometrika* **74**, 703-716.

[71] Stone A. A. and Shiffman S. (2002). Capturing momentary, self-report data: A proposal for reporting guidelines. *Annals of Behavioral Medicine* **24**, 236-243.

[72] Stute, W. and Wang, J.L. (1993). The strong law under random censorship. *The Annals of Statistics* **21**, 1591-1607.

[73] Taylor J. M. G., Cumberland W. G. and Sy J.P. (1994). A stochastic model for analysis of longitudinal data. *Journal of American Statistical Association* **89**, 727-776.

[74] Tierney L. and Kadane J. B. (1986). Accurate approximation for posterior moments and marginal densities. *Journal of American Statistical Association* **81**, 82-86.

[75] Tseng Y. K., Hsieh F. and Wang J. L. (2005). Joint modeling of accelerated failure time and longitudinal data. *Biometrika* **92**, 587-603.

[76] Tsiatis A. A. (1981). A large sample study of Cox's regression model. *The Annals of Statistics* **9**, 93-108.

[77] Tsiatis A. A. and Davidian M. (2001). A semiparametric estimator for the proportional hazards model with longitudinal covariates measured with error. *Biometrika* **88**, 447-458.

[78] Tsiatis A. A. and Davidian M. (2004). Joint modeling of longitudinal and time-to-event data: An overview. *Statistica Sinica* **14**, 809-834.

[79] Tsiatis A. A., DeGruttola V. and Wulfsohn M. S. (1995). Modeling the relationship of survival to longitudinal data measured with error: Applications to survival data and CD4 counts in patients with AIDS. *Journal of the American Statistical Association* **90**, 27-37.

[80] Turnbull B. W. (1974). Nonparametric estimation of a survivorship function with doubly censored data. *Journal of the American Statistical Association* **69**, 169-173.

[81] Van der Laan M. J. and Robins J.M. (2003). *Unified methods for censored longitudinal data and causality.* Springer-Verlag, New York.

[82] Walls T. A. and Schafer J. L. (2006). *Intensive Longitudinal Data.* Edited by T. A. Walls and J. L. Schafer. Oxford University Press. Cambridge.

[83] Wang Y. and Taylor J. M. G. (2001). Jointly modeling longitudinal and event time data with application to acquired immunodeficiency syndrome. *Journal of the American Statistical Association* **96**, 895-905.

[84] Wellner J. A. and Zhan Y. (1997). A hybrid algorithm for computation of the NPMLE for censored data. *Journal of the American Statistical Association* **92**, 945-959.

[85] Wulfsohn M. S. and Tsiatis A. A. (1997). A joint model for survival and longitudinal data measured with error. *Biometrics*, **53**, 330-339.

[86] Xu J. and Zeger S. L. (2001). The evaluation of multiple surrogate end points. *Biometrics* **57**, 81-87.

[87] Yao F., Müller H. G. and Wang J. L. (2005). Functional data analysis for sparse longitudinal analysis. *Journal of the American Statistical Association*, **100**, 577-590.

[88] Yao F., Müller H. G., Clifford A. J., Dueker S. R., Follett J., Lin Y., Buchholz B. A. and Vogel J. S. (2003). Shrinkage estimation for functional principle component scores with application to the population kinetics of plasma folate. *Biometrics* **59**, 676-685.

[89] Zeger S. L. and Diggle P. J. (1994). Semiparametric models for longitudinal data with application to CD4 cell numbers in HIV seroconverters. *Biometrics* **50**, 689-699.

[90] Zeng D. and Cai J. (2005) Asymptotic results for maximum likelihood estimators in joint analysis of repeated measurements and survival time. *The Annals of Statistics* **33**, 2132-2163.

[91] Zhang C. H. and Li X. (1996). Linear regression with doubly censored data. *The Annals of Statistics* **24**, 2720-2743.

Part V

Analysis of Survival and Longitudinal Data

Chapter 8

Survival Analysis with High-Dimensional Covariates [*]

Bin Nan [†]

Abstract

Recent interest in the application of microarray technology focuses on relating gene expression profiles to censored survival outcome such as patients' overall survival time or time to cancer relapse. Due to the high-dimensional nature of the gene expression data, regularization becomes an effective approach for such analyses. In this chapter, we review several aspects of the recent development of penalized regression models for censored survival data with high-dimensional covariates, e.g. gene expressions. We first discuss the Cox proportional hazards model (Cox 1972) as the primary example and then the accelerated failure time model (Kalbfleisch and Prentice 2002) for further consideration.

Keywords: Accelerated failure time model, adaptive Lasso, censoring, Cox model, elastic net, grouped variables, hierarchical penalty, Lasso, partial likelihood, regularization, SCAD.

1 Introduction

Censored survival data arise commonly in health science research, such as cancer studies, when the outcome of interest is time to some event, i.e. death or cancer relapse, that may be subject to censoring due to limited follow-up time. Although many methods and corresponding theory on analyzing survival data have been well developed in the past two to three decades, new issues emerge, for example, survival time prediction and variable selection in microarray gene expression studies, and there is a great demand in developing new methods for such emerging research problems.

---

[*]The author is partly supported by the NSF grant DMS-0706700.

[†]Department of Biostatistics, University of Michigan, Ann Arbor, MI 48109, USA, E-mail: bnan@umich.edu

1.1 The Cox model

Among many statistical models for censored survival data, the Cox model gained
its popularity due to its meaningful interpretation of relative risk, nice numerical properties of the partial likelihood approach for fitting the Cox model, and
its semiparametric feature that assumes an unspecified baseline hazard function,
which has motivated the development of the whole new statistical area of semiparametric models (see e.g. Bickel, Klaassen, Ritov, and Wellner 1993). To be
specific, let $Y = \min(T, C)$ be the observed time and $\Delta = I(T \leqslant C)$ the failure
indicator, where T is the failure time and C the censoring time; let $\mathbf{X}$ be a set of
covariates. The hazard function give $\mathbf{X}$ is defined as

$$h(t|\mathbf{X}) = \lim_{\delta \downarrow 0} \frac{1}{\delta} \Pr(t < T \leqslant t + \delta | T \geqslant t, \mathbf{X}).$$

In the Cox model, $h(t|\mathbf{X})$ has the following proportional hazards structure:

$$h(t|\mathbf{X}) = h_0(t) \exp(\boldsymbol{\beta}'\mathbf{X}), \tag{1.1}$$

where the unspecified function h_0 is called the baseline hazard function and $\boldsymbol{\beta}$ is
the vector of regression parameters that are interpreted as log relative hazards.

Assuming T and C are independent given $\mathbf{X}$, Cox (1972, 1975) proposed
a partial likelihood approach to estimate $\boldsymbol{\beta}$ in (1.1) from a random sample of n
observations, which solves the following optimization problem:

$$\max_{\boldsymbol{\beta}} \prod_{i=1}^{n} \left\{ \frac{\exp(\boldsymbol{\beta}'\mathbf{X}_i)}{\sum_{j=1}^{n} I(Y_j \geqslant Y_i) \exp(\boldsymbol{\beta}'\mathbf{X}_j)} \right\}^{\Delta_i}. \tag{1.2}$$

There is a huge amount of literature on the partial likelihood estimation, among
which, Andersen and Gill (1982) provided an elegant prove of its asymptotic properties using the counting process martingale theory.

1.2 The accelerated failure time model

An alternative strategy of modeling censored survival data, which has draw much
attention recently, is the so-called accelerated failure time (AFT) model that postulates a linear relationship between the monotonically transformed survival time,
e.g. by log transformation, and a set of covariates. If we still use the same letters
Y, T, and C to denote their transformed counterparts, we have

$$T_i = \boldsymbol{\beta}'\mathbf{X} + \epsilon_i, \quad i = 1, \ldots, n, \tag{1.3}$$

where ϵ_i, $i = 1, \ldots, n$, are independent and identically distributed random errors
that follow an unspecified distribution. Thus the model is also semiparametric.
Because T_i are not always observed due to censoring, the classical least squares
approach can not be applied directly to the linear model (1.3) for the estimation
of the slope parameters $\boldsymbol{\beta}$.

Commonly used methods for the estimation of β include the Buckley-James approach (Buckley and James 1979) and the rank based estimating equation approach (see e.g. Tsiatis 1990; Wei, Ying, and Lin 1990). The Buckley-James method is basically the least squares method for imputed data, thus provides a loss function that can be regularized for high-dimensional covariates. For censored data, the key idea of the Buckley-James method is to recover those censored T_i by their conditional expectations given corresponding censoring times and covariates. This is the same idea as the single imputation of Little and Rubin (2002). Define the "imputed" failure time Y_i^* as

$$Y_i^* = \begin{cases} Y_i & \Delta_i = 1, \\ \mathrm{E}(T_i|T_i > Y_i, \mathbf{X}_i) & \Delta_i = 0. \end{cases} \tag{1.4}$$

Absorbing the unknown intercept into ϵ_i in model (1.3), the quantity $E(T_i|T_i > Y_i, \mathbf{X}_i)$ for a censored subject i can be calculated by

$$\mathrm{E}(T_i|T_i > Y_i, \mathbf{X}_i) = \boldsymbol{\beta}'\mathbf{X}_i + \mathrm{E}(\epsilon_i|\epsilon_i > Y_i - \boldsymbol{\beta}'\mathbf{X}_i)$$
$$= \boldsymbol{\beta}'\mathbf{X}_i + \int_{Y_i - \boldsymbol{\beta}'\mathbf{X}_i}^{\infty} \frac{t dF(t)}{1 - F(Y_i - \boldsymbol{\beta}'\mathbf{X}_i)}, \tag{1.5}$$

where F is the distribution function of $\epsilon = T - \boldsymbol{\beta}'\mathbf{X}$ in which the intercept is absorbed. That $\mathbf{X}_i$ disappears from the conditional expectation of ϵ is due to a common assumption of independence between the error term and covariates in linear regression. Buckley and James (1979) substituted the above F by its Kaplan-Meier estimator $\hat{F}$ for a given value of β. Then the least squares method can be applied to the following regression model

$$Y_i^* = \boldsymbol{\beta}'\mathbf{X}_i + \epsilon_i^*, \tag{1.6}$$

iteratively between $\hat{F}$ and β, where the intercept parameter is absorbed when each component of $\mathbf{X}$ is centered around its sample average.

The rank based estimating equation approach proceeds by solving the following equation

$$\Psi_n(\boldsymbol{\beta}) = \frac{1}{n} \sum_{i=1}^{n} \rho_n(e_i(\boldsymbol{\beta}); \boldsymbol{\beta}) \left\{ \mathbf{X}_i - \frac{\sum_{j=1}^{n} \mathbf{X}_i I(e_j(\boldsymbol{\beta}) \geqslant e_i(\boldsymbol{\beta}))}{\sum_{j=1}^{n} I(e_j(\boldsymbol{\beta}) \geqslant e_i(\boldsymbol{\beta}))} \right\} \Delta_i$$
$$= o_p(n^{-1/2}), \tag{1.7}$$

where ρ_n is a weight function and $e_i(\boldsymbol{\beta}) = Y_i - \boldsymbol{\beta}'\mathbf{X}_i$. When Gehan weight is used, i.e.

$$\rho_n(t; \boldsymbol{\beta}) = \frac{1}{n} \sum_{i=1}^{n} I(e_i(\boldsymbol{\beta}) \geqslant t),$$

the estimating function $\Psi_n(\boldsymbol{\beta})$ is monotone (see Fygenson and Ritov 1994) with the following form

$$\Psi_n(\boldsymbol{\beta}) = \frac{1}{n^2} \sum_{i=1}^{n} \sum_{j=1}^{n} (\mathbf{X}_i - \mathbf{X}_j) I(e_j(\boldsymbol{\beta}) \geqslant e_i(\boldsymbol{\beta})) \Delta_i,$$

which can be viewed as the gradient of the following convex function

$$\ell_n(\boldsymbol{\beta}) = \frac{1}{n^2} \sum_{i=1}^{n} \sum_{j=1}^{n} \{e_j(\boldsymbol{\beta}) - e_i(\boldsymbol{\beta})\} I(e_j(\boldsymbol{\beta}) \geq e_i(\boldsymbol{\beta})) \Delta_i. \tag{1.8}$$

Therefor, a valid estimator of $\boldsymbol{\beta}$ can be obtained by minimizing $\ell_n(\boldsymbol{\beta})$ through linear programming (see e.g. Jin et al. 2003).

1.3   Regularization methods for high-dimensional covariates

Denote the dimension of $\boldsymbol{\beta}$ by p. Let $\ell_n(\boldsymbol{\beta})$ be a loss function, e.g. the negative log partial likelihood function for the Cox model or the residual sum of squares for the linear model. A desirable estimation of $\boldsymbol{\beta}$ can be obtained by minimizing $\ell_n(\boldsymbol{\beta})$ in a classical setting where p is a small fixed integer. When p is large, the estimation problem can be ill-posed. To prevent overfitting, one can apply the regularization method by minimizing penalized loss function:

$$\min_{\boldsymbol{\beta}} \ \ell_n(\boldsymbol{\beta}) + \text{Pen}_{\lambda_n}(\boldsymbol{\beta}), \tag{1.9}$$

where λ_n is a positive tuning parameter.

Equation (1.9) is a general form of the regularization methods. We briefly review the least squares regression in this section. The most classical example of regularization method for a linear regression model is the ridge regression that has

$$\text{Pen}_{\lambda_n}(\boldsymbol{\beta}) = \lambda_n \sum_{j=1}^{p} \beta_j^2,$$

see e.g. Hoerl (1962), Hoerl and Kennard (1970a,b). The ridge regression, however, does not shrink the regression coefficients to exactly zero, thus can not be used for variable selection.

The least absolute shrinkage and selection operator (Lasso) method has drawn much attention since it was proposed by Tibshirani (1996), which has the L_1-norm penalty

$$\text{Pen}_{\lambda_n}(\boldsymbol{\beta}) = \lambda_n \sum_{j=1}^{p} |\beta_j|.$$

The popularity of Lasso is mainly gained from its shrinkage property, i.e. some of the regression coefficients are shrunk to exactly zero, thus Lasso does variable selection while obtaining parameter estimates simultaneously. Several fast algorithms for the Lasso method have been developed, see e.g. the least angle regression method (LARS) of Efron et al. (2004) and the shooting algorithm of Fu (1998) and Friedman et al. (2007). Theoretical results on the consistency of variable selection in linear regression models have be established by Zhao and Yu (2006), Yuan and Lin (2007), Meinshausen and Yu (2009), among others, and issues of choosing the tuning parameter λ_n have been discussed by, e.g., Leng et al. (2006). The parameter estimation as well as prediction error bounds for linear regression models have been shown to satisfy the so-called oracle inequalities (i.e., the non-asymptotic error bounds have the same order as if the set of underlying non-zero coefficients were given ahead by an *oracle*), by Bunea et al. (2007) and Bickel et al. (2008), among others.

To reduce the bias of the Lasso method, Fan and Li (2001) proposed the smoothly clipped absolute deviation (SCAD) penalty

$$\text{Pen}_{\lambda_n}(\boldsymbol{\beta}) = \sum_{j=1}^{p} p_{\lambda_n}(|\beta_j|)$$

satisfying

$$p'_{\lambda_n}(|\beta_j|) = I(|\beta_j| \leqslant \lambda_n) + \frac{(a\lambda_n - |\beta_j|)_+}{(a-1)\lambda_n} I(|\beta_j| > \lambda_n), \quad a > 2, \quad (1.10)$$

where $a = 3.7$ is recommended by Fan and Li (2001). They then applied the SCAD method to the Cox regression model (see Fan and Li 2002). Note that the SCAD penalty is non-convex. Fan and Li (2001, 2002) proved that the SCAD method yields an estimation of $\boldsymbol{\beta}$ that achieves the asymptotic oracle property, i.e., the method performs as well as if the correct underlying model were provided in advance by an oracle.

Another approach that improves the Lasso method is the adaptive Lasso method proposed by Zou (2006) and Zhang and Lu (2007), which is a weighted Lasso penalty:

$$\text{Pen}_{\lambda_n}(\boldsymbol{\beta}) = \lambda_n \sum_{j=1}^{p} w_j |\beta_j|.$$

When $w_j = 1/\tilde{\beta}_j$, where $\tilde{\boldsymbol{\beta}}$ is a consistent estimator, the adaptive Lasso estimator achieves the same oracle property as that in Fan and Li (2001, 2002).

Many different penalized methods have been developed in recent literature to address different aspects of variable selection in regressions settings, particularly when covariates are naturally grouped, e.g. each gene pathway contains multiple genes in a microarray data analysis. We do not attempt to provide an exhaustive review of all the published or to be published methods in this chapter. Instead, we will discuss the above commonly used methods for the Cox regression in §2 and a new method considering the group structure of variables for the Cox regression in §3. Much detail will be provided in §3. In section §4, we will provide a brief review of recent developments in fitting the AFT model with high-dimensional covariates.

2 Regularized Cox regression

2.1 The Lasso method

The Lasso method (see Tibshirani 1996) was initially proposed as a variable selection tool for linear regression models. In biomedical research, however, event time that may subject to right censoring is a common outcome measure, and the partial likelihood method for the Cox model is widely used to fit such data. From (1.2), we obtain the negative log partial likelihood function as follows

$$\ell_n(\boldsymbol{\beta}) = -\frac{1}{n}\sum_{i=1}^{n}\Delta_i\left[\boldsymbol{\beta}'\mathbf{X}_i - \log\left\{\sum_{j=1}^{n}I(Y_j \geqslant Y_i)\exp(\boldsymbol{\beta}'\mathbf{X}_j)\right\}\right]. \tag{2.1}$$

Tibshirani (1997) extended the Lasso method to the Cox model by solving the following minimization problem:

$$\min_{\boldsymbol{\beta}}\ \ell_n(\boldsymbol{\beta}) + \lambda_n\sum_{j=1}^{p}|\beta_j|,$$

or equivalent,

$$\min_{\boldsymbol{\beta}}\ \ell_n(\boldsymbol{\beta}),\quad \text{subject to } \sum_{j-1}^{p}|\beta_j| \leqslant s_n, \tag{2.2}$$

where λ_n and s_n are tuning parameters that have a one-to-one correspondence. See Osborne et al. (2000) for a discussion of the relationship between λ_n and s_n. Tibshirani (1997) proposed an iterative procedure to reformulate the constrained optimization problem (2.2) as a Lasso problem for linear regression models. Specifically, let $\mathbb{X}$ be the design matrix of covariates and $\boldsymbol{\eta} = \mathbb{X}\boldsymbol{\beta}$. Define $\boldsymbol{\mu} = \partial\ell_n/\partial\boldsymbol{\eta}$, $\mathbf{A} = -\partial^2\ell_n/\partial\boldsymbol{\eta}\boldsymbol{\eta}'$, and $\mathbf{z} = \boldsymbol{\eta} + \mathbf{A}^{-}\boldsymbol{\mu}$. When $p > n$, $\mathbf{A}$ is clearly singular, then the generalized inverse can be used (see Gui and Li 2005). Then a one-term Taylor series expansion for $\ell_n(\boldsymbol{\beta})$ has the form

$$(\mathbf{z} - \boldsymbol{\eta})'\mathbf{A}(\mathbf{z} - \boldsymbol{\eta}),$$

and problem (2.2) can be solved by the following procedure:

Step 1. Fix s_n and initialize $\hat{\boldsymbol{\beta}} = 0$.
Step 2. Compute $\boldsymbol{\eta}$, $\boldsymbol{\mu}$, $\mathbf{A}$, and $\mathbf{z}$ based on the current value of $\hat{\boldsymbol{\beta}}$.
Step 3. Minimize $(\mathbf{z} - \mathbb{X}\boldsymbol{\beta})'\mathbf{A}(\mathbf{z} - \mathbb{X}\boldsymbol{\beta})$ subject to $\sum_{j=1}^{p}|\beta_j| \leqslant s_n$.
Step 4. Repeat Steps 2 and 3 until $\hat{\boldsymbol{\beta}}$ does not change.

Tibshirani (1997) proposed to solve Step 3 by using a quadratic programming procedure. However, Gui and Li (2005) pointed out that the quadratic programming can not be directly applied to the setting $p > n$. They proposed to convert Step 3 to a standard Lasso problem and then use what was then newly developed LARS algorithm by Efron et al. (2004) to solve Step 3. To be specific, first apply the Cholesky decomposition to obtain $\mathbf{T} = \mathbf{A}^{1/2}$ such that $\mathbf{T}'\mathbf{T} = \mathbf{A}$, and define $\mathbf{y} = \mathbf{T}\mathbf{z}$ and $\widetilde{\mathbb{X}} = \mathbf{T}\mathbb{X}$; then Step 3 of the above iterative procedure can be rewritten as

Step 3': Minimize $(\mathbf{y} - \widetilde{\mathbb{X}}\boldsymbol{\beta})'(\mathbf{y} - \widetilde{\mathbb{X}})$ subject to $\sum_{j=1}^{p}|\beta_j| \leqslant s_n$, which can be efficiently solved by the LARS algorithm for a given s_n. Gui and Li (2005) named this approach the LARS-Cox regression.

Since Step 3' is a standard Lasso problem, the LARS algorithm can be improved further by the shooting algorithm of Fu (1998) that was shown by Friedman et al. (2007) to be a much faster algorithm than its competitors for solving the Lasso problem. Consider the following Lasso problem for a linear regression model:

$$\min_{\beta} \sum_{i=1}^{n} \left(y_i - \sum_{j=1}^{p} x_{ij} \beta_j \right)^2 + \lambda_n \sum_{j=1}^{p} |\beta_j|,$$

where x_{ij} are standardized so that $\sum_{i=1}^{n} x_{ij} = 0$ and $\sum_{i=1}^{n} x_{ij}^2 = 1$. With a single predictor, the Lasso solution is very simple, and is a soft-thresholding (Donoho and Johnstone 1995) of the least squares estimation $\hat{\beta}$:

$$\hat{\beta}^{\text{lasso}}(\lambda_n) = S(\hat{\beta}, \lambda_n) \equiv \text{sign}(\hat{\beta})(|\hat{\beta}| - \lambda_n/2)_+. \tag{2.3}$$

For multiple predictors, consider a simple iterative algorithm that applies soft-thresholding with a "partial residual" as a response variable. With all the values of β_k for $k \neq j$ being held at values $\tilde{\beta}_k(\lambda_n)$, minimizing

$$\sum_{i=1}^{n} \left(y_i - \sum_{k \neq j} x_{ik} \beta_k - x_{ij} \beta_j \right)^2 + \lambda_n \sum_{k \neq j} |\beta_k| + \lambda_n |\beta_j|$$

with respect to β_j, we obtain

$$\tilde{\beta}_j(\lambda_n) \leftarrow S\left(\sum_{i=1}^{n} x_{ij}(y_i - \tilde{y}_i^{(j)}), \lambda_n \right), \tag{2.4}$$

where $\tilde{y}_i^{(j)} = \sum_{k \neq j} x_{ik} \tilde{\beta}_k(\lambda_n)$ and the function S is defined in (2.3). This is simply the univariate regression coefficient of the partial residual $y_i - \tilde{y}_i^{(j)}$ on the j-th variable. The update in (2.4) is repeated for $j = 1, 2, \ldots, p, 1, 2, \ldots$ until convergence. This shooting algorithm (or called the pathwise coordinate descent algorithm by Friedman et al 2007) has been widely used in solving similar optimization problems.

2.2 The SCAD method

Fan and Li (2002) Applied the SCAD penalty to the Cox model (1.1). In particular, they proposed to solve the following optimization problem

$$\min_{\beta} \ell_n(\beta) + \sum_{j=1}^{p} p_{\lambda_n}(|\beta_j|), \tag{2.5}$$

where $\ell_n(\beta)$ is the negative log partial likelihood given in (2.1) and $p_{\lambda_n}(|\beta_j|)$ satisfies (1.10). The SCAD method for the Cox regression also has the oracle property of Fan and Li (2001) for a linear model, which will be described in detail later.

To solve the SCAD problem, Fan and Li (2001, 2002) proposed to use the local quadratic approximation of $p_{\lambda_n}(|\beta_j|)$ as follows. Given an initial value β_0 that is close to the minimizer of (2.5), when β_j is not very close to 0, the derivative of $p_{\lambda_n}(|\beta_j|)$ to β_j can be locally approximated by:

$$\partial p_{\lambda_n}(|\beta_j|)/\partial \beta_j = p'_{\lambda_n}(|\beta_j|)\mathrm{sign}(\beta_j) \approx [p'_{\lambda_n}(|\beta_{j0}|)/|\beta_{j0}|]\beta_j,$$

otherwise, set $\beta_j = 0$. In other words,

$$p_{\lambda_n}(|\beta_j|) \approx p_{\lambda_n}(|\beta_{j0}|) + \frac{1}{2}p'_{\lambda_n}(|\beta_{j0}|)(\beta_j^2 - \beta_{j0}^2) \quad \text{for } \beta_j \approx \beta_{j0}.$$

The negative log partial likelihood function $\ell_n(\beta)$ can also be approximated by the Taylor expansion. Then the optimization in (2.5) is approximated to a local quadratic minimization problem and can be solved by the Newton-Raphson algorithm. With a good initial value, a one-step penalized partial likelihood estimator can be as efficient as the fully iterative one.

We now consider the asymptotic performance, i.e. the oracle property, of the SCAD method for the Cox model. Note that here we only discuss the case that p is fixed while n approaches to infinity. The results can be extended to the case that p approaches to infinity as n goes to infinity, but at a slower rate (i.e. $p/n \to 0$), see Fan and Peng (2004). Assume the regularity conditions A–D of Andersen and Gill (1982) throughout this subsection. Let the true parameter values be β^0. Let

$$\mathcal{A} = \{j : \beta_j^0 \neq 0, 1 \leqslant j \leqslant p\}.$$

Denote

$$Q_n(\beta) = \ell_n(\beta) + \sum_{j=1}^{p} p_{\lambda_n}(|\beta_j|)$$

and

$$a_n = \max_j \{p'_{\lambda_n}(|\beta_j^0|) : \beta_j^0 \neq 0\},$$
$$b_n = \max_j \{|p''_{\lambda_n}(|\beta_j^0|)| : \beta_j^0 \neq 0\}.$$

Theorem 2.1. *If $a_n = O_p(n^{-1/2})$ and $b_n \to 0$, then for fixed p, there exists a local minimizer $\hat{\beta}_n$ of $Q_n(\beta)$ such that $\|\hat{\beta}_n - \beta^0\| = O_p(n^{-1/2})$.*

Theorem 2.2. (Oracle property). *Let $\hat{\beta}_n$ be a root-n consistent local minimizer of $Q_n(\beta)$. Under the conditions in Theorem 2.1, we have:*

(i) *(Sparsity) if $n^{1/2}p'_{\lambda_n}(|\hat{\beta}_{n,j}|) \to \infty$ as $n \to \infty$ for all $j \in \mathcal{A}^c$, then $\hat{\beta}_{n,\mathcal{A}^c} = 0$ with probability approaching 1.*

(ii) *(Asymptotic normality) if further $n^{1/2}a_n \to 0$ as $n \to \infty$, then under (i), $n^{1/2}(\hat{\beta}_{n,\mathcal{A}} - \beta_{\mathcal{A}}^0)$ converges in distribution to a zero-mean normal random variable with covariance matrix $I_{\mathcal{A}}(\beta_{\mathcal{A}}^0)^{-1}$. Here $I_{\mathcal{A}}$ is the information matrix for $\beta_{\mathcal{A}}$ knowing that $\beta_{\mathcal{A}^c}^0 = 0$.*

The proofs of Theorems 2.1 and 2.2 have the same sprit of Fan and Li (2001, 2002) and are special cases of the general problem setting in the next Section 3, i.e. variables are in groups. We will provided detailed proofs in Section 2.3 for grouped variables, which easily reduce to the current setting with a group size of one.

Corollary 2.3. *If $\lambda_n \to 0$ and $n^{1/2}\lambda_n \to \infty$, then there exists a root-n consistent local minimizer $\hat{\beta}_n$ of $Q_n(\beta)$ which satisfies the oracle property in Theorem 2.2.*

Proof. Since $\lambda_n \to 0$, when $\beta_j \neq 0$, for large enough n we have $I(|\beta_j| \leqslant \lambda_n) = 0$ and $(a\lambda_n - |\beta_j|)_+ = 0$. Thus $p'_{\lambda_n}(|\beta_j|) = 0$ when n is large enough, which implies $a_n = 0$ and $b_n = 0$ when n is large enough. So we have $n^{1/2}a_n \to 0$ and $b_n \to 0$. By Theorem 2.1 we know that there exists a root-n consistent local minimizer $\hat{\beta}_n$. When $\beta_j = 0$, we know that $n^{1/2}\hat{\beta}_j = O_p(1)$, so $I(|\hat{\beta}_j| \leqslant \lambda_n) = I(|n^{1/2}\hat{\beta}_j| \leqslant n^{1/2}\lambda_n) = 1$ with probability approaching 1 because $n^{1/2}\lambda_n \to \infty$. Thus $n^{1/2}p'_{\lambda_n}(|\hat{\beta}_j|) \geqslant n^{1/2}\lambda_n I(|\hat{\beta}_j| \leqslant \lambda_n) \to \infty$ in probability. Together with $n^{1/2}a_n \to 0$ and $b_n \to 0$, we obtain the oracle property of $\hat{\beta}_n$ by Theorem 2.2. $\square$

2.3 The adaptive Lasso method

Zhang and Lu (2007) proposed the adaptive Lasso method for the Cox regression, which solves the following optimization problem

$$\min_{\beta} \; \ell_n(\beta) + \lambda_n \sum_{j=1}^{p} w_j|\beta_j|, \tag{2.6}$$

where $\ell_n(\beta)$ is the negative log partial likelihood function given in (2.1) and the positive weights $\mathbf{w} = (w_1, \ldots, w_p)'$ are chosen adaptively by data. The values of $\mathbf{w}$ are crucial for guaranteeing the optimality of the solution. Zhang and Lu (2007) proposed to use $w_j = 1/|\tilde{\beta}_j|$, where $\tilde{\beta} = (\tilde{\beta}_1, \ldots, \tilde{\beta}_p)'$ is the maximum partial likelihood estimator that is consistent and asymptotically normal under a set of general regularization conditions (see Andersen and Gill 1982).

Once $\mathbf{w}$ is given, the optimization in (2.6) is slightly more general than but almost identical to the Lasso method, hence can be solved by a numerical procedure provided in previous Section 2.1. Using the same notation as in Section 2.1, the algorithm is given as follows.

1. Obtain $\tilde{\beta}$ by minimizing the negative log partial likelihood function $\ell_n(\beta)$.
2. Fix λ_n and initialize $\hat{\beta} = 0$.
3. Compute $\boldsymbol{\eta}$, $\boldsymbol{\mu}$, $\mathbf{A}$, $\mathbf{z}$, $\mathbf{y}$ and $\tilde{\mathbb{X}}$ based on the current value of $\hat{\beta}$.
4. Minimize $(\mathbf{y} - \tilde{\mathbb{X}}\beta)'(\mathbf{y} - \tilde{\mathbb{X}}\beta) + \lambda_n \sum_{j=1}^{p} |\beta_j|/|\tilde{\beta}_j|$ by the shooting algorithm that updates $\hat{\beta}_j$ for $j = 1, 2, \ldots, p, 1, 2, \ldots$ until convergence:

$$\hat{\beta}_j(\lambda_n) \leftarrow S\left(\sum_{i=1}^{n} \tilde{x}_{ij}(y_i - \tilde{y}_i^{(j)}), \lambda_{n,j} \right),$$

where $\lambda_{n,j} = \lambda_n/|\tilde{\beta}_j|$.

5. Repeat Steps 3 and 4 until $\hat{\boldsymbol{\beta}}$ does not change.

Zhang and Lu (2007) also showed that the adaptive Lasso estimator has the oracle property that is given in the following corollaries for the case the p is fixed. We refer to Huang et al. (2008) for the adaptive Lasso estimation in a linear model with a divergent p.

Corollary 2.4. *Assume the regularity conditions A–D of Andersen and Gill* (1982). *If $n^{1/2}\lambda_n = O_p(1)$, then the adaptive Lasso estimator satisfies $\|\hat{\boldsymbol{\beta}}_n - \boldsymbol{\beta}^0\| = O_p(n^{-1/2})$.*

Proof. Since $p'_{\lambda_n}(|\beta_j|) = \lambda_n/|\tilde{\beta}_j|$, we have

$$n^{1/2}a_n = n^{1/2}\lambda_n \max_j\{|\tilde{\beta}_j|^{-1} : \beta_j \neq 0\}.$$

Thus $n^{1/2}a_n = O_p(1)$ and clearly $b_n = 0$, so by Theorem 2.1 we know that there exists a root-n consistent $\hat{\beta}_n$. □

Corollary 2.5. *Assume that $n^{1/2}\lambda_n \to 0$ and $n\lambda_n \to \infty$. Then under the conditions of Corollary 2.4, with probability tending to 1, the root-n consistent adaptive Lasso estimator $\hat{\boldsymbol{\beta}}_n$ satisfy:*

(i) *(Sparsity) $\hat{\boldsymbol{\beta}}_{n,\mathcal{A}^c} = 0$.*

(ii) *(Asymptotic normality) $n^{1/2}(\hat{\boldsymbol{\beta}}_{n,\mathcal{A}} - \boldsymbol{\beta}_{\mathcal{A}}^0)$ converges in distribution to a zero-mean normal random variable with covariance matrix $I_{\mathcal{A}}(\boldsymbol{\beta}_{\mathcal{A}}^0)^{-1}$. Here $I_{\mathcal{A}}$ is the information matrix for $\boldsymbol{\beta}_{\mathcal{A}}$ knowing that $\boldsymbol{\beta}_{\mathcal{A}^c}^0 = 0$.*

Proof. When $\beta_j = 0$, we have

$$P(n^{1/2}\lambda_n/|\tilde{\beta}_j| > K) = P\left(\frac{n\lambda_n}{|n^{1/2}\tilde{\beta}_{n,j}|} > K\right) \to 1 \quad \forall K > 0$$

since $n\lambda_n \to \infty$. Hence $n^{1/2}p'_{\lambda_n}(|\hat{\beta}_{n,j}|) \to \infty$ with probability approaching 1. Together with $n^{1/2}a_n \to 0$ and $b_n = 0$, we obtain the oracle property of $\hat{\beta}_n$ by Theorem 2.2. □

3   Hierarchically penalized Cox regression with grouped variables

In this section, we introduce a new method developed by Wang et al. (2009) for the variable selection problem in the Cox model when predictors can be naturally grouped. The problem of interest is to select important groups as well as important individual variables within identified groups. One motivation of considering the group variable selection problem arises in genomic research. Genomic data can often be naturally divided into small sets based on biological knowledge.

For example, when analyzing microarray gene expression data, one can group genes into functionally similar sets as in The Gene Ontology Consortium (2000) or into known biological pathways such as the KEGG pathways (Kanehisa and Goto 2002). Usually, making use of the group information in modeling can be helpful for identifying pathways as well as genes within the pathways related to the phenotypes, hence improves understanding of biological processes.

By shrinking some of the regression coefficients to be exact zero, the methods described in the previous Section 2 automatically remove unimportant variables. However, when the predictors are grouped, one drawback of these methods is that they treat variables individually or "flatly," i.e., ignoring the group structure. Consequently, they tend to perform variable selection based on the strength of the individual variables rather than the strength of the group, often resulting in selecting more variables than necessary.

The variable selection problem with grouped predictors has been tackled recently by several authors. Antoniadis and Fan (2001) and Cai (2001) discussed block-wise thresholding, Zhao et al.(2009) proposed a method that penalizes the L_∞-norm of the coefficients within each group, Yuan and Lin (2006) extended the lasso method to penalize the L_2-norm of the coefficients within each group for linear regression models, and Wang and Leng (2008) considered an adaptive grouped lasso method. Based on the boosting technique, Luan and Li (2007) and Wei and Li (2007), respectively, proposed a group additive regression model and a nonparametric pathway-based regression model to identify groups of genomic features that are related to several clinical phenotypes including the survival outcome. All of these useful group variable selection methods have a common limitation: they select variables in an "all-in-all-out" fashion, i.e., when one variable in a group is selected, all other variables in the same group are also selected. In other words, these methods do not do variable selection within an identified group. The reality, however, may not be the case. For example, some genes in a pathway may not be related to the phenotype, though the pathway as a whole is involved in the biological process.

3.1   Regularization via the hierarchical penalty for the Cox model

Assume that the total of p variables in $\mathbf{X}$ can be divided into K groups. Let the kth group have p_k variables, denoted by $\mathbf{X}_{(k)} = (X_{k1}, \ldots, X_{kp_k})'$, and $\boldsymbol{\beta}_{(k)} = (\beta_{k1}, \ldots, \beta_{kp_k})'$ represent the corresponding regression coefficients. We first assume that the K groups do not overlap, i.e., each variable belongs to only one group. Later we will allow variables to belong to multiple groups.

Let $\mathbf{X}_{i,(k)} = (X_{i,k1}, \ldots, X_{i,kp_k})'$ denote the p_k variables in the kth group for the ith subject, and $\mathbf{X}_i = (\mathbf{X}'_{i,(1)}, \ldots, \mathbf{X}'_{i,(K)})'$ denote the total of p variables for the ith subject. We assume that T_i and C_i are conditionally independent given $\mathbf{X}_i$ and that the censoring mechanism is noninformative.

The proportional hazards model (1.1) now can be written as

$$h(t|\mathbf{X}) = h_0(t) \exp\left(\sum_{k=1}^{K} \sum_{j=1}^{p_k} \beta_{kj} X_{kj} \right)$$
$$= h_0(t) \exp(\boldsymbol{\beta}'_{(1)} \mathbf{X}_{(1)} + \cdots + \boldsymbol{\beta}'_{(K)} \mathbf{X}_{(K)}).$$

We consider continuous failure times and assume that there are no ties in the observed times for simplicity. The partial likelihood is then given by

$$L_n(\boldsymbol{\beta}) = \prod_{i \in D} \frac{\exp\left(\sum_{k=1}^{K} \boldsymbol{\beta}'_{(k)} \mathbf{X}_{i,(k)} \right)}{\sum_{l \in R_i} \exp\left(\sum_{k=1}^{K} \boldsymbol{\beta}'_{(k)} \mathbf{X}_{l,(k)} \right)},$$

where D is the set of indices of observed failures and R_i is the set of indices of the subjects who are at risk at time Y_i.

Let $\ell_n(\boldsymbol{\beta})$ be the negative log partial likelihood defined in (2.1). Variable selection can be realized via minimizing the penalized negative log partial likelihood function

$$\ell_n(\boldsymbol{\beta}) + \sum_{k=1}^{K} \sum_{j=1}^{p_k} p_{\lambda_n}(\beta_{kj}),$$

where $p_{\lambda_n}(\beta_{kj})$ denotes a general penalty function. When $p_{\lambda_n}(\beta_{kj}) = \lambda_n |\beta_{kj}|$, the above optimization corresponds the Lasso method discussed in Section 2.1; when p_{λ_n} satisfies (1.10), it corresponds to the SCAD method discussed in Section 2.2; when $p_{\lambda_n}(\beta_{kj}) = \lambda_n w_{kj} |\beta_{kj}|$, the above optimization corresponds the adaptive Lasso method discussed in Section 2.3.

To make use of the group structure among the predictors, we reparameterize β_{kj} as

$$\beta_{kj} = \gamma_k \theta_{kj}, \quad k = 1, \ldots, K; \ j = 1, \ldots, p_k, \qquad (3.1)$$

where $\gamma_k \geqslant 0$ for identifiability. This decomposition reflects the information that all β_{kj}, $j = 1, \ldots, p_k$, belong to the kth group by treating each β_{kj} hierarchically. Parameter γ_k controls all β_{kj}, $j = 1, \ldots, p_k$, as a group at the first level of the hierarchy; θ_{kj}'s reflect differences within the kth group at the second level of the hierarchy. Let $\boldsymbol{\theta}_{(k)} = (\theta_{k1}, \ldots, \theta_{kp_k})'$, $k = 1, \ldots, K$, then we have $\boldsymbol{\beta}_{(k)} = \gamma_k \cdot \boldsymbol{\theta}_{(k)}$. Now the partial likelihood function can be written as

$$L_n(\boldsymbol{\gamma}, \boldsymbol{\theta}) = \prod_{i \in D} \frac{\exp\left(\sum_{k=1}^{K} \gamma_k \boldsymbol{\theta}'_{(k)} \mathbf{X}_{i,(k)} \right)}{\sum_{l \in R_i} \exp\left(\sum_{k=1}^{K} \gamma_k \boldsymbol{\theta}'_{(k)} \mathbf{X}_{l,(k)} \right)}.$$

Let $\ell_n(\boldsymbol{\gamma}, \boldsymbol{\theta}) = -(1/n) \log\{L_n(\boldsymbol{\gamma}, \boldsymbol{\theta})\}$. For the purpose of variable selection, we consider the following penalized log partial likelihood with a hierarchical penalty:

$$\min_{\gamma_k, \theta_{kj}} \ell_n(\boldsymbol{\gamma}, \boldsymbol{\theta}) + \lambda_\gamma \sum_{k=1}^{K} \gamma_k + \lambda_\theta \sum_{k=1}^{K} \sum_{j=1}^{p_k} |\theta_{kj}| \qquad (3.2)$$

subject to $\gamma_k \geqslant 0$, $k = 1, \ldots, K$,

where $\lambda_\gamma \geq 0$ and $\lambda_\theta \geq 0$ are two tuning parameters. Parameter λ_γ controls the estimators at the group level and can effectively remove unimportant groups: if γ_k is shrunken to zero, all β_{kj} in the kth group will be equal to zero. Parameter λ_θ controls the estimators at the variable-specific level: if γ_k is not equal to zero, some θ_{kj} and hence the corresponding β_{kj}, $j = 1, \ldots, p_k$, are still possible to be shrunken to zero. It is clearly seen that for fixed β and given values of λ_γ and λ_θ, the maximizer of (3.2), where $\ell_n(\gamma, \theta)$ is constant, is unique.

Since there are two tuning parameters in (3.2), it may be complicated to tune their values in practice. However, it turns out that the two tuning parameters λ_γ and λ_θ can be simplified into one. Specifically, let $\lambda = \lambda_\gamma \cdot \lambda_\theta$, we can show that (3.2) is equivalent to

$$\min_{\gamma_k, \theta_{kj}} \ell_n(\gamma, \theta) + \sum_{k=1}^{K} \gamma_k + \lambda \sum_{k=1}^{K} \sum_{j=1}^{p_k} |\theta_{kj}| \qquad (3.3)$$

$$\text{subject to } \gamma_k \geq 0, \; k = 1, \ldots, K.$$

The meaning of "equivalence" is illustrated in the following lemma:

Lemma 3.1. *Let $(\hat{\gamma}^*, \hat{\theta}^*)$ be a local minimizer of (3.2), then there exists a local minimizer $(\hat{\gamma}^\dagger, \hat{\theta}^\dagger)$ of (3.3) such that $\hat{\gamma}_k^* \hat{\theta}_{kj}^* = \hat{\gamma}_k^\dagger \hat{\theta}_{kj}^\dagger$. Similarly, if $(\hat{\gamma}^\dagger, \hat{\theta}^\dagger)$ is a local minimizer of (3.3), then there exists a local minimizer $(\hat{\gamma}^*, \hat{\theta}^*)$ of (3.2) such that $\hat{\gamma}_k^* \hat{\theta}_{kj}^* = \hat{\gamma}_k^\dagger \hat{\theta}_{kj}^\dagger$.*

Proof. In the proof, we use $|a|$ to denote the l_1-norm of a. Let $Q^*(\lambda_\gamma, \lambda_\theta, \gamma, \theta)$ denote the criterion that we would like to minimize in Equation (3.2), let $Q^\dagger(\lambda, \gamma, \theta)$ denote the corresponding criterion in Equation (3.3), and let $(\hat{\gamma}^*, \hat{\theta}^*)$ denote a local minimizer of $Q^*(\lambda_\gamma, \lambda_\theta, \gamma, \theta)$. We will prove that $(\hat{\gamma}^\dagger = \lambda_\gamma \hat{\gamma}^*, \hat{\theta}^\dagger = \hat{\theta}^* / \lambda_\gamma)$ is a local minimizer of $Q^\dagger(\lambda, \gamma, \theta)$.

We immediately have $Q^*(\lambda_\gamma, \lambda_\theta, \gamma, \theta) = Q^\dagger(\lambda, \lambda_\gamma \gamma, \theta / \lambda_\gamma)$. Since $(\hat{\gamma}^*, \hat{\theta}^*)$ is a local minimizer of $Q^*(\lambda_\gamma, \lambda_\theta, \gamma, \theta)$, there exists $\delta > 0$ such that if (γ', θ') satisfies $|\gamma' - \hat{\gamma}^*| + |\theta' - \hat{\theta}^*| < \delta$, then $Q^*(\lambda_\gamma, \lambda_\theta, \gamma', \theta') \geq Q^*(\lambda_\gamma, \lambda_\theta, \hat{\gamma}^*, \hat{\theta}^*)$.

Choose δ' such that $\delta' / \min(\lambda_\gamma, 1/\lambda_\gamma) \leq \delta$. Then for any (γ'', θ'') satisfying $|\gamma'' - \hat{\gamma}^\dagger| + |\theta'' - \hat{\theta}^\dagger| < \delta'$, we have

$$\left| \frac{\gamma''}{\lambda_\gamma} - \hat{\gamma}^* \right| + |\lambda_\gamma \theta'' - \hat{\theta}^*| \leq \frac{\lambda_\gamma |\frac{\gamma''}{\lambda_\gamma} - \hat{\gamma}^*| + \frac{1}{\lambda_\gamma}|\lambda_\gamma \theta'' - \hat{\theta}^*|}{\min(\lambda_\gamma, \frac{1}{\lambda_\gamma})} < \frac{\delta'}{\min(\lambda_\gamma, \frac{1}{\lambda_\gamma})} \leq \delta.$$

Hence

$$Q^\dagger(\lambda, \hat{\gamma}'', \hat{\theta}'') = Q^*(\lambda_\gamma, \lambda_\theta, \hat{\gamma}'' / \lambda_\gamma, \lambda_\gamma \hat{\theta}'') \geq Q^*(\lambda_\gamma, \lambda_\theta, \hat{\gamma}^*, \hat{\theta}^*) = Q^\dagger(\lambda, \hat{\gamma}^\dagger, \hat{\theta}^\dagger).$$

Therefore, $(\hat{\gamma}^\dagger = \lambda_\gamma \hat{\gamma}^*, \hat{\theta}^\dagger = \hat{\theta}^* / \lambda_\gamma)$ is a local minimizer of $Q^\dagger(\lambda, \gamma, \theta)$.

Similarly we can prove that for any local minimizer $(\hat{\gamma}^\dagger, \hat{\theta}^\dagger)$ of $Q^\dagger(\lambda, \gamma, \theta)$, there is a corresponding local minimizer $(\hat{\gamma}^*, \hat{\theta}^*)$ of $Q^*(\lambda_\gamma, \lambda_\theta, \gamma, \theta)$ such that $\hat{\gamma}_k^* \hat{\theta}_{kj}^* = \hat{\gamma}_k^\dagger \hat{\theta}_{kj}^\dagger$. $\qquad \square$

Lemma 3.1 indicates that we only need to tune one parameter $\lambda = \lambda_\gamma \cdot \lambda_\theta$ in (3.3). Furthermore, criterion (3.3) for the hierarchically penalized Cox regression using γ_k and θ_{kj} can be written in an equivalent form using the original regression coefficients β_{kj}:

Lemma 3.2. *If* $(\hat{\gamma}, \hat{\theta})$ *is a local minimizer of* (3.3), *then* $\hat{\beta}$, *where* $\hat{\beta}_{kj} = \hat{\gamma}_k \hat{\theta}_{kj}$, *is a local minimizer of*

$$\min_{\beta_{kj}} \ \ell_n(\beta) + 2\lambda^{1/2} \sum_{k=1}^{K} \left(\sum_{j=1}^{p_k} |\beta_{kj}| \right)^{1/2}. \tag{3.4}$$

On the other hand, if $\hat{\beta}$ *is a local minimizer of* (3.4), *then* $(\hat{\gamma}, \hat{\theta})$ *is a local minimizer of* (3.3), *where* $\hat{\gamma}_k = \left(\lambda \sum_{j=1}^{p_k} |\hat{\beta}_{kj}| \right)^{1/2}$ *and* $\hat{\theta}_{kj} = \hat{\beta}_{kj}/\hat{\gamma}_k$ *if* $\hat{\gamma}_k \neq 0$ *and zero otherwise.*

Proof. Suppose $(\hat{\gamma}, \hat{\theta})$ is a local minimizer of (3.3). Let $\hat{\beta}$ satisfy $\hat{\beta}_{kj} = \hat{\gamma}_k \hat{\theta}_{kj}$. It is trivial that $\hat{\gamma}_k = 0$ if and only if $\hat{\theta}_{(k)} = 0$. Hence if $\hat{\gamma}_k \neq 0$, then $|\hat{\beta}_{(k)}| \neq 0$, and we further have $\hat{\gamma}_k = \left(\lambda |\hat{\beta}_{(k)}| \right)^{1/2}$ and $\hat{\theta}_{(k)} = \hat{\beta}_{(k)}/\hat{\gamma}_k$ that can be shown in the following.

Let β be fixed at $\hat{\beta}$. Then $Q^\dagger(\lambda, \gamma, \theta)$ only depends on the penalty. For some k with $|\hat{\beta}_{(k)}| \neq 0$, the corresponding penalty term is $\gamma_k + \lambda \sum_{j=1}^{p_k} |\hat{\beta}_{kj}|/\gamma_k$, which is minimized at $\hat{\gamma}_k = \left(\lambda |\hat{\beta}_{(k)}| \right)^{1/2}$.

Let $Q(\lambda, \beta)$ be the corresponding criterion in Equation (3.4). We first show that $\hat{\beta}$ is a local minimizer of $Q(\lambda, \beta)$, i.e., there exists a $\delta' > 0$ such that if $|\Delta\beta| < \delta'$, then $Q(\lambda, \hat{\beta} + \Delta\beta) \geqslant Q(\lambda, \hat{\beta})$. Denote $\Delta\beta = \Delta\beta^{(1)} + \Delta\beta^{(2)}$, where $\Delta\beta_{(k)}^{(1)} = 0$ if $|\hat{\beta}_{(k)}| = 0$ and $\Delta\beta_{(k)}^{(2)} = 0$ if $|\hat{\beta}_{(k)}| \neq 0$. We thus have $|\Delta\beta| = |\Delta\beta^{(1)}| + |\Delta\beta^{(2)}|$.

We first show $Q(\lambda, \hat{\beta} + \Delta\beta^{(1)}) \geqslant Q(\lambda, \hat{\beta})$ for some δ'. By the argument given at the beginning of the proof, we have $\hat{\gamma}_k = \left(\lambda |\hat{\beta}_{(k)}| \right)^{1/2}$ and $\hat{\theta}_{(k)} = \hat{\beta}_{(k)}/\hat{\gamma}_k$ if $|\hat{\gamma}_k| \neq 0$, and $\hat{\theta}_{(k)} = 0$ if $|\hat{\gamma}_k| = 0$. Furthermore, let $\hat{\gamma}'_k = \left(\lambda |\hat{\beta}_{(k)} + \Delta\beta_{(k)}^{(1)}| \right)^{1/2}$ and $\hat{\theta}'_{(k)} = (\hat{\beta}_{(k)} + \Delta\beta_{(k)}^{(1)})/\hat{\gamma}'_k$ if $|\hat{\gamma}_k| \neq 0$, and let $\hat{\gamma}'_k = 0$ and $\hat{\theta}'_{(k)} = 0$ if $|\hat{\gamma}_k| = 0$. Then we have $Q^\dagger(\lambda, \hat{\gamma}', \hat{\theta}') = Q(\lambda, \hat{\beta} + \Delta\beta^{(1)})$ and $Q^\dagger(\lambda, \hat{\gamma}, \hat{\theta}) = Q(\lambda, \hat{\beta})$. Hence we only need to show $Q^\dagger(\lambda, \hat{\gamma}', \hat{\theta}') \geqslant Q^\dagger(\lambda, \hat{\gamma}, \hat{\theta})$. As $(\hat{\gamma}, \hat{\theta})$ is a local minimizer of $Q^\dagger(\lambda, \gamma, \theta)$, there exists a δ such that for any (γ', θ') satisfying $|\gamma' - \hat{\gamma}| + |\theta' - \hat{\theta}| < \delta$, we have $Q^\dagger(\lambda, \gamma', \theta') \geqslant Q^\dagger(\lambda, \hat{\gamma}, \hat{\theta})$. Straightforward calculation shows that, for $a = \min\{|\hat{\beta}_{(k)}| : |\hat{\beta}_{(k)}| \neq 0, k = 1, \ldots, K\}$, $b = \max\{|\hat{\beta}_{(k)}| : |\hat{\beta}_{(k)}| \neq 0, k = 1, \ldots, K\}$, and $\delta' < a/2$, we have

$$|\hat{\gamma}'_k - \hat{\gamma}_k| \leqslant \frac{\lambda |\Delta\beta_{(k)}^{(1)}|}{(2\lambda a)^{1/2}}, \quad |\hat{\theta}'_{(k)} - \hat{\theta}_{(k)}| \leqslant |\Delta\beta_{(k)}^{(1)}| \left\{ \frac{1}{(\lambda a/2)^{1/2}} + \frac{b}{a(\lambda a)^{1/2}} \right\}.$$

Therefore, we are able to choose a δ' satisfying $\delta' < a/2$ such that $|\hat{\gamma}' - \hat{\gamma}| + |\hat{\theta}' - \hat{\theta}| < \delta$ when $|\triangle\beta^{(1)}| < \delta'$. Hence we have $Q^\dagger(\lambda, \hat{\gamma}', \hat{\theta}') \geqslant Q^\dagger(\lambda, \hat{\gamma}, \hat{\theta})$ due to the local minimality. Hence $Q(\lambda, \hat{\beta} + \triangle\beta^{(1)}) \geqslant Q(\lambda, \hat{\beta})$.

Next we show $Q(\lambda, \hat{\beta} + \triangle\beta^{(1)} + \triangle\beta^{(2)}) \geqslant Q(\lambda, \hat{\beta} + \triangle\beta^{(1)})$. This is trivial when $\triangle\beta^{(2)} = 0$. If $\triangle\beta^{(2)} \neq 0$, then $\triangle\beta^{(1)} = 0$ and we have

$$Q(\lambda, \hat{\beta} + \triangle\beta^{(1)} + \triangle\beta^{(2)}) - Q(\lambda, \hat{\beta} + \triangle\beta^{(1)})$$

$$= \left(\triangle\beta^{(2)}\right)' n^{-1} \frac{\partial \ell_n(\beta^*)}{\partial \beta} + 2\sum_{k=1}^{K} \left(\lambda |\triangle\beta^{(2)}_{(k)}|\right)^{1/2},$$

where β^* is a vector between $\hat{\beta} + \triangle\beta^{(1)} + \triangle\beta^{(2)}$ and $\hat{\beta} + \triangle\beta^{(1)}$. Since $|\triangle\beta^{(2)}| < \delta'$, for a small enough δ', the second term on the right-hand-side of the above equality dominates the first term, hence we have $Q(\lambda, \hat{\beta} + \triangle\beta^{(1)} + \triangle\beta^{(2)}) \geqslant Q(\lambda, \hat{\beta} + \triangle\beta^{(1)})$. Thus we have shown that there exists a $\delta' > 0$ such that if $|\triangle\beta| < \delta'$, then $Q(\lambda, \hat{\beta} + \triangle\beta) \geqslant Q(\lambda, \hat{\beta})$, which implies that $\hat{\beta}$ is a local minimizer of $Q(\lambda, \beta)$.

Similarly, we can prove that if $\hat{\beta}$ is a local minimizer of $Q(\lambda, \beta)$, then $(\hat{\gamma}, \hat{\theta})$ is a local minimizer of $Q^\dagger(\lambda, \gamma, \theta)$, and where $\hat{\gamma}_k = \left(\lambda |\hat{\beta}_{(k)}|\right)^{1/2}$ and $\hat{\theta}_{(k)} = \hat{\beta}_{(k)}/\hat{\gamma}_k$ if $|\hat{\beta}_{(k)}| \neq 0$, and $\hat{\gamma}_k = 0$ and $\hat{\theta}_{(k)} = 0$ if $|\hat{\beta}_{(k)}| = 0$. $\square$

The equivalence of optimizations in (3.2), (3.3) and (3.4) is a very useful property. As we will see in the following, the numerical computation is based on (3.3) while the proof of asymptotic properties is based on (3.4). Note that the penalty in (3.4) reduces to a bridge penalty (see Knight and Fu 2000) when $p_k = 1$ for all k.

3.2 Computational algorithm

To estimate γ_k and θ_{kj} in (3.3), we can use an iterative approach, i.e., we first fix γ_k and estimate θ_{kj}, then fix θ_{kj} and estimate γ_k, and we iterate between these two steps until numerical convergence is achieved. Specifically, the algorithm proceeds as the following:

1. Center and normalize X_{kj}, and obtain an initial value $\gamma_k^{(0)}$ for each γ_k, for example, $\gamma_k^{(0)} = 1$. Let $m = 1$.

2. At the mth iteration, let $\tilde{X}_{i,kj} = \gamma_k^{(m-1)} X_{i,kj}$, $k = 1, \ldots, K$, $j = 1, \ldots, p_k$, and estimate $\theta_{kj}^{(m)}$ by

$$\theta_{kj}^{(m)} = \arg\min_{\theta_{kj}} -\frac{1}{n} \log\left(\prod_{i \in D} \frac{\exp\left(\sum_{k=1}^{K} \sum_{j=1}^{p_k} \theta_{kj} \tilde{X}_{i,kj}\right)}{\sum_{l \in R_i} \exp\left(\sum_{k=1}^{K} \sum_{j=1}^{p_k} \theta_{kj} \tilde{X}_{l,kj}\right)}\right)$$

$$+ \lambda \sum_{k=1}^{K} \sum_{j=1}^{p_k} |\theta_{kj}|.$$

3. Let $\tilde{X}_{i,k} = \sum_{j=1}^{p_k} \theta_{kj}^{(m)} X_{i,kj}$, $k = 1, \ldots, K$, and estimate $\gamma_k^{(m)}$ by

$$\gamma_k^{(m)} = \arg\min_{\gamma_k \geq 0} -\frac{1}{n} \log \left(\prod_{i \in D} \frac{\exp\left(\sum_{k=1}^K \gamma_k \tilde{X}_{i,k}\right)}{\sum_{l \in R_i} \exp\left(\sum_{k=1}^K \gamma_k \tilde{X}_{l,k}\right)} \right) + \sum_{k=1}^K \gamma_k.$$

4. Repeat Steps 2 and 3 until $\gamma_k^{(m)}$ and $\theta_{kj}^{(m)}$ converge numerically. Let $\beta_{kj} = \gamma_k^{(m)} \theta_{kj}^{(m)}$ be the final solution.

Since at each step, the value of objective function (3.3) is non-increasing, the algorithm always converges. Step 2 is a lasso-type problem, and we can use the shooting algorithm in Section 2.1 or Section 2.3 to efficiently solve for θ_{kj}. Step 3 is a non-negative garrote algorithm, and we can use one of the algorithms in Fan and Li (2002), Yuan and Lin (2007), or in particular the shooting algorithm of Friedman et al. (2007) for the quadratic approximation of the negative log likelihood by the Taylor expansion to efficiently solve for γ_k.

3.3   Asymptotic theory: general penalty

We first extend Theorems 2.1 and 2.2 to the setting with grouped variables, then show the asymptotic property of the solution of (3.4). For the ease of notation in the proofs, instead of negative log likelihood function we have been considering so far, we consider the penalized log partial likelihood function in this subsection with a general penalty function:

$$Q_n^*(\boldsymbol{\beta}) = \ell_n^*(\boldsymbol{\beta}) - \sum_{k=1}^K p_{\lambda_n}^{(k)}(|\boldsymbol{\beta}_{(k)}|), \tag{3.5}$$

where $\ell_n^*(\boldsymbol{\beta}) = \log L_n(\boldsymbol{\beta})$ and $p_{\lambda_n}^{(k)}(|\boldsymbol{\beta}_{(k)}|) = p_{\lambda_n}^{(k)}(|\beta_{k1}|, \ldots, |\beta_{kp_k}|)$ is a general p_k-variate penalty function for parameters in the kth group satisfying the following two conditions:

(i) $p_{\lambda_n}^{(k)}(|\boldsymbol{\beta}_{(k)}|) \geq 0$ for all $\boldsymbol{\beta}_{(k)} \in R^{p_k}$, and $p_{\lambda_n}^{(k)}(0) = 0$. (3.6)

(ii) $p_{\lambda_n}^{(k)}(|\boldsymbol{\beta}_{(k)}|) \geq p_{\lambda_n}^{(k)}(|\boldsymbol{\beta}_{(k)}^*|)$ if $|\beta_{kj}| \geq |\beta_{kj}^*|$ for all $j = 1, \ldots, p_k$. (3.7)

The problem now becomes to maximize (3.5).

The penalty functions $p_{\lambda_n}^{(k)}(\cdot)$, $k = 1, \ldots, K$, in (3.5) are not necessarily the same for all groups. We also allow $p_{\lambda_n}^{(k)}(\cdot)$ to depend on the tuning parameter λ_n that varies with n. Similar to Anderson and Gill (1982), we consider a finite time interval. The regularity conditions A–D in Anderson and Gill (1982) are also assumed here.

We write the true parameter vector as $\boldsymbol{\beta}^0 = (\boldsymbol{\beta}_\mathcal{A}^{0\,\prime}, \boldsymbol{\beta}_\mathcal{B}^{0\,\prime}, \boldsymbol{\beta}_\mathcal{C}^{0\,\prime})'$, where

$$\mathcal{A} = \{(k, j) : \beta_{kj}^0 \neq 0\},$$
$$\mathcal{B} = \{(k, j) : \beta_{kj}^0 = 0, \ \boldsymbol{\beta}_{(k)}^0 \neq 0\},$$
$$\mathcal{C} = \{(k, j) : \boldsymbol{\beta}_{(k)}^0 = 0\}.$$

Here $\mathcal{A}$ contains the indices of nonzero coefficients, $\mathcal{B}$ contains the indices of zero coefficients that belong to nonzero groups, and $\mathcal{C}$ contains the indices of zero coefficients that belong to zero groups. So $\mathcal{A}$, $\mathcal{B}$ and $\mathcal{C}$ are disjoint and form a partition of the set of all indices of coefficients. We denote $\mathcal{D} = \mathcal{B} \cup \mathcal{C}$, which contains the indices of all zero coefficients.

Let $I(\boldsymbol{\beta}^0)$ be the Fisher information matrix based on the log partial likelihood function for all $\boldsymbol{\beta} = \boldsymbol{\beta}^0$, and $I_{\mathcal{A}}(\boldsymbol{\beta}_{\mathcal{A}}^0)$ the Fisher information matrix for $\boldsymbol{\beta}_{\mathcal{A}} = \boldsymbol{\beta}_{\mathcal{A}}^0$ knowing that $\boldsymbol{\beta}_{\mathcal{D}}^0 = 0$. Similarly to that in §2.2, we also define

$$a_n = \max_{(k,j)} \left\{ \frac{\partial p_{\lambda_n}^{(k)}(|\beta_{k1}^0|,\ldots,|\beta_{kp_k}^0|)}{\partial |\beta_{kj}|} : \beta_{kj}^0 \neq 0 \right\},$$

$$b_n = \max_{(k,j)} \left\{ \left| \frac{\partial^2 p_{\lambda_n}^{(k)}(|\beta_{k1}^0|,\ldots,|\beta_{kp_k}^0|)}{\partial |\beta_{kj}|^2} \right| : \beta_{kj}^0 \neq 0 \right\}.$$

The following two theorems state main asymptotic results for the penalized partial likelihood with a general penalty function, which contain Theorems 2.1 and 2.2 as special examples. Their proofs follow the same sprit of Fan and Li (2001, 2002).

Theorem 3.3. *Under regularity conditions A–D in Anderson and Gill (1982), if $a_n = O_p(n^{-1/2})$ and $b_n \to 0$, then there exists a local maximizer $\hat{\boldsymbol{\beta}}_n$ of $Q_n^*(\boldsymbol{\beta})$ in (3.5) such that $\|\hat{\boldsymbol{\beta}}_n - \boldsymbol{\beta}^0\| = O_p(n^{-1/2})$.*

Proof. Let s be the number of nonzero groups. Without loss of generality, we assume that $\boldsymbol{\beta}_{(k)}^0 \neq 0$ $(k = 1,\ldots,s)$ and $\boldsymbol{\beta}_{(k)}^0 = 0$ $(k = s+1,\ldots,K)$. Let s_k be the number of nonzero coefficients in group k $(k = 1,\ldots,s)$. Again, without loss of generality, we assume that $\beta_{kj}^0 \neq 0$ $(k = 1,\ldots,s; \; j = 1,\ldots,s_k)$ and $\beta_{kj}^0 = 0$ $(k = 1,\ldots,s; \; j = s_k+1,\ldots,p_k)$.

To prove the consistency, it is sufficient to show that for any given $\epsilon > 0$, there exists a constant C such that

$$Pr\left\{ \sup_{\|u\|=C} Q_n^*(\boldsymbol{\beta}^0 + n^{-1/2}u) < Q_n^*(\boldsymbol{\beta}^0) \right\} \geqslant 1 - \epsilon. \qquad (3.8)$$

This implies that with a probability of at least $1 - \epsilon$, there exists a local maximum in the ball $\left\{ \boldsymbol{\beta}^0 + n^{-1/2}u : \|u\| \leqslant C \right\}$. Hence, there exists a local maximizer $\hat{\boldsymbol{\beta}}_n$ such that $\|\hat{\boldsymbol{\beta}}_n - \boldsymbol{\beta}^0\| = O_p(n^{-1/2})$. Since p_{λ_n} satisfies conditions (3.6) and (3.7), we have

$$\begin{aligned}
D_n(u) &= Q_n^*(\boldsymbol{\beta}^0 + n^{-1/2}u) - Q_n^*(\boldsymbol{\beta}^0) \\
&\leqslant n^{-1}\{\ell_n^*(\boldsymbol{\beta}^0 + n^{-1/2}u) - \ell_n^*(\boldsymbol{\beta}^0)\} \\
&\quad - \sum_{k=1}^{s}\left\{ p_{\lambda_n}^{(k)}(|\beta_{k1}^0 + n^{-1/2}u_{k1}|,\ldots,|\beta_{kp_k}^0 + n^{-1/2}u_{ks_k}|, 0) \right. \\
&\quad \left. - p_{\lambda_n}^{(k)}(|\beta_{k1}^0|,\ldots,|\beta_{ks_k}^0|, 0) \right\} \\
&= A - B.
\end{aligned}$$

By Taylor expansion of the negative partial likelihood function, we have

$$
\begin{aligned}
A &= n^{-1/2}\left\{n^{-1/2}\frac{\partial \ell_n^*(\beta^0)}{\partial \beta}\right\}' n^{-1/2}u \\
&\quad - \frac{1}{2}n^{-1}u'\left\{-n^{-1}\frac{\partial^2 \ell_n^*(\beta^0)}{\partial \beta^2}\right\}u + n^{-1}o_p(n^{-1}\|u\|^2) \\
&\leqslant n^{-1}O_p(1)|u| - \frac{1}{2}n^{-1}u'\left\{I(\beta^0) + o_p(1)\right\}u + n^{-1}o_p(n^{-1}\|u\|^2) \\
&\leqslant p^{1/2}n^{-1}\|u\|O_p(1) - \frac{1}{2}n^{-1}u'I(\beta^0)u + o_p(n^{-1}\|u\|^2) \\
&= A_1 + A_2 + A_3.
\end{aligned}
$$

By Taylor expansion of the penalty function, we have

$$
\begin{aligned}
B &= \sum_{k=1}^{s}\left\{\sum_{j=1}^{s_k}\frac{\partial p_{\lambda_n}^{(k)}(|\beta_{k1}^0|,\ldots,|\beta_{kp_k}^0|)}{\partial|\beta_{kj}|}\mathrm{sgn}(\beta_{kj}^0)n^{-1/2}u_{kj}\right. \\
&\quad \left. + \frac{1}{2}\sum_{i=1}^{s_k}\sum_{j=1}^{s_k}\frac{\partial^2 p_{\lambda_n}^{(k)}(|\beta_{k1}^0|,\ldots,|\beta_{kp_k}^0|)}{\partial|\beta_{ki}|\partial|\beta_{kj}|}n^{-1}u_{ki}u_{kj}\right\} + o_p\{n^{-1}(u_{k1}^2 + \cdots + u_{ks_k}^2)\} \\
&\leqslant q_1^{1/2}n^{-1/2}a_n\|u\| + \frac{1}{2}n^{-1}b_n\|u\|^2 + o_p(n^{-1}\|u\|^2) \\
&= q_1^{1/2}\|u\|O_p(n^{-1}) + o_p(n^{-1}\|u\|^2) \\
&= B_1 + B_2,
\end{aligned}
$$

where $q_1 = \sum_{k=1}^{s}p_k$. We can see that, by choosing a sufficiently large C, A_2 dominates A_1, A_3, B_1, B_2 uniformly in $\|u\| = C$. Hence, (3.8) holds. This completes the proof. $\qquad\square$

Theorem 3.4. *Let $\hat{\beta}_n = (\hat{\beta}'_{n,\mathcal{A}}, \hat{\beta}'_{n,\mathcal{B}}, \hat{\beta}'_{n,\mathcal{C}})'$ be a root-n consistent local maximizer of $Q_n^*(\beta)$ in (3.5). Assume regularity conditions A–D in Anderson and Gill (1982) hold.*

(i) *(Sparsity) For $(k,j) \in \mathcal{D}$, i.e., $\beta_{kj}^0 = 0$, if*

$$
n^{1/2}\partial p_{\lambda_n}^{(k)}(|\hat{\beta}_{n,k1}|,\ldots,|\hat{\beta}_{n,kp_k}|)/\partial|\beta_{kj}| \to \infty
$$

as $n \to \infty$, then we have $\hat{\beta}_{n,kj} = 0$ with probability approaching to 1.

(ii) *(Asymptotic normality) For $(k,j) \in \mathcal{A}$, i.e., $\beta_{kj}^0 \neq 0$, if $b_n \to 0$ and*

$$
n^{1/2}\partial p_{\lambda_n}^{(k)}(|\beta_{k1}^0|,\ldots,|\beta_{kp_k}^0|)/\partial|\beta_{kj}| \to 0,
$$

then under (i) we have that $n^{1/2}(\hat{\beta}_{n,\mathcal{A}} - \beta_{\mathcal{A}}^0)$ converges in distribution to a zero-mean normal random variable with covariance matrix $I_{\mathcal{A}}(\beta_{\mathcal{A}}^0)^{-1}$.

Proof. First we prove the sparsity: $Pr(\hat{\beta}_{n,kj} = 0) \to 1$ as $n \to \infty$ if $\beta_{kj}^0 = 0$. Using Taylor expansion, we have

$$\frac{\partial Q_n^*(\hat{\beta}_n)}{\partial \beta_{kj}} = n^{-1} \frac{\partial \ell_n^*(\beta^0)}{\partial \beta_{kj}} + \sum_{k',j'} n^{-1} \frac{\partial^2 \ell_n^*(\beta^*)}{\partial \beta_{k'j'} \partial \beta_{kj}} (\hat{\beta}_{n,k'j'} - \beta_{k'j'}^0)$$
$$- \frac{\partial p_{\lambda_n}^{(k)}(|\hat{\beta}_{n,k1}|, \ldots, |\hat{\beta}_{n,kp_k}|)}{\partial |\beta_{kj}|} \operatorname{sgn}(\hat{\beta}_{n,kj}), \qquad (3.9)$$

where β^* lies between $\hat{\beta}_n$ and β^0. By Theorem 3.2 in Andersen and Gill (1982) and the fact that $\hat{\beta}_n$ is a root-n consistent estimator, the first two terms at the right hand side of (3.9) are both $O_p(n^{-1/2})$. Hence we have

$$\frac{\partial Q_n^*(\hat{\beta}_n)}{\partial \beta_{kj}} = n^{-1/2} \left\{ O_p(1) - n^{1/2} \frac{\partial p_{\lambda_n}^{(k)}(|\hat{\beta}_{n,k1}|, \ldots, |\hat{\beta}_{n,kp_k}|)}{\partial |\beta_{kj}|} \operatorname{sgn}(\hat{\beta}_{n,kj}) \right\}.$$

If $n^{1/2} \partial p_{\lambda_n}^{(k)}(|\hat{\beta}_{n,k1}|, \ldots, |\hat{\beta}_{n,kp_k}|)/\partial |\beta_{kj}| \to \infty$ with probability tending to 1 as $n \to \infty$, then for an arbitrary $\epsilon > 0$ and $k = s+1, \ldots, K$, when n is large we have

$$\frac{\partial Q_n^*(\hat{\beta}_n)}{\partial \beta_{kj}} < 0, \ 0 < \hat{\beta}_{n,kj} < \epsilon, \qquad \frac{\partial Q_n^*(\hat{\beta}_n)}{\partial \beta_{kj}} > 0, \ -\epsilon < \hat{\beta}_{n,kj} < 0.$$

Therefore, $Pr(\hat{\beta}_{n,kj} = 0) \to 1$ as $n \to \infty$.

Secondly, we prove the asymptotic normality. Following Theorem 3.3 and the sparsity property that we just have showed, there exists a root-n consistent estimator $\hat{\beta}_n = (\hat{\beta}_{n,\mathcal{A}}, 0)'$ that satisfies the equation

$$\frac{\partial Q_n^*(\hat{\beta}_n)}{\partial \beta_{kj}} = 0, \ (k, j) \in \mathcal{A}.$$

Hence we have

$$0 = n^{-1} \frac{\partial \ell_n^*(\hat{\beta}_{n,\mathcal{A}}, \mathbf{0})}{\partial \beta_\mathcal{A}} - \sum_{k=1}^{s} \left\{ \frac{\partial p_{\lambda_n}^{(k)}(|\hat{\beta}_{n,k1}|, \ldots, |\hat{\beta}_{n,ks_k}|, 0)}{\partial |\beta_\mathcal{A}|} \operatorname{sgn}(\hat{\beta}_{n,\mathcal{A}}) \right\}$$

$$= n^{-1} \frac{\partial \ell_n^*(\beta_\mathcal{A}^0, \mathbf{0})}{\partial \beta_\mathcal{A}} + n^{-1} \frac{\partial^2 \ell_n^*(\beta_\mathcal{A}^*, \mathbf{0})}{\partial \beta_\mathcal{A}^2} (\hat{\beta}_{n,\mathcal{A}} - \beta_\mathcal{A}^0)$$

$$- \sum_{k=1}^{s} \left\{ \frac{\partial p_{\lambda_n}^{(k)}(|\beta_{k1}^0|, \ldots, |\beta_{ks_k}^0|, 0)}{\partial |\beta_\mathcal{A}|} \operatorname{sgn}(\beta_\mathcal{A}^0) \right.$$

$$\left. + \frac{\partial^2 p_{\lambda_n}^{(k)}(|\beta_{k1}^0|, \ldots, |\beta_{ks_k}^0|, 0)}{\partial |\beta_\mathcal{A}|^2} (\hat{\beta}_{n,\mathcal{A}} - \beta_\mathcal{A}^0) \right\} + o_p(n^{-1/2}),$$

where $\beta_\mathcal{A}^*$ lies between $\hat{\beta}_{n,\mathcal{A}}$ and $\beta_\mathcal{A}^0$. If $b_n \to 0$ and

$$n^{1/2} \partial p_{\lambda_n}^{(k)}(|\beta_{k1}^0|, \ldots, |\beta_{kp_k}^0|)/\partial |\beta_{kj}| \to 0$$

as $n \to \infty$, it follows by Theorem 3.2 in Andersen and Gill (1982) and Slutsky's lemma that $n^{1/2}(\hat{\boldsymbol{\beta}}_{\mathcal{A}} - \boldsymbol{\beta}_{\mathcal{A}}^0)$ converges in distribution to a normal random variable with mean zero and variance $I_{\mathcal{A}}(\boldsymbol{\beta}_{\mathcal{A}}^0)^{-1}$. $\square$

Theorem 3.3 indicates that by choosing proper penalty functions $p_{\lambda_n}^{(k)}$ and a proper λ_n, there exists a root-n consistent penalized partial likelihood estimator. Theorem 3.4 implies that by choosing proper penalty functions $p_{\lambda_n}^{(k)}$ and a proper λ_n, the corresponding penalized partial likelihood estimator can possess the sparse property that $\hat{\beta}_{n,\mathcal{D}} = 0$ with probability tending to 1. Furthermore, the estimator for the nonzero coefficients, $\hat{\beta}_{n,\mathcal{A}}$, can have the same asymptotic distribution as they would have if the zero coefficients were known in advance. Therefore we can say that asymptotically, the penalized partial likelihood estimator can perform as well as if the true underlying model were provided in advance, i.e., it possess the oracle property of Fan and Li (2001, 2002).

3.4   Asymptotic theory: hierarchical penalty

We will show the asymptotic results for the hierarchically penalized Cox regression based on criterion (3.4). If we write $2\lambda^{1/2}$ in (3.4) as λ_n, then based on Theorems 3.3 and 3.4 we have

Corollary 3.5. *If $\lambda_n = O_p(n^{-1/2})$, then there exists a root-n consistent local minimizer $\hat{\beta}_n = (\hat{\beta}'_{n,\mathcal{A}}, \hat{\beta}'_{n,\mathcal{B}}, \hat{\beta}'_{n,\mathcal{C}})'$ for the hierarchically penalized Cox regression (3.4); if further $n^{3/4}\lambda_n \to \infty$ as $n \to \infty$, then $\hat{\beta}_{n,\mathcal{C}} = 0$ with probability tending to 1.*

Proof. Minimizer of (3.4) is the maximizer of the negative objective function in (3.4) that corresponds to the situation of maximizing (3.5). It is then straightforward to check that the corresponding conditions in Theorems 3.3 and 3.4(i) hold for the penalty function $p_{\lambda_n}^{(k)}(|\boldsymbol{\beta}_{(k)}|) = \lambda_n(|\beta_{k1}| + \cdots + |\beta_{kp_k}|)^{1/2}$. The details are omitted. $\square$

The above Corollary 3.5 implies that the hierarchically penalized Cox regression can effectively remove unimportant groups. For the above root-n consistent estimator, if $\mathcal{B}$ is not empty, however, then the sparse property may not hold, i.e., there is no guarantee that $\hat{\beta}_{n,\mathcal{B}} = 0$ with probability approaching to 1. This means that although the hierarchically penalized Cox regression can effectively remove unimportant groups, it cannot effectively remove unimportant variables within the important groups.

To tackle this limitation, we apply the adaptive idea in Section 2.3 that has also been used in Breiman (1995), Shen and Ye (2002), Wang et al. (2006), Zhao and Yu (2006), Zou (2006, 2008), among others, which is to penalize different coefficients differently. Specifically, we consider

$$\min_{\beta_{kj}} Q_n^W(\boldsymbol{\beta}) = \ell_n(\boldsymbol{\beta}) + \lambda_n \sum_{k=1}^{K} \left(\sum_{j=1}^{p_k} w_{n,kj} |\beta_{kj}| \right)^{1/2}, \qquad (3.10)$$

where $\ell_n(\boldsymbol{\beta})$ takes the form in (2.1) and $w_{n,kj}$ are pre-specified non-negative weights. The intuition is that if the effect of a variable is strong, we would like the corresponding weight to be small, hence the corresponding coefficient is lightly penalized. If the effect of a variable is not strong, we would like the corresponding weight to be large, hence the corresponding coefficient is heavily penalized. The next theorem shows that, by controlling weights properly, the adaptive hierarchically penalized Cox regression (3.10) possesses the oracle property as stated in Theorem 3.4.

Theorem 3.6. *Define*

$$w_{n,max}^{\mathcal{A}} = \max\{w_{n,kj} : (k,j) \in \mathcal{A}\}, \; w_{n,min}^{\mathcal{A}} = \min\{w_{n,kj} : (k,j) \in \mathcal{A}\};$$
$$w_{n,max}^{\mathcal{D}} = \max\{w_{n,kj} : (k,j) \in \mathcal{D}\}, \; w_{n,min}^{\mathcal{D}} = \min\{w_{n,kj} : (k,j) \in \mathcal{D}\}.$$

Under regularity conditions A–D in Anderson and Gill (1982), if

$$n^{1/2}\lambda_n w_{n,max}^{\mathcal{A}} w_{n,min}^{\mathcal{A}}{}^{-1/2} \to 0,$$
$$\lambda_n w_{n,max}^{\mathcal{A}}{}^2 w_{n,min}^{\mathcal{A}}{}^{-3/2} \to 0, \quad and$$
$$n^{1/2}\lambda_n w_{n,min}^{\mathcal{D}}(w_{n,max}^{\mathcal{D}} + w_{n,max}^{\mathcal{A}})^{-1/2} \to \infty$$

as $n \to \infty$, then there exists a root-n consistent local minimizer $\hat{\boldsymbol{\beta}}_n$ of $Q_n^W(\boldsymbol{\beta})$ such that $\hat{\boldsymbol{\beta}}_{n,\mathcal{D}} = 0$ with probability tending to 1 and $n^{1/2}(\hat{\boldsymbol{\beta}}_{n,\mathcal{A}} - \boldsymbol{\beta}_{\mathcal{A}}^0) \to N\left(0, I_{\mathcal{A}}(\boldsymbol{\beta}_{\mathcal{A}}^0)^{-1}\right)$ in distribution.

Proof. We only need to check that the conditions in Theorems 3.3 and 3.4 hold for the penalty function $p_{\lambda_n}^{(k)}(|\boldsymbol{\beta}_{(k)}|) = \lambda_n(w_{n,k1}|\beta_{k1}| + \cdots + w_{n,kp_k}|\beta_{kp_k}|)^{1/2}$.

First, we prove the root-n consistency. For $\beta_{kj} \in \mathcal{A}$, i.e., $\beta_{kj}^0 \neq 0$, we have

$$\begin{aligned}
a_n &= \max_{(k,j)\in\mathcal{A}} \frac{\partial p_{\lambda_n}(|\beta_{k1}^0|, \ldots, |\beta_{kp_k}^0|)}{\partial|\beta_{kj}|} \\
&= \max_{(k,j)\in\mathcal{A}} \frac{1}{2}\lambda_n w_{n,kj}(w_{n,k1}|\beta_{k1}^0| + \cdots + w_{n,kp_k}|\beta_{kp_k}^0|)^{-1/2} \\
&\leqslant \frac{1}{2}\lambda_n w_{n,max}^{\mathcal{A}} w_{n,min}^{\mathcal{A}}{}^{-1/2}M^{-1/2},
\end{aligned}$$

$$\begin{aligned}
b_n &= \max_{(k,j)\in\mathcal{A}} \left|\frac{\partial^2 p_{\lambda_n}(|\beta_{k1}^0|, \ldots, |\beta_{kp_k}^0|)}{\partial|\beta_{kj}|^2}\right| \\
&= \frac{1}{4}\lambda_n w_{n,kj}^2(w_{n,k1}|\beta_{k1}^0| + \cdots + w_{n,kp_k}|\beta_{kp_k}^0|)^{-3/2} \\
&\leqslant \frac{1}{4}\lambda_n w_{n,max}^{\mathcal{A}}{}^2 w_{n,min}^{\mathcal{A}}{}^{-3/2}M^{-3/2},
\end{aligned}$$

where $M = \min_k(|\beta_{k1}^0| + \cdots + |\beta_{ks_k}^0|)$ is a finite constant. Then the consistency follows from Theorem 3.3.

Secondly, we prove the sparsity. Assume $\hat{\beta}_{n,kj}$ is any root-n consistent local minimizer of $Q_n^W(\boldsymbol{\beta})$, then we can find a constant M^*, such that $|\hat{\beta}_{n,kj}| \leqslant M^*$ for all (k,j) with probability tending to 1. Then for $(k,j) \in \mathcal{D}$, i.e., $\beta_{kj}^0 = 0$, we have

$$n^{1/2}\frac{\partial p_{\lambda_n}(|\hat{\beta}_{n,k1}|, \ldots, |\hat{\beta}_{n,kp_k}|)}{\partial |\beta_{kj}|} = \frac{n^{1/2}\lambda_n w_{n,kj}}{2(w_{n,k1}|\hat{\beta}_{k1}| + \cdots + w_{n,kp_k}|\hat{\beta}_{kp_k}|)^{1/2}}$$

$$\geqslant \frac{n^{1/2}\lambda_n w_{n,min}^{\mathcal{D}}}{2M^{*1/2}(w_{n,max}^{\mathcal{A}} + w_{n,max}^{\mathcal{D}})^{1/2}}.$$

Therefore, when $n^{1/2}\lambda_n w_{n,min}^{\mathcal{D}}/(w_{n,max}^{\mathcal{A}} + w_{n,max}^{\mathcal{D}})^{1/2} \to \infty$, by Theorem 3.4(i) we have $Pr(\hat{\boldsymbol{\beta}}_{n,\mathcal{D}} = 0) \to 1$.

Finally, the asymptotic normality follows from Theorem 3.4(ii) in a manner similar to the proof of consistency. $\square$

Now the remaining question is how we specify λ_n and the weights $w_{n,kj}$ so that the conditions in Theorem 3.6 are satisfied. This is answered by the following corollary.

Corollary 3.7. *Let $\tilde{\boldsymbol{\beta}}_n$ be a n^α-consistent estimator, i.e., $\|\tilde{\boldsymbol{\beta}}_n - \boldsymbol{\beta}^0\| = O_p(n^{-\alpha})$ with $0 < \alpha \leqslant 1/2$. If we choose $\lambda_n = n^{-1/2}/\log(n)$ and $w_{n,kj} = 1/|\tilde{\beta}_{n,kj}|^r$, where $r > 0$, then there exists a root-n consistent local minimizer $\hat{\boldsymbol{\beta}}_n$ of $Q_n^W(\boldsymbol{\beta})$ such that $\hat{\boldsymbol{\beta}}_{n,\mathcal{D}} = 0$ with probability tending to 1 and $n^{1/2}(\hat{\boldsymbol{\beta}}_{n,\mathcal{A}} - \boldsymbol{\beta}_{\mathcal{A}}^0) \to N\left(0, I_{\mathcal{A}}(\boldsymbol{\beta}_{\mathcal{A}}^0)^{-1}\right)$ in distribution as $n \to \infty$.*

Proof. We only need to verify that $w_{n,kj} = |\tilde{\beta}_{n,kj}|^r$ satisfy the conditions in Theorem 3.6. Let $A = \max\{\beta_{kj}^0\}$ and $B = \min\{\beta_{kj}^0 : \beta_{kj}^0 \neq 0\}$. Then by the consistency of $\tilde{\boldsymbol{\beta}}_n$, it is easy to show that $w_{n,max}^{\mathcal{A}} \to B^{-r}$ and $w_{n,min}^{\mathcal{A}} \to A^{-r}$. Thus, when taking $\lambda_n = n^{-1/2}/\log(n)$, we have $n^{1/2}\lambda_n w_{n,max}^{\mathcal{A}}/w_{n,min}^{\mathcal{A}}{}^{1/2} \to 0$ and $\lambda_n w_{n,max}^{\mathcal{A}}{}^2/w_{n,min}^{\mathcal{A}}{}^{3/2} \to 0$.

For each (k,j) with $\beta_{kj}^0 = 0$, we have $\tilde{\beta}_{n,kj} = O_p(n^{-\alpha})$ with $0 < \alpha \leqslant 1/2$. Therefore, $w_{n,min}^{\mathcal{D}}/(w_{n,max}^{\mathcal{D}} + w_{n,max}^{\mathcal{A}})^{1/2} = O_p(n^{\alpha/2})$. When taking $\lambda_n = n^{-1/2}/\log(n)$, we have $n^{1/2}\lambda_n w_{n,min}^{\mathcal{D}}/(w_{n,max}^{\mathcal{D}} + w_{n,max}^{\mathcal{A}})^{1/2} \to \infty$. $\square$

3.5 Overlapped variables

The group structure we have considered in previous subsections does not have overlaps, i.e., each variable belongs to only one group. In many practical problems, however, a variable can belong to multiple groups. For example, one gene can be shared by many different pathways. In this section, we extend the proposed method for problems with overlaps.

With slightly different notation, we reparameterize each β_j as

$$\beta_j = \left(\sum_{k \in G_j} \gamma_k\right) \cdot \theta_j, \quad 1 \leqslant j \leqslant p, \tag{3.11}$$

where $\gamma_k \geqslant 0, 1 \leqslant k \leqslant K$, and G_j is the index set of groups to which variable X_j belongs. This is a natural generalization of the decomposition (3.1) for the non-overlap case. We still treat each β_j hierarchically. Parameter γ_k at the first level controls the contribution of β_j to group k, while θ_j at the second level of the hierarchy reflects the specific-effect of variable X_j. One can see that when each variable belongs to only one group, the decomposition (3.11) is the same as the decomposition (3.1).

Under this decomposition, the corresponding partial likelihood function can be written as

$$L_n^{OL}(\boldsymbol{\gamma}, \boldsymbol{\theta}) = \prod_{i \in D} \frac{\exp\left\{\sum_{j=1}^p \left(\sum_{k \in G_j} \gamma_k\right)\theta_j X_{ij}\right\}}{\sum_{l \in R_i} \exp\left\{\sum_{j=1}^p \left(\sum_{k \in G_j} \gamma_k\right)\theta_j X_{lj}\right\}}.$$

Let $\ell_n^{OL}(\boldsymbol{\gamma}, \boldsymbol{\theta}) = -(1/n)\log\{L_n^{OL}(\boldsymbol{\gamma}, \boldsymbol{\theta})\}$. For variable selection, we mimic (3.3) and consider

$$\min_{\gamma_k \geqslant 0, \theta_j} \ell_n^{OL}(\boldsymbol{\gamma}, \boldsymbol{\theta}) + \sum_{k=1}^K \gamma_k + \lambda \sum_{j=1}^p |\theta_j|. \tag{3.12}$$

The iterative algorithm given in §3.2 can then be used to estimate γ_k and θ_j in (3.12).

3.6 Real data example

Simulations in Wang et al. (2009) showed that the (adaptive) hierarchically penalized Cox regression has better finite sample performs in both variable selection and prediction than Lasso, adaptive Lasso and SCAD. Here we review the real data analysis example in Wang et al. (2009).

The data set was presented in Miller et al. (2005), where the gene expression levels were profiled on 251 frozen primary breast cancer tissues resected in Uppsala County, Sweden, from January 1, 1987 to December 31, 1989 using Affymetrix Chip HG-133A (GEO Accession No. GSE3494). Among these patients, 236 had follow-up information in terms of time and event of disease-specific survival. In our analysis, we use the same pathways as in Wei and Li (2007) and Luan and Li (2007), which were obtained by merging the Affymetrix data with the cancer-related pathways provided by SuperArray (http://superarray.com/). There are 245 genes in 33 cancer-related sub-pathways. We list all the pathways in Table 3.1 for ease of reference. Note that some genes belong to multiple pathways. Our goal is to identify the pathways that are related to survival time in breast cancer patients.

The final model is selected based on the generalized cross validation criterion (Fan and Li 2002, Zhang and Lu 2007), and pathways are identified. The three pathways are: regulation of cell cycle (# 12 in Table 3.1) in which 48 out of 75 genes are selected, cell growth and maintenance (# 14 in Table 3.1) in which 40 out of 62 genes are selected, and small GTPase-medicated signal transduction (# 32 in Table 3.1) in which 6 out of 11 genes are selected. The first two pathways were also identified by Wei and Li (2007), Luan and Li (2008) and Miller et al.

(2005). In total, 82 genes are selected, among which 12 genes are shared by two pathways.

Table 3.1   **Thirty-three pathways and their descriptions considered in the breast cancer data analysis. "# of Genes" indicates the number of genes in the corresponding pathway**

Pathway	# of Genes	Description
1	17	Anti-apoptosis
2	3	VHLCaspase activation
3	3	DNA damage response
4	22	Factors involved in other aspects of apoptosis
5	6	Induction of apoptosis
6	9	Induction of apoptosis by signals
7	5	Regulation of apoptosis
8	2	Apoptosis others
9	13	Cell cycle arrest
10	4	Cell cycle checkpoint
11	29	Factors involved in other aspects of cell cycle
12	75	Regulation of cell cycle
13	6	Cell differentiation/cell fate determinations
14	62	Cell growth and/or maintenance
15	38	Cell proliferation
16	11	Growth factors
17	45	Regulation of cell proliferation/differentiation
18	10	Cell migration and motility
19	2	Cell-cell adhesion
20	5	Cell-matrix adhesion
21	10	MMPs/MMP inhibitors
22	13	Cell surface receptor-linked signal transduction
23	9	Frizzled and frizzled-2 signaling pathways
24	17	G-protein-coupled receptor signaling pathway
25	2	Insulin receptor signaling pathway
26	4	Integrin-mediated signaling pathway
27	29	Intracellular signaling cascade
28	6	JAK-STAT cascade
29	2	Notch signaling pathway
30	3	RAS protein signal transduction
31	4	Rho protein signal transduction
32	11	Small GTPase-mediated signal transduction
33	16	Wnt receptor signaling pathway

We then use the model with the selected three pathways to predict the survival for subjects in an independent breast cancer data set that was reported by Sotiriou et al. (2006) (GEO Accession No. GSE2990). The data set contains gene expressions obtained by the same microarray platform for 94 patients from the John Radcliffe Hospital (Oxford, UK). First, we estimate the cumulative baseline hazard function by the Breslow estimator. Then we compute the risk score $\mathbf{X}'\hat{\beta}_n$ that yields a 50% survival probability at five years for subjects in Miller's data set, which is chosen to be the threshold for the high and low risk groups. Finally, we compute the risk scores for subjects in Sotiriou's data set using $\hat{\beta}_n$ obtained from Miller's data set, and assign each subject in Sotiriou's data set into the high or low risk group based on the comparison to the threshold. Among the total of 94 subjects, 26 are in the high risk group, and 68 are in the low risk group. Kaplan-Meier curves of these two groups are plotted in Figure 3.1. We can see that the two curves are well separated. The p-value of the log rank test is less than 0.001.

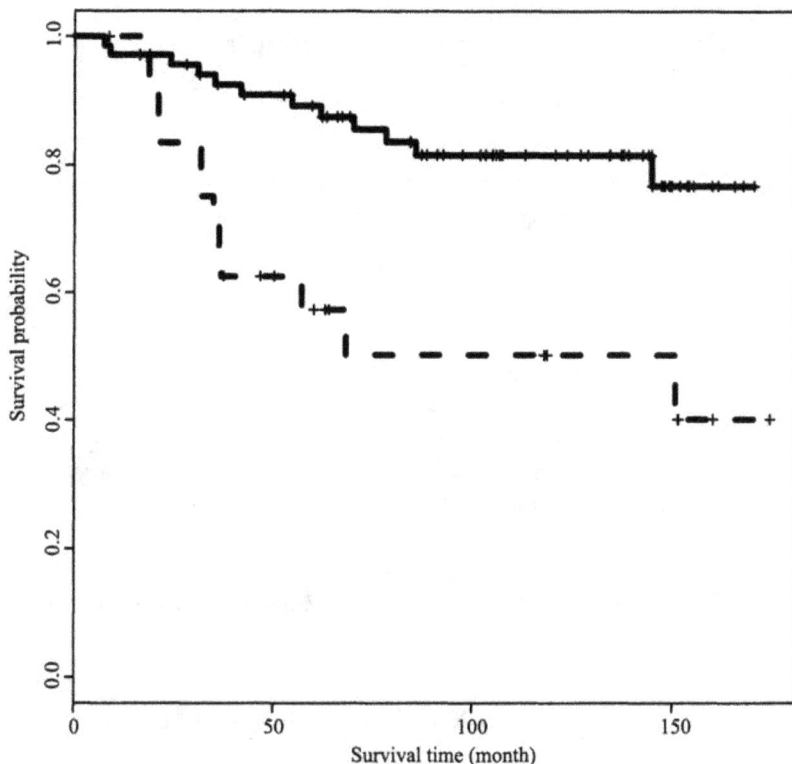

Figure 3.1   Kaplan-Meier curves for the breast cancer survival of the high and low risk groups in Sotiriou's data set. Solid line represents the low risk group and dashed line represents the high risk group. The p-value of the log rank test is less than 0.001.

4   Regularized methods for the accelerated failure time model

Though most of the work on variable selection for censored survival outcome focuses on the Cox model, there has been some attempts to use the accelerated failure time (AFT) model, an important alternative method for modeling survival data.

The AFT model (1.3) postulates a simple linear relationship between the transformed survival time and a set of covariates. Commonly used estimating methods include the Buckley-James method and the rank based method. Recently, motivated by cancer microarray studies, Wang et al. (2008) applied the elastic net penalty of Zou and Hastie (2005) to the Buckley-James estimation, which is able to select (or remove) highly correlated covariates, and Cai et al. (2009) applied the Lasso penalty to the Gehan weighted estimating method based on ranks. We provide a review of these methods in this section.

4.1   Buckley-James method with elastic net penalty

In general, the Lasso method suffers from two drawbacks (Zou and Hastie 2005):

1. When there are several genes that share one biological pathway, it is possible that their expression levels are highly correlated. The Lasso penalty, however, can usually only select one gene. The ideal method should be able to automatically select the whole group of relevant and yet highly correlated genes while eliminating trivial ones.

2. As shown in Rosset et al. (2004), the Lasso penalty can select at most n input variables. But for microarray data, the sample size n is usually on the order of 10s or 100s, while the number of attributes p is typically on the order of 10,000s. So claiming that no more than n genes are involved in a complicated biological process seems to be unrealistic for many biomedical studies. The ideal method should be able to select an arbitrary number of genes relevant to the clinical outcome.

To overcome these drawbacks, an elastic net penalty (Zou and Hastie 2005) can be applied to the AFT model. To be specific, consider the following minimization problem

$$\min_{\beta} \frac{1}{2}\sum_{i=1}^{n}(Y_i^* - \beta'\mathbf{X}_i)^2 + \lambda_1 \sum_{j=1}^{p}|\beta_j| + \lambda_2 \sum_{j=1}^{p}\beta_j^2, \tag{4.1}$$

where Y_i^* are given in (1.4) and every covariate is centered and also scaled by its sample standard deviation in order to make the numerical implementation more stable.

This type of regularization method with double penalties was originally developed by Zou and Hastie (2005) for linear models with uncensored data. They called it the elastic net regression. By using the mixture of the L_1-norm and the L_2-norm penalties, it combines good features of the two. Similar to the regression with the L_1-norm penalty, the elastic net method simultaneously performs automatic variable selection and continuous shrinkage. The added advantages by including the L_2-norm penalty are that groups of correlated variables now can be selected together and the number of selected variables is no longer limited by n. The doubly penalized Buckley-James method extends these good features to the linear regression with censored data. Following are the major steps of the algorithm for a given pair of (λ_1, λ_2).

1. Let $\beta^{(0)}$ be the initial value of β.
2. At the m-th iteration,

(a) compute

$$Y_i^* = \delta_i Y_i + (1 - \delta_i) \left\{ \mathbf{X}_i' \boldsymbol{\beta}^{(m-1)} + \int_{Y_i - \mathbf{x}_i' \boldsymbol{\beta}^{(m-1)}}^{\infty} \frac{t d\hat{F}(t)}{1 - \hat{F}(Y_i - \mathbf{X}_i' \boldsymbol{\beta}^{(m-1)})} \right\};$$

(b) compute $\boldsymbol{\beta}^{(m)}$ by

$$\min_{\boldsymbol{\beta}} \frac{1}{2} \sum_{i=1}^{n} (Y_i^* - \mathbf{X}_i' \boldsymbol{\beta})^2 + \lambda_1 \sum_{j=1}^{p} |\beta_j| + \lambda_2 \sum_{j=1}^{p} \beta_j^2; \qquad (4.2)$$

(c) stop the iteration if $|\boldsymbol{\beta}^{(m)} - \boldsymbol{\beta}^{(k)}| < d$ for some $k \in \{0, 1, \ldots, m-1\}$, here d is a prespecified precision.

3. When convergence is claimed, rescale $\hat{\boldsymbol{\beta}}$ obtained from the last iteration to be $(1 + \lambda_2)\hat{\boldsymbol{\beta}}$, and compute $\hat{\alpha} = \bar{Y}^* - \sum_{i=1}^{p} \bar{X}_i' \hat{\beta}$.

Note that in the m-th iteration, all the remaining mass is placed at the last $Y_i - \mathbf{X}_i' \boldsymbol{\beta}^{(m-1)}$ when calculating the conditional expectation of residuals. The optimization in Step 2(b) is a standard elastic net problem and can be carried out by the method of Zou and Hastie (2005) or the following shooting algorithm (Friedman et al. 2007) by rescaling (2.4):

$$\tilde{\beta}_j \leftarrow \frac{S\left(\sum_{i=1}^{n} x_{ij} (y_i^* - \tilde{y}_i^{*(j)}), \lambda_1 \right)}{1 + \lambda_2}.$$

The stopping rule given in Step 2(c) considers possible oscillation among iterations, a common phenomenon for a discrete estimating function for which there is no clearly defined root except a region where the estimating function changes sign. Oscillation occurs when the numerical procedure reiterates among a few points in that region, so numerical convergence can be claimed and the current solution can be chosen as the final solution. Yu and Nan (2006) provided a detailed discussion on numerical convergence of the rank based estimating method that has the similar problem as the Buckley-James method.

The final rescale step in the algorithm is very important. We can see in Step 2(b) that β is doubly shrunken by both the L_1-norm and the L_2-norm penalties. This double shrinkage actually introduces unnecessary extra bias comparing to using either the L_1-norm or the L_2-norm penalty only. Following Zou and Hastie (2005), we rescale $\hat{\beta}$ by multiplying the amplifying factor $1 + \lambda_2$.

Similar to the elastic-net method for the linear regression, the doubly penalized Buckly-James model can select correlated genes, and the number of selected genes can exceed the sample size.

The elastic net method may be further improved by replacing the Lasso penalty in (4.1) by the weighted Lasso penalty:

$$\min_{\boldsymbol{\beta}} \frac{1}{2} \sum_{i=1}^{n} (Y_i^* - \boldsymbol{\beta}' \mathbf{X}_i)^2 + \lambda_1 \sum_{j=1}^{p} w_j |\beta_j| + \lambda_2 \sum_{j=1}^{p} \beta_j^2,$$

a method called the adaptive elastic net by Zou and Zhang (2009). The asymptotic oracle property of Fan and Li (2001) has also been established for the adaptive elastic net method when weights are given by

$$w_j = |\hat{\beta}_j(\text{enet})|^{-r}, \quad j = 1, \ldots, p,$$

where $\hat{\beta}(\text{enet})$ is an elastic net estimator and r is a positive number (see Zou and Zhang 2009).

4.2   Rank based method with Lasso penalty

The convex function $\ell_n(\boldsymbol{\beta})$ in (1.8) can be used as a loss function in selecting variables for the AFT model. When Lasso penalty is used, one can solve the following optimization problem

$$\min_{\beta} \frac{1}{n^2} \sum_{i=1}^{n} \sum_{j=1}^{n} \{e_j(\boldsymbol{\beta}) - e_i(\boldsymbol{\beta})\} I(e_j(\boldsymbol{\beta}) \geqslant e_i(\boldsymbol{\beta})) \Delta_i + \lambda_n \sum_{j=1}^{p} |\beta_j|,$$

which is equivalent to

$$\min_{\beta} \frac{1}{n^2} \sum_{i=1}^{n} \sum_{j=1}^{n} \gamma_{ij} \Delta_i, \quad \text{subject to} \tag{4.3}$$

$$\gamma_{ij} \geqslant 0,$$
$$\gamma_{ij} \geqslant (Y_j - \boldsymbol{\beta}'\mathbf{X}_j) - (Y_i - \boldsymbol{\beta}'\mathbf{X}_i),$$
$$\sum_{j=1}^{p} |\beta_j| \leqslant s_n.$$

Here the last constraint corresponds the Lasso penalty and s_n is a tuning parameter. This is clearly a linear programming problem.

Cai et al. (2009) also applied the adaptive Lasso penalty to the above problem. Specifically, the last constraint in (4.3) is replaced by

$$\sum_{j=1}^{p} |\beta_j|/|\tilde{\beta}_j| \leqslant s_n,$$

where $\tilde{\boldsymbol{\beta}}$ is an estimator obtained by minimizing $\ell_n(\boldsymbol{\beta})$ given in (1.8), or equivalently, by solving the linear programming problem (4.3) without imposing the last Lasso constraint. The adaptive Lasso estimation for the AFT model was shown to possesses the oracle property of Fan and Li (2001).

To overcome the computational burden of the linear programming especially when the optimal tuning parameter is selected by a data-driven procedure, Cai et al. (2009) proposed an efficient numerical algorithm that computes the exact entire Lasso regularization path. Detailed algorithm can be found in the associated Web Appendix posted on the *Biometrics* web site.

5 Tuning parameter selection and a concluding remark

Tuning parameters in the regularization problems discussed in this chapter are usually determined by one of the commonly used data-driven procedures, including V-fold cross-validation (CV, Stone 1974), generalized cross-validation (GCV, Wahba 1990), Akaike information criterion (AIC, Akaike 1974), and Bayesian information criterion (BIC, Schwarz 1978). Due to the heavy computational burden of high-dimensional problems, GCV is often preferred over CV. For censored survival data, the approximation ideas of O'Sullivan (1988) and Nan et al. (2005) for GCV can be useful. Though there are some interesting theoretical results for model selection criteria primarily for linear regression models, see e.g. Shao (1997) and Yang (2005), not much concrete theoretical investigation exits in the literature particularly for survival models.

The problem becomes more challenging when p increases with n, especially when $p > n$. The asymptotic results discussed in this chapter are derived under the situation that p is fixed. For the case that p grows with n but less than n, similar results may be obtained following Fan and Peng (2004). Though simulations in cited publications usually show that the proposed methods work reasonable well for the case that $p > n$, theoretical properties for survival models, either asymptotic or non-asymptotic, need further exploration.

References

[1] Akaike H.(1974). A new look at the statistical model identification. *IEEE Transactions on Automatic Control* **19**, 716-723.

[2] Andersen P. K. and Gill R. D.(1982). Cox's regression model for counting processes: a large sample study. *Ann. Statist.* **10**, 1100-1120.

[3] Antoniadis A. and Fan. J.(2001). Regularization of wavelet approximations (with discussions). *Journal of the American Statistical Association* **96**, 939-967.

[4] Bickel P. J., Klaassen C. A. J., Ritov Y. and Wellner J. A.(1993). *Efficient and Adaptive Estimation for Semiparametric Models.* Johns Hopkins University Press, Baltimore.

[5] Bickel P. J., Ritov Y. and Tsybakov A. (2008). Simultaneous analysis of Lasso and Dantzig selector. *Annals of Statistics*, to appear.

[6] Breiman L.(1995). Better subset regression using the non-negative garrote. *Technometrics* **37**, 373-384.

[7] Bunea F., Tsybakov A. B. and Wegkamp M. H. (2007). Sparsity oracle inequalities for the Lasso. *Electronic Journal of Statistics* **1**, 169-194.

[8] Cai T.(2001). Discussion of "Regularization of wavelet approximations" by Antoniadis and Fan. *Journal of the American Statistical Association* **96**, 960-962.

[9] Cai T., Huang J. and Tian L.(2009). Regularized estimation for the accelerated failure time model. *Biometrics* **65**, 394-404.

[10] Cox D. R.(1972). Regression models and life tables (with discussion). *J. Roy. Statist. Soc. Ser. B* **34**, 187-220.

[11] Cox D. R.(1975). Partial likelihood. *Biometrika* **62**, 269-276.

[12] Donoho D. and Johnstone I. (1995). Adapting to unknown smoothness via wavelet shrinkage. *J. Amer. Statist. Assoc.* **90**, 1200-1224.

[13] Efron B., Hastie T., Johnstone I. and Tibshirani R. (2004). Least angle regression (with discussion). *The Annals of Statistics* **32**, 407-499.

[14] Fan J. and Li R.(2001). Variable selection via nonconcave penalized likelihood and its oracle properties. *Journal of the American Statistical Association* **96**, 1348-1360.

[15] Fan J. and Li R.(2002). Variable selection for Cox's proportional hazards model and frailty model. *Annals of Statistics* **30**, 74-99.

[16] Fan J. and Peng H.(2004) . Nonconcave penalized likelihood with a diverging number of parameters. *Annals of Statistics* **32**, 928-961.

[17] Friedman J., Hastie T., Hoefling H. and Tibshirani R.(2007). Pathwise coordinate optimization. *Annals of Applied Statistics* **2**, 302-332.

[18] Fu W. J.(1998). Penalized regressions: The bridge versus the lasso. *Journal of Computational and Graphical Statistics* **7**, 397-416.

[19] Fygenson M. and Ritov Y.(1994). Monotone estimating equations for censored data. *Annals of Statistics* **22**, 732-746.

[20] Gui J. and Li H.(2005). Penalized Cox regression analysis in the high-dimensional and low-smaple size settings, with applications to microarray gene expression data. *Bioinformatics* **21**, 3001-3008.

[21] Hoerl A. E.(1962). Application of ridge analysis to regression problems. *Chemical Engineering Progress* **58**, 54-59.

[22] Hoerl A. E. and Kennard R. W.(1970a). Ridge regression: biased estimation for nonorthogonal problems. *Technometrics* **12**, 55-67.

[23] Hoerl A. E. and Kennard R. W.(1970b). Ridge regression: applications to nonorthogonal problems. *Technometrics* **12**, 69-82.

[24] Huang J., Ma S. and Zhang C.-H.(2008). Adaptive Lasso for sparse high-dimensional regression models. *Statistica Sinica* **18**, 1603-1618.

[25] Jin Z., Lin D. Y., Wei L. J., and Ying Z.(2003). Rank-based inference for the accelerated failure time model. *Biometrika* **90**, 341-353.

[26] Kalbfleisch J. D. and Prentice R. L.(2002). *The Statistical Analysis of Failure Time Data, 2nd Ed.* John Wiley & Sons, Inc., New York.

[27] Kanehisa M. and KEGG S. Goto.(2002). Kyoto encyclopedia of genes and genomes. *Nucl. Acids Res.* **28**, 27-30.

[28] Knight K. and Fu W.(2000). Asymptotics for Lasso-type estimators. *The Annals of Statistics* **28**, 1356-1378.

[29] Leng C., Lin Y. and Wahba G.(2006). A note on the Lasso and related procedures in model selection. *Statistica Sinica* **16**, 1273-1284.

[30] Little R. J. A. and Rubin D. B.(2002). *Statistical Analysis with Missing Data.* John Wiley & Sons, New Jersey.

[31] Luan Y. and Li H.(2008). Group additive regression models for genomic data analysis. *Biostatistics* **9**, 100-113.

[32] Meinshausen N. and Yu B.(2009). Lasso-type recovery of sparse representa-

tions for high-dimensional data. *Annals of Statistics* **37**, 246-270.

[33] Miller R. G.(1976). Least squares regression with censored data. *Biometrika* **63**, 449-464.

[34] Nan B., Lin X., Lisabeth L. D. and Harlow S. D.(2005). A varying-coefficient Cox model for the effect of age at a marker event on age at menopause. *Biometrics* **61**, 576-583.

[35] Osborne M. R., Presnell B. and Turlach B. A.(2000). On the Lasso and its dual. *Journal of Computational and Graphical Statistics* **9**, 319-337.

[36] O'Sullivan F.(1988). Nonparametric estimation of relative risk using splines and cross-validation. *SIAM Journal on Scientific and Statistical Computing* **9**, 531-542.

[37] Rosset S., Zhu J. and Hastie T. (2004). Boosting as a regularized path to a maximum margin classifier. *Journal of Machine Learning Research* **5**, 941-973.

[38] Schwarz G. E.(1978). Estimating the dimension of a model. *Annals of Statistics* **6** , 461-464.

[39] Shao J.(1997). An asymptotic theory for linear model selection. *Statistica Sinica* **7**, 221-264.

[40] Shen X. and Ye J. (2002). Adaptive model selection. *J. Am. Statist. Assoc.* **97**, 210-221.

[41] Sotiriou C., Wirapati P., Loi S., Harris A., Fox S., Smeds J., Nordgren H., Farmer P., Praz V., Haibe-Kains B., Desmedt C., Larsimont D., Cardoso F., Peterse H., Nuyten D., Buyse M., Van de Vijver M. J., Bergh J., Piccart M. and Delorenzi M. (2006). Gene expression profiling in breast cancer: understanding the molecular basis of histologic grade to improve prognosis. *J. Nat. Cancer Inst.* **98**, 262-272.

[42] Stone M.(1974). Cross-validation choice and assessment of statistical predictions (with discussion). *Journal of the Royal Statistical Society, Series B* **36**, 111-147.

[43] The Gene Ontology Consortium. Gene ontology: tool for the unification of biology. Nat. Genet. 25 (2000), 25-29.

[44] Tibshirani R.(1996). Regression shrinkage and selection via the lasso. *Journal of the Royal Statistical Society, Series B* **58**, 267-288.

[45] Tibshirani R.(1997). The Lasso method for variable selection in the Cox model. *Statistics in Medicine* **16**, 385-395.

[46] Wahba G.(1990). *Spline Models for Observational Data.* CBMS-NSF Regional Conferences Series in Applied Mathematics, SIAM.

[47] Wang H. and Leng C.(2008). A note on adaptive group lasso. *Computational Statistics and Data Analysis* **52**, 5277-5286.

[48] Wang H., Li G. and Tsai C.-L.(2007). Regression coefficient and autoregressive order shrinkage and selection via the lasso. *J. R. Statist. Soc. B* **69**, 63-78.

[49] Wei Z. and Li H.(2007). Nonparametric pathway-based regression models for analysis of genomic data. *Biostatistics* **8**, 265-284.

[50] Wang S., Nan B., Zhu J. and Beer D. G.(2008). Doubly penalized Buckley-James method for survival data with high-dimensional covariates. *Biometrics* **64**, 132-140.

[51] Wang S., Nan B., Zhou N. and Zhu J.(2009). Hierarchically penalized Cox regression for censored data with grouped variables and its oracle property. *Biometrika*, to appear.

[52] Yang Y.(2005). Can the strengths of AIC and BIC be shared? A conflict between model indentification and regression estimation. *Biometrika* **92**, 937-950.

[53] Yu M. and Nan B.(2006). A hybrid Newton-type method for censored survival data using double weights in linear models. *Lifetime Data Analysis* **12**, 345-364.

[54] Yuan M. and Lin Y.(2006). Model selection and estimation in regression with grouped variables. *Journal of the Royal Statistical Society, Series B* **68**, 49-67.

[55] Yuan M. and Lin Y.(2007). On the non-negative garrotte estimator. *Journal of the Royal Statistical Society, Series B* **69**, 143-161.

[56] Zhang H. H. and Lu W.(2005). Adaptive Lasso for Cox's proportional hazards model. *Biometrika* **94**, 691-703.

[57] Zhao P., Rocha G. and Yu B. (2009). The composite absolute penalties family for grouped and hierarchical variable selection. *Annals of Statistics*, to appear.

[58] Zhao P. and Yu B. (2006). On model selection consistency of Lasso. *Journal of Machine Learning Research* **7**, 2541-2563.

[59] Zou H.(2006). The adaptive lasso and its oracle properties. *Journal of the American Statistical Association* **101** 1418-1429.

[60] Zou H. and Hastie T.(2005). Regularization and variable selection via the elastic net. *Journal of the Royal Statistical Society, Series B* **67**, 301-320.

[61] Zou H. and Zhang H. H.(2009). On The Adaptive Elastic-Net With A Diverging Number of Parameters.*Annals of Statistics*, to appear.

Part VI

Sufficient Dimension Reduction in Regression

Chapter 9

Sufficient Dimension Reduction in Regression [*]

Xiangrong Yin [†]

Abstract

Technological advances allow scientists to collect high-dimensional data sets in which the number of variables is considerably large. However, the analysis of such kinds of data sets challenges the traditional statistical methods. Dimension reduction then becomes an attractive approach in reducing variables without loss of data information. In this chapter, I give a selective review of sufficient dimension reduction in regression. I first introduce some basic sufficient dimension reduction concepts and theory in regression. The critical feature of sufficient dimension reduction in regression is that it reduces the variables without loss of regression information. Methodologically, I describe three different approaches of sufficient dimension reduction in regression: inverse regression approach, forward regression approach and joint association approach. Along this line, I also discuss sufficient variable selection in regression, mainly focusing on ideas that combine sufficient dimension reduction and variable selection procedure in regression. An appealing feature of sufficient variable selection is that it provides the interpretation of the informative variables for the data. Finally, I review some advances of sufficient dimension reduction in regression on highly correlated and large-p-small-n data. A number of topics worthy of further study are given throughout this chapter.

Keywords: Central subspace, central mean subspace, central k-th moment subspace, dimension reduction subspace, nonparametric methods, SIR, sufficient dimension reduction, variable selection.

1 Introduction

For the past 20 years or so, computing technology development profoundly changed all the fields allowing scientists to collect large data and many variables such as biotech and genetic data, financial data, environmental data, and satellite imagery and hyper-spectral imagery. This new development raises significant statistical challenges for data analysis. Due to the "curse of dimensionality" (Bellman, 1961),

---

[*]The author is partly supported by NSF grants DMS-0806120.

[†]Department of Statistics, University of Georgia, Athens, GA 30677, USA, E-mail: xryin@stat.uga.edu

257

many classical models derived from oversimplified assumptions and nonparametric methods are no longer reliable. However, dimension reduction that reduces the dimension(s) but retains (sufficient) important information then can play a critical role in high-dimensional data analysis. Indeed, sufficient dimension reduction has become a frontier and cutting edge area of research in statistics. With dimension reduction as a pre-process, often the number of reduced dimensions is small. Hence, parametric and nonparametric modeling methods then can be readily applied to the reduced data.

In this chapter, I focus on sufficient dimension reduction in regression. Throughout this chapter, I assume that Y is a univariate response and $\mathbf{X}$ is a p-dimensional vector of continuous predictors. Section 2 reviews the basic concepts of sufficient dimension reduction in regression and three broad estimation approaches for sufficient dimension reduction in regression. Section 3 describes sufficient variable selection which is mainly on the combined ideas of sufficient dimension reduction and variable selection through penalization in regression. Section 4 discusses sufficient dimension reduction for correlated data as well as data set with large-p-small-n. Section 5 concludes the chapter with additional discussions.

2   Sufficient dimension reduction in regression

In full generality, the goal of regression is to infer about the conditional distribution of Y given $\mathbf{X}$. Often time, the primary focus is to infer about the regression mean function of Y given $\mathbf{X}$, while in other cases the interest of regression may be extended to conditional variance of Y given $\mathbf{X}$ or even its higher conditional moments. I will discuss sufficient dimension reduction for these different regression purposes and their respective estimation methods with respect to their advantages and disadvantages.

2.1   Basic concepts of sufficient dimension reduction

The basic idea of *sufficient dimension reduction* (SDR; Cook, 1994, 1996, 1998a) is to replace the predictor vector by its projection on to a low-dimensional subspace without losing information on the conditional distribution of $Y|\mathbf{X}$, and without assuming any specific model for $Y|\mathbf{X}$.

In mathematical terms, a dimension reduction subspace (Li, 1991; Cook, 1994, 1996, 1998a) for regression of Y on $\mathbf{X}$ is any subspace S of $\mathbb{R}^p$ such that Y and $\mathbf{X}$ are independent conditioning on $\mathbf{P}_S\mathbf{X}$, where $\mathbf{P}_S$ stands for orthogonal projection on to S with respect to the standard inner product. That is,

$$Y \perp\!\!\!\perp \mathbf{X}|\mathbf{P}_S\mathbf{X}. \tag{2.1}$$

The intersection of all such S is called the *central subspace* if itself satisfies (2.1), and is denoted by $S_{Y|\mathbf{X}}$ (Cook, 1994, 1996, 1998a). As shown in Cook (1998a), and more recently in Yin, Li and Cook (2008), under very general conditions, the central subspace exists and is uniquely the smallest possible dimension reduction subspace.

Let $\Sigma_{\mathbf{X}} = \text{Var}(\mathbf{X})$, which is assumed to be non-singular, and let $\mathbf{Z} = \Sigma_{\mathbf{X}}^{-\frac{1}{2}}(\mathbf{X} - \text{E}(\mathbf{X}))$ be the standardized predictor. Then $S_{Y|\mathbf{X}} = \Sigma_{\mathbf{X}}^{-\frac{1}{2}} S_{Y|\mathbf{Z}}$ (Cook 1998a, proposition 6.1). If $\boldsymbol{\eta}$ is a $p \times d$ matrix whose columns span $S_{Y|\mathbf{X}}$, then the columns of the matrix $\boldsymbol{\gamma} = \Sigma_{\mathbf{X}}^{\frac{1}{2}}\boldsymbol{\eta}$ form a basis of $S_{Y|\mathbf{Z}}$, the central subspace for the regression of Y on $\mathbf{Z}$. Thus there is no loss of generality working on the $\mathbf{Z}$ scale because any basis for $S_{Y|\mathbf{Z}}$ can be back-transformed to basis for $S_{Y|\mathbf{X}}$, as long as the inverse of $\Sigma_{\mathbf{X}}$ exists.

When the goal of regression is to infer about the regression mean function, a different SDR subspace is proposed by Cook and Li (2002). That is, a sufficient dimension reduction subspace is any subspace S of $\mathbb{R}^p$ such that Y and $\text{E}(Y|\mathbf{X})$ are independent conditioning on $\mathbf{P}_S\mathbf{X}$, or

$$Y \perp\!\!\!\perp \text{E}(Y|\mathbf{X})|\mathbf{P}_S\mathbf{X}. \tag{2.2}$$

The intersection of all such S is itself satisfying (2.2), then it is called the *central mean subspace*, and is denoted by $S_{\text{E}(Y|\mathbf{X})}$. Similarly to the central subspace, one can establish the same conditions for the existence of the central mean subspace, and that $S_{\text{E}(Y|\mathbf{X})} = \Sigma_{\mathbf{X}}^{-\frac{1}{2}} S_{\text{E}(Y|\mathbf{Z})}$.

The central subspace and central mean subspace are designed to capture the entire conditional distribution of $Y|\mathbf{X}$ and the conditional mean distribution of $\text{E}(Y|\mathbf{X})$, respectively. Yin and Cook (2002) suggested the *central k-th moment DRSs*, and study their properties. The construction of a central k-th moment subspace is similar to that of the central subspace and that of the central mean subspace, but dimension reduction is aimed at reducing the regression mean function, variance function and up to the k-th moment function, leaving the rest of $Y|\mathbf{X}$ as the "nuisance parameter".

When considering conditional moments, dimension reduction hinges on finding a S so that the random vector $\mathbf{P}_S\mathbf{X}$ contains all the information about Y that is available from $\text{E}(Y|\mathbf{X}), \text{Var}(Y|\mathbf{X}), \ldots, M^{(k)}(Y|\mathbf{X})$, where $M^{(k)}(Y|\mathbf{X}) = \text{E}[(Y - \text{E}(Y|\mathbf{X}))^k|\mathbf{X}]$ for $k \geqslant 2$. For notational convenience, $M^{(1)}(Y|\mathbf{X})$ stands for $\text{E}(Y|\mathbf{X})$. That is,

$$Y \perp\!\!\!\perp \{M^{(1)}(Y|\mathbf{X}), \ldots, M^{(k)}(Y|\mathbf{X})\}|\mathbf{P}_S\mathbf{X}, \tag{2.3}$$

then S is a k-th moment sufficient dimension reduction subspace for the regression of Y on $\mathbf{X}$. The intersection of all such S is itself satisfying (2.3), then it is called the *central k-th moment subspace*, and is denoted by $S_{Y|\mathbf{X}}^{(k)}$. See Yin and Cook (2002) for more details about the relations among those subspaces. More recently, Zhu and Zhu (2009a) proposed a concept of central variance subspace that focused on dimension reduction for the conditional variance function only. Again one can establish the same existence conditions for the central k-th moment subspace, and the relation that $S_{Y|\mathbf{X}}^{(k)} = \Sigma_{\mathbf{X}}^{-\frac{1}{2}} S_{Y|\mathbf{Z}}^{(k)}$.

These different concepts of central SDR subspaces (as central subspace, central mean subspace, central k-th moment subspace and central variance subspace) suggest different scopes of the interests for SDR in regression. Throughout this chapter I assume the existences of the respective central SDR subspaces. The dimension of the respective central SDR subspace is called the structural dimension and is denoted by d without confusion.

The following illustrative example shall clarify the relations among those different SDR subspaces.

Illustrative Example: Let e_i be the p-dimensional vector whose ith element is 1 and all other elements are 0. Suppose that $\epsilon \perp\!\!\!\perp \mathbf{X}, \epsilon \sim N(0,1)$, the model is

$$Y = 1 + \sin(X_1 + X_2) + |X_1 - X_2|\epsilon.$$

Then by the definitions of the respective SDR subspaces, we have that

$\mathcal{S}_{Y|\mathbf{X}} = \mathrm{Span}(e_1, e_2) = \mathcal{S}_{Y|\mathbf{X}}^{(k)}$, for $k \geqslant 2$, where $d = 2$;

$\mathcal{S}_{\mathrm{E}(Y|\mathbf{X})} = \mathrm{Span}(e_1 + e_2) = \mathcal{S}_{Y|\mathbf{X}}^{(1)}$, where $d = 1$;

While the central variance subspace is spanned by $e_1 - e_2$, where $d = 1$.

The goal of SDR is to estimate d and a $p \times d$ basis matrix of the respective central SDR subspace, say, $\boldsymbol{\eta}$. The reduction is then from p-dimensional vector $\mathbf{X}$ to d-dimensional vector $\boldsymbol{\eta}^T\mathbf{X}$. The different central SDR subspaces defined previously in turn may suggest that different estimation methods should be used when a regression has a specific aim whether it is the regression mean function, variance function, or the entire conditional distribution. For instance, in the illustrative example, if the main focus of this regression is to infer the regression mean function, then the recovering of the direction, $e_1 + e_2$, is the main goal of SDR. Hence, estimation methods focusing on $\mathcal{S}_{\mathrm{E}(Y|\mathbf{X})}$ should be preferred because these methods may be more powerful in estimating this direction. However, if the main focus of the regression is to infer the entire conditional distribution of $Y|\mathbf{X}$, then methods that can recover the central subspace perhaps should be used, because they may estimate the central subspace more accurate.

In the next section, I will discuss three general approaches of estimation methods for the respective central SDR subspaces as well as their advantages and disadvantages.

2.2 Estimation methods

Considering the entire conditional distribution, the central subspace can be equivalently described as one of the following statements (Cook, 2007):

    (i) $\mathbf{X}|(Y, \mathbf{P}_S\mathbf{X})$ is distributed as $\mathbf{X}|\mathbf{P}_S\mathbf{X}$

    (ii) $Y|\mathbf{X}$ is distributed as $Y|\mathbf{P}_S\mathbf{X}$

    (iii) $Y \perp\!\!\!\perp \mathbf{X}|\mathbf{P}_S\mathbf{X}$

Statement (i) corresponds to inverse regression, statement (ii) corresponds to forward regression, while statement (iii) requires a joint distribution of Y and $\mathbf{X}$. Thus different estimation methods can be constructed based on these equivalent statements. In fact, SDR methods can be roughly classified into one of these three approaches: inverse regression, forward regression and joint association. Similar equivalent statements can be made on the central mean subspace, central kth moment subspace as well as central variance subspace.

2.2.1 Inverse approach

This main class of estimators for the central subspace used the regression of $\mathbf{X}|Y$, rather than $Y|\mathbf{X}$ in the traditional modeling, although the goal is to recover the sufficient dimensions in the forward regression. The class includes the most well-known method of sliced inverse regression (SIR) (Li, 1991) which used the inverse mean, $E(\mathbf{X}|Y)$. The advantage of SIR doesn't require a model assumption on $Y|\mathbf{X}$ as in traditional regression. However, it does require a condition on the distribution of $\mathbf{X}$: the linearity condition on $\mathbf{X}$. That is, $E(\mathbf{X}|P_S\mathbf{X})$ is a linear function of $\mathbf{X}$. Generally in practice the condition is satisfied by requiring elliptic-contoured distribution of $\mathbf{X}$. For instance, the normality of $\mathbf{X}$ will be sufficient to satisfy the linearity condition. Under this condition, the space spanned by $E(\mathbf{X}|Y)$ can then be used to estimate the $S_{Y|\mathbf{X}}$, because $E(\mathbf{X}|Y) \in \Sigma_\mathbf{X}S_{Y|\mathbf{X}}$. Practically with a data set, first standardize the predictor variables, and then the mean $E(\mathbf{Z}|Y)$ can be obtained by slicing on the range of Y, calculating the mean within each slice and finally forming a SIR dimension reduction matrix such as $\text{Var}(E(\mathbf{Z}|Y))$. The d eigenvectors corresponding to the largest d eigenvalues of this matrix can be used as the estimated directions, which then can be transformed to $\mathbf{X}$-scale (Li, 1991). While SIR is very popular, it does require the linearity condition. Although this condition may be achieved by transformation of the variables or re-weighted methods (Cook and Nachtsheim, 1994). And moreover, Hall and Li (1993) argued that this may not be a restrictive assumption, since it holds to a reasonable approximation as p increases. Nevertheless, it is well-known that SIR is powerful in recovering linear structures, but not to recover nonlinear structure, in particular, when the model is symmetric. For example, suppose that $\mathbf{X} \sim N(0, I_p)$, where I_p is a $p \times p$ identity matrix, and ϵ is standard normal random variable, and $\mathbf{X}$ and ϵ are independent. Hence the linearity condition is satisfied. If the model is

$$y = x_1^2 + \epsilon,$$

then the SIR dimension reduction matrix is a null matrix. Although the subspace spanned by the columns of the SIR matrix belongs to the central subspace, it fails to recover the true dimension, $\mathbf{e}_1$. To overcome this drawback, inverse methods beyond $E(\mathbf{X}|Y)$ have been developed. For instance, sliced average variance estimation (SAVE) (Cook and Weisberg, 1991) used $\text{Var}(\mathbf{X}|Y)$, which requires linearity and constant variance conditions (that is, $\text{Var}(\mathbf{X}|P_S\mathbf{X})$ is not random) on $\mathbf{X}$, but SAVE is more comprehensive than SIR. Inverse methods can be implemented easily with simple computation. However, exhaustiveness of the estimation for the central subspace may be an issues unless strong condition are imposed (Cook and Yin 2001).

One way or another, many other inverse methods trying to fix one or all these issues have been proposed. For instance, SAVE could miss directions that found by SIR practically, while on the other hand, SIR can easily detect linear structure, but nonlinearity such as symmetric could be missed by SIR. Hence, Hybrid methods of the first two inverse moments (Ye and Weiss, 2003; Zhu, Ohtaki and Li, 2007) have been suggested to take both of the advantages of the first and second inverse moments. Bura and Cook (2001a) developed a parametric inverse regression. Sliced average third moment (SAT) (Yin and Cook, 2003) used inverse third moment to capture more general structure, for which Zhu and Zhu (2006) developed kernel estimation. Contour regression (Li, Zha and Chiaromonte, 2005) and directional regression (Li and Wang, 2007) tried to improve the efficiency of the estimation as well as the exhaustiveness of the estimation for the central subspace using inverse second moment. Cook and Ni (2005) proposed a class of inverse family estimators via minimum discrepancy approach, and Cook and Ni (2006) suggested a method using more information for the estimation of inverse regression. Wang, Ni and Tsai (2008) used contour-projection to improve SIR and SAVE.

SIR and its sibling inverse methods in fact have generated intense interests since its introduction (Li, 1991). Many extensions and refinements ensued. Hsing and Carroll (1992), Zhu and Ng (1995), Zhu and Fang (1996), Gannon and Saracco (2003) studied the asymptotic properties of the SIR estimator and its variations. Schott (1994), Velilla (1998), Ferré (1998), Bura and Cook (2001b) and Cook (2004) introduced different asymptotic inference procedures to determine the dimension of the subspace estimated by SIR. Yin and Seymour (2005) obtained asymptotic test for determining the structure dimension for SIR-II. Shao, Cook and Weisberg (2007) developed general tests for SAVE. Permutation tests for determining the dimensionality were proposed by Cook and Weisberg (1991), Cook and Yin (2001) and Yin and Cook (2002). Chen and Li (1998) studied the relation between SIR and maximal correlation. Hsing (1999) used nearest-neighbor method to develop a variation of SIR. Naik and Tsai (2000) compared the performance of SIR with partial least squares in the context of a single-index model.

Gather, Hilker and Becker (2001) developed a robust version of SIR. Fung, He, Liu, and Shi (2002), and Bura (2003) further expanded the scope of SIR by replacing inverse conditional mean $E(\mathbf{X}|Y)$ with basis expansion or smoother. Zhu, Miao, and Peng (2006) studied asymptotic behavior for SIR when the number of covariates increases with increasing sample size. Zhu and Zhu (2007) proposed kernel estimation for SAVE while Yu, Zhu and Zhu (2009) proposed spline method to estimate SAVE. Li and Zhu (2007) studied the asymptotics for SAVE which posed a different behavior than that of SIR. Li and Dong (2008) extended SDR methods for non-elliptic distributed predictors.

Recently Cook (2007) nicely linked principal component analysis (PCA) and inverse regression approach, and proposed a principal fitted component method which is further developed by Cook and Forzani (2008). More recently, Cook and Forzani (2009) proposed likelihood approach SDR which is better than methods such as SIR, SAVE, etc. under the conditional normality of $\mathbf{X}|Y$, and it seems quite robust against the conditional normal assumption. This type of inverse regression is model-based approach, which brings new ideas and differs from the classical inverse regression aforementioned previously.

2.2.2 Forward approach

This class of estimators used traditional view of forward regression of $Y|\mathbf{X}$. For instance, even if the model via the ordinary least squares (OLS) is not true, OLS estimator can be used as a link-free direction in the central (mean) subspace under the linearity condition (Li and Duan, 1989; Duan and Li, 1991). Nonparametric methods without requiring specific conditions such as linearity on $\mathbf{X}$ have been developed for SDR: Samarov (1993) extended the average derivatives estimator for single-index model that was developed by Härdle and Stoker (1989) to multi-index, while Hristache, Juditsky, Polzehl, and Spokoiny (2001) proposed structure adaptive approach for dimension reduction. Generally nonparametric methods of forward approach in dimension reduction may gain accuracy of the estimators, while they could lose computational power relatively to inverse regression methods that without model assumptions.

Xia, Tong, Li and Zhu (2002) proposed local polynomial method for dimension reduction in the central mean subspace. Their approach, Minimum Average Variance Estimation (MAVE) seems computationally efficient in addition to the accuracy. Moreover, without requiring any specific conditions such as linearity condition on the predictor $\mathbf{X}$, MAVE can exhaustively estimate the central mean subspace. Extending the MAVE approach to condition density, Xia (2007) developed a new method of density MAVE which can exhaustively estimate the central subspace. An improved method over density MAVE is developed by Wang and Xia (2008) who combined the slicing idea and MAVE idea to take both of the advantages. Along this line, Xia (2008) further developed a method to identify a specific linear structure in the central mean subspace, in addition to estimate the central mean subspace exhaustively. Fukumizu, Bach and Jordan (2009) developed kernel dimension reduction (KDR) which involves the use of conditional covariance operators on reproducing kernel Hilbert spaces (RKHSs) and then used these operators to measure departures from conditional independence to recover the entire central subspace.

2.2.3 Joint approach

This class of methods can be flexibly regarded as forward approach or inverse approach. For instance, OLS estimator can be reformulated as a correlation coefficient index, thus it may be regarded as a joint approach. The correlation coefficient then can be rewritten as an inverse OLS form of $\mathbf{X}|Y$, which then is an inverse approach. In fact, Xia, Tong, Li and Zhu (2002) developed an inverse MAVE. Under linearity and constant variance conditions on $\mathbf{X}$, principal Hessian directions (PHD) (Li 1992; Cook 1998b), as a joint moment method, can be used for estimating the central mean subspace. Cook and Li (2004) further developed iterative Hessian Transformation (IHT) method under weaker conditions. Yin and Cook (2004) proposed fourth moments method which can recover more complicated structures in the central mean subspace but under more restrictive conditions on $\mathbf{X}$.

Joint approaches that go beyond moments have been developed as well. Yin, Li and Cook (2008) developed a SDR method using information index that can successively and exhaustively estimate the central subspace for normal predictors $\mathbf{X}$ or conditional normal distribution of $\mathbf{X}|Y$. Their method extended the method for single-index proposed by Yin and Cook (2005) where no specific distribution on $\mathbf{X}$ is imposed. Zhu and Zeng (2006) proposed two Fourier transform methods without requiring restrictive distributions on $\mathbf{X}$, and in particular, they developed two methods for normal predictors. These two Fourier transform methods can also exhaustively estimate the central mean subspace and central subspace, respectively. Zeng (2008) further developed a testing method of estimating the dimensionality for the Fourier transform approach.

Broadly speaking, most of the SDR methods in regression can be classified into one of these three classes. However, the usefulness or advantage for the method is the method itself rather than what class the method belongs to. For instance, under normality condition of $\mathbf{X}$, $\mathbf{T}_2$ (Samarov, 1993) is equivalent to PHD (Li, 1992). However, PHD still estimates the central mean subspace under weaker conditions: linearity and constant variance of $\mathbf{X}$ (Cook, 1998b). The trade between $\mathbf{T}_2$ and PHD is that one is more general, the other is computationally much simpler but under stronger conditions of the predictors. While $\mathbf{T}_2$ can be viewed as forward approach from its formulation, PHD as a joint (correlation) method may also be naturally viewed from inverse approach. Nevertheless, both approaches will miss linear structure in the model. Furthermore, different algorithms for the same dimension reduction method can bring different aspects: asymptotics, computational cost and accuracy of the results. For example, $\mathbf{T}_1$ (Samarov, 1993) and the Outer Product of Gradients (OPG) (Xia, Tong, Li and Zhu, 2002), are in fact the same methods with different algorithms. And both the approaches can exhaustively recover the central mean subspace.

For estimating the respective central SDR subspaces, an important issue is the estimation of the structure dimensions, d. Roughly speaking, there seem also three classes of estimations: asymptotic approach, computation extensive approach and information approach. Asymptotic test includes Li (1991), Schott (1994), Velilla (1998), Ferré (1998), Bura and Cook (2001b) and Cook (2004) for SIR; Li (1992) and Cook (1998b) for PHD; Cook (2004), Yin and Seymour (2005), Shao, Cook and Weisberg (2007) for SAVE or SIR-II; Zeng (2008) for a Fourier transform method. Those asymptotic distributions of the test statistics are mostly χ^2 or weighted χ^2's, depending on their respective set ups. Computation extensive approach includes permutation test developed by Cook and Weisberg (1991), Cook and Yin (2001) and Yin and Cook (2002); Bootstrap method (Ye and Weiss 2003); and cross-validation method by Xia, Tong, Li and Zhu (2002); Information criterion includes modified BIC (Zhu, Miao and Peng 2006).

If the respective central SDR subspaces and the corresponding structure dimension (d) are accurately estimated, then the goal of SDR in regression is achieved. On the other hand, even so, the reduced d variables, $\boldsymbol{\eta}^T\mathbf{X}$, are the linear combinations of all the original variables in $\mathbf{X}$. Although d is generally much smaller than p, it may still be difficult to interpret the effects of the original variables. Selection of the important and informative variables then becomes critical, in particular when the model is sparse. That is, only a few variables are important and related to the response variable. In such a case, then the selection of these variables is critical for the interpretation of the data. I will discuss this issue in the next section.

3 Sufficient variable selection (SVS)

One way to select informative variables is to perform hypothesis tests, much like the way that is used in classical multiple regression model. Cook (2004) proposed three general hypothesis tests in the content for SDR methods, particularly using SIR as the example by reformulating the SIR dimension reduction matrix as an OLS estimator. His methods can then be used to test the significant variables in the reduced variables.

Li, Cook and Nachtsheim (2005) proposed a method of variable selection assuming no model, combining sufficient dimension reduction and variable selection via penalization. Their work has brought much attention. Ni, Cook and Tsai (2005) proposed a shrinkage SIR; Li and Nachtsheim (2006) suggested Sparse SIR; and Li (2007) proposed a general sparse method for sufficient dimension reduction matrices. Zhou and He (2008) proposed Constrained Canonical Correlation (C^3), which uses CANCOR (Fung, He, Liu and Shi, 2002) with an L^1 norm constraint. In addition, a variable filtering and re-estimation procedure was added to enhance sparsity and accuracy. Note that Fung, He, Liu and Shi (2002) showed that CAN-COR is equivalent to the SIR matrix, thus C^3 may be regarded as an alternative approach to that of Li (2007).

Generally, the basic idea in this approach is that instead of minimizing a loss function say, $L(Y, X, \eta)$, one uses $L(Y, X, \eta) + p_\lambda(\eta)$, where λ is a tuning parameter which is often selected by some information criteria, such as AIC (Akaike, 1973), BIC (Schwarz, 1978) and RIC (Shi and Tsai, 2002). There is a huge literature on this topic in regression. For instance, if $L(Y, X, \eta)$ is the OLS form and the penalty term is L^1 norm, then it is LASSO (Tibshirani, 1996). The content in SDR is that $L(Y, X, \eta)$ is an L^2 loss by reformulating various SDR methods and a penalty term of $p_\lambda(\eta)$ such as L^1 term is often used. For instance, Wang and Yin (2008) developed a sparse MAVE (SMAVE) combining MAVE with L^1 penalty, and Bondell and Li (2009) developed sparse version of inverse regression estimators (IRE) by combining IRE (Cook and Ni, 2005) and L^1 penalty. Though many other penalty terms such as SCAD (Fan and Li, 2001) may be used.

Wang and Yin (2008) argued that sufficient variable selection method resulting from the combination of SDR method and variable selection procedure is sufficient, if the SDR method can exhaustively estimate the aimed central SDR subspace. Otherwise, dimension reduction and variable selection can be achieved but informative variables still could be lost. Nevertheless, the following example shows the improvement of sufficient variable selection over SDR method.

Sufficient Variable Selection Example: Let $p=10$, $x_1, x_2, ..., x_{10}$, and ϵ are iid standard normal distributed.

$$y = x_1/\{0.5 + (x_2 + 1.5)^2\} + 0.5\epsilon. \tag{3.1}$$

Here, the central subspace is spanned by $\{\mathbf{e}_1, \mathbf{e}_2\}$.

Assuming the dimension $d = 2$ is known, for one simulated data set I apply SIR, MAVE and SAMVE to the data set, respectively. Table 3.1 below gives typical results for two different sample sizes ($n = 100, 200$). As expected, the estimated directions from SIR are not as accurate as those from MAVE. Although some of the estimated coefficients in MAVE directions are very small, all original variables are still in those directions. Only the directions found by SMAVE perfectly recover the true central subspace, shrinking the noninformative variables to zeros. Results for large simulations and more details can be found in Wang and Yin (2008). In general, SVS methods that combine the SDR methods with variable selection procedure not only can select the informative variables, but also can improve the accuracy of the estimations for the respective SDR subspaces over those corresponding SDR methods forming in the SVS procedures.

Table 3.1   Estimated directions for the central subspace

		n = 100					n = 200		
SIR		MAVE		sMAVE		SIR		MAVE		sMAVE
0.922	−0.289	0.963	0.057	0.991	−0.136	0.960	−0.081	0.976	0.101	1	0
0.185	0.766	0.023	−0.883	0.136	0.991	0.029	−0.807	−0.087	0.946	0	1
0.108	0.040	0.076	0.025	0.000	0.000	0.042	0.344	−0.001	−0.044	0	0
−0.029	0.027	−0.035	0.149	0.000	0.000	−0.016	0.062	0.074	−0.009	0	0
0.062	−0.329	0.213	0.034	0.000	0.000	−0.041	0.050	−0.018	0.146	0	0
−0.030	−0.025	−0.070	0.287	0.000	0.000	−0.012	0.010	0.073	−0.192	0	0
0.138	−0.267	−0.032	0.041	0.000	0.000	−0.010	−0.259	−0.124	−0.034	0	0
−0.178	−0.123	0.052	−0.167	0.000	0.000	−0.229	−0.352	0.008	−0.071	0	0
0.219	0.211	0.072	0.155	0.000	0.000	−0.012	0.153	−0.113	0.001	0	0
−0.015	0.294	−0.081	0.239	0.000	0.000	0.147	0.056	0.017	−0.168	0	0

Although the development for sufficient variable selection and SDR methods has greatly progressed, up to this point of my discussion, they all require one critical condition. That is, the covariance matrix of $\mathbf{X}$, $\mathbf{\Sigma_X}$, is nonsingular. This is nothing wrong in the population sense, and perhaps needed for the existence of the respective central SDR subspaces. However, practically it may be a trouble when it is estimated by a sample version, say, $\hat{\mathbf{\Sigma}}_\mathbf{X}$, for the cases of $n << p$, or highly correlated data. Because in the two latter cases, $\hat{\mathbf{\Sigma}}_\mathbf{X}$ may not be invertible or unstable. I will discuss some of the developments in this direction in the next section.

4   SDR for correlated data and large-p-small-n

One way to partly circumvent the problem of inversion of $\hat{\mathbf{\Sigma}}_\mathbf{X}$ is to apply principal component analysis on $\hat{\mathbf{\Sigma}}_\mathbf{X}$ first, and then reduce $\mathbf{X}$ to eigenvectors corresponding to positive eigenvalues only. Finally SDR methods such as SIR are applied to the reduced principal variables (Chiaromonte and Martinelli, 2002; Li and Li, 2004). However, such methods mainly focused on building a predictive model, and it is difficult to perform individual predictor selection.

Second type of approach is to adopt the ridge regression. Zhong, et al. (2005) developed a modified version of SIR to incorporate the singularity of $\hat{\Sigma}_\mathbf{X}$, which was shown by Li and Yin (2008b) that Zhong, et al. (2005) in fact used a general L^2 penalty. Li and Yin (2008a) used L^2 and L^1 penalty terms sequentially to deal with the correlated data as well as for variable selection. Although the methods developed by Zhong, et al. (2005), and Li and Yin (2008a), respectively to avoid the singularity of $\hat{\Sigma}_\mathbf{X}$, still one has to invert a $p \times p$ matrix. This is because their approaches adopt L^2 penalty, a ridge regression method. This approach may still be a trouble when p is huge. Hence, techniques for inverting a large matrix could be very useful in SDR content.

Although the above two approaches have been proved useful in practice either in prediction or variable selection, theoretically they are remain to be seen that the reduced directions are indeed contained in the respective central SDR subspaces. However, the reduced directions for the next approach using partial least squares idea in fact can be shown in the respective central SDR subspaces.

When data are highly correlated, for single-index model Naik and Tsai (2000) showed that SIR performed poorly while partial least squares (PLS) performed much better. This may be expected since SIR requires the inversion of $\hat{\Sigma}_\mathbf{X}$, and in such a case the calculation of $\hat{\Sigma}_\mathbf{X}^{-1}$ is not stable. To overcome this, Li, Cook and Tsai (2007) then combined PLS and SIR to develop a partial SIR for single-index model, which greatly improved SIR. And they showed that the direction found by partial SIR indeed is in the central subspace. Cook, Li and Chiaromonte (2007) further developed a more general approach for sufficient dimension reduction matrices without the inversion of $\hat{\Sigma}_\mathbf{X}$, by using partial least squares idea with SDR matrices. Their method is quite general and will only compute the inverse of a relative small matrix. However, their method still could be in trouble for large-p-small-n regressions in which d increases as p grows. Note that the definitions of SDR subspaces have nothing to do with the sample size n, but only related to d and p. However, practically n plays an important and perhaps critical role in developing SDR methods. More recently, Zhu and Zhu (2009b) combined SDR method and partial least squares approach to develop a method for highly correlated data and study the properties with diverging number of predictors.

5 Further discussion

This chapter reviews topics of SDR in regression only. I have thus by-passed SDR topics in some other closely related areas. Those include SDR in functional data analysis and in regression with categorical predictors, to name a few. For instance, functional SDR methods developed by Ferré and Yao, (2003, 2005), and more recently investigation by Cook, Forzani and Yao (2010), are closely related to that in Section 4, since the covariance matrix in functional data is singular. On the other hand, Carroll and Li (1995) studied the use of SDR methods when there are binary regressors. Along this line, Chiaromonte, Cook and Li (2002), and Li, Cook and Chiaromonte (2003) developed new SDR theory in regression with categorical predictors, among others.

To summarize, I have reviewed three related SDR topics in regression. The research on SDR methods in regression as those described in Section 2 seems rich already, but it's still growing. Inverse methods started with no model assumption so that the algorithms are very simple comparing with forward nonparametric approach. However, current on-going research with model-based inverse regression seems very interesting, bringing some fresh comparisons with forward approaches in generality, computational cost and accuracy. On the other hand, it seems that SVS methods that combine SDR methods and variable selection methods as discussed in section 3 have a lot to be explored. Not only different penalty terms can be used, but how those penalty methods differs, if any, from penalized modeling approach. Moreover, great efforts seem to be needed for developing SDR and SVS methods for large-p-small-n and correlated data, but this seems a very challenging area.

With on-going methodology development of SDR in regression, the applications of these methods are also picking up the pace. SIR has found wide applications in diverse fields such as computer vision (Ling, Yin and Bhandarkar, 2003,; Ling, Bhandarkar, Yin and Lu, 2005), and biological science (Chiaromonte and Martinelli, 2002; Bura and Pfeiffer, 2003; Li and Li, 2004). More recently, Li and Yin (2009) extended partial OLS (Li, Cook and Chiaromonte, 2003) to longitudinal data analysis. Li and Lu (2008) extended SDR method for missing data. While applications of SDR methods in some areas may be straightforward without much efforts, others may pose new challenges. Often time, these new challenges will stimulate and advance SDR research into a new hight.

References

[1] Akaike H.(1973). Information theory and an extension of the maximum likelihood principle. *Second International Symposium on Information Theory*, 267-281. Akademiai Kiado, Budapest.

[2] Bellman R. E.(1961). *Adaptive Control Processes*. Princeton University Press, Princeton.

[3] Bondell H. D. and L. Li.(2009). Shrinkage inverse regression estimation for model free variable selection. *Journal of the Royal Statistical Society B* **71**, 287-299.

[4] Bura E.(2003). Using linear smoothers to assess the structural dimension of regressions. *Statistica Sinica* **13**, 143-162.

[5] Bura E. and Cook R. D.(2001a). Estimating the Structural Dimension of Regressions via Parametric Inverse Regression. *Journal of the Royal Statistical Society Series B* **63**, 393-410.

[6] Bura E. and Cook R. D.(2001b). Extending sliced inverse regression: the weighted chi-squared test. *Journal of American Statistical Association* **96**, 996-1003.

[7] Bura E., and Pfeiffer R. M.(2003). Graphical methods for class prediction usi-

ng dimension reduction techniques on DNA microarray data. *Bioinformatics* **19**, 1252-1258.

[8] Carroll R. J. and Li K. C.(1995). Binary regressors in dimension reduction models: A new look at treatment comparisons. *Statistica Sinica* **5**, 367-388.

[9] Chen C. H. and Li K. C.(1998). Can SIR be as popular as multiple linear regression? *Statistica Sinica* **8**, 289-316.

[10] Chiaromonte F., Cook R. D. and Li B.(2002). Sufficient dimension reduction in regression with categorical predictors. *The Annals of Statistics* **30**, 475-497.

[11] Chiaromonte F. and Martinelli J.(2002). Dimension reduction strategies for analyzing global gene expression data with a response. *Mathematical Biosciences* **176**, 123-144.

[12] Cook R. D.(1994). On the interpretation of regression plots. *Journal of the American Statistical Association* **89**, 177-190.

[13] Cook R. D.(1996). Graphics for regressions with a binary response. *Journal of the American Statistical Association* **91**, 983-992.

[14] Cook R. D.(1998a). *Regression Graphics*. Wiley, New York.

[15] Cook R. D.(1998b). Principal Hessian directions revisited (with discussion). *Journal of the American Statistical Association* **93**, 84-100.

[16] Cook R. D.(2004). Testing predictor contributions in sufficient dimension reduction. *The Annals of Statistics* **32**, 1062-1092.

[17] Cook R. D.(2007). Fisher lecture: Dimension reduction in regression. *Statistical Science* **22**, 1-26.

[18] Cook R. D. and Forzani L.(2008). Principal fitted components for dimension reduction in regression. *Statistical Science* **23**, 485-501.

[19] Cook R. D. and Forzani L.(2009). Likelihood-based sufficient dimension reduction. *Journal of the American Statistical Association* **104**, 197-208.

[20] Cook R. D., Forzani L. and Yao A. F.(2010). Necessary and sufficient conditions for sufficiency of a method for smoothed functional inverse regression. *Statistica Sincia* **20**, 235-238.

[21] Cook R. D., Li B. and Chiaromonte F.(2007). Dimension reduction in regression without matrix inversion. *Biometrika* **94**, 569-584.

[22] Cook R. D. and Li B.(2002). Dimension reduction for the Conditional mean in regression. *The Annals of Statistics* **30**, 455-474.

[23] Cook R. D. and Li B.(2004). Determining the dimension of iterative Hessian transformation.*The Annals of Statistics* **32**, 2501-2531.

[24] Cook R. D., Li B. and Chiaromonte F.(2007). Dimension reduction without matrix inversion. *Biometrika* **94**, 569-584.

[25] Cook R. D. and Nachtsheim C. J.(1994). Re-weighting to achieve elliptically contoured covariates in regression. *Journal of the American Statistical Association* **89**, 592-600.

[26] Cook R. D.and Ni L.(2005). Sufficient dimension reduction via inverse regression: a minimum discrepancy approach. *Journal of the American Statistical Association* **100**, 410-428.

[27] Cook R. D. and Ni L.(2006). Using Intra-slice Covariances for Improved Estimation of the Central Subspace in Regression. *Biometrika* **93**, 65-74.

[28] Cook R.D.and Weisberg S.(1991). Discussion of "Sliced inverse regression

for dimension reduction," by K. C. Li. *Journal of the American Statistical Association* **86** , 328-332.

[29] Cook R. D. and Yin X.(2001). Dimension reduction and visualization in discriminant analysis (With discussion). *The Australia-New Zealand Journal of Statistics* **43**, 147-199.

[30] Duan N. and Li K-C.(1991). Slicing regression: a link free regression method. *The Annals of Statistics* **19**, 505-530.

[31] Fan J. and Li R.(2001). Variable selection via nonconcave penalized likelihood and Its Oracle Properties. *Journal of the American Statistical Association* **96**, 1348-1360.

[32] Ferré L.(1998). Determing the dimension in sliced inverse regression and related methods. *Journal of the American Statistical Association* **93**, 132-140.

[33] Ferré L. and Yao A. F.(2003). Functional sliced inverse regression analysis. *Statistics* **37**, 475-488.

[34] Ferré L. and Yao A. F.(2005). Smoothed functional inverse regression. *Statistica Sinica* **15**, 665-683.

[35] Fukumizu K., Bach F. R. and Jordan M. I.(2009). Kernel dimension reduction in regression. *The Annals of Statistics* **37**, 1871-1905.

[36] Fung W. K., He X., Liu L. and Shi P.(2002). Dimension reduction based on canonical correlation. *Statistica Sinica* **12**, 1093-1113.

[37] Gather U., Hilker T. and Becker C.(2001). A Robustified Version of Sliced Inverse Regression. Statistics in Genetics and in the Environmental Sciences, *Proceedings of the Workshop on Statistical Methodology for the Sciences: Environmetrics and Genetics held in Ascona* from May 23 to 28, eds. L. T. Fernholz, S. Morgenthaler, W. Stahel. Basel: Birkhäuser, 147-157.

[38] Gannoun A. and Saracco J.(2003). Asymptotic theory for SIR_α method. *Statistica Sinica* **13**, 297-310.

[39] Hall P. and Li K-C.(1993). On almost linearity of low dimensional projections from high dimensional data. *The Annals of Statistics* **21**, 867-889.

[40] Härdle W. and Stoker T.(1989). Investigating smooth multiple regression by the method of average derivatives. *Journal of the American Statistical Association* **84**, 986-995.

[41] Hristache M., Juditsky A., Polzehl J. and Spokoiny V.(2001). Structure adaptive approach for dimension reduction. *The Annals of Statistics* **29**, 1537-1566.

[42] Hsing T.(1999). Nearest neighbor inverse regression. *Annals of Statistics* **27**, 697-731.

[43] Hsing T. and Carroll R. J.(1992). An asymptotic theory of sliced inverse regression. *The Annals of Statistics* **20**, 1040-1061.

[44] Li B. and Dong Y.(2009). Dimension reduction for nonelliptically distributed predictors. *The Annals of Statistics* **37**, 1272-1298.

[45] Li B. and Wang S.(2007). On directional regression for dimension reduction. *Journal of American Statistical Association* **102**, 997-1008.

[46] Li B., Zha H. and Chiaromonte F. (2005). Contour regression: a general approach to dimension reduction. *Annals of Statististics* **33**, 1580-1616.

[47] Li K.-C.(1991). Sliced inverse regression for dimension reduction. *Journal of the American Statistical Association* **86**, 316-342.

[48] Li K.-C.(1992). On principal Hessian directions for data visualization and dimension reduction: Another application of Stein's lemma. *Journal of the American Statistical Association* **87**, 1025-1039.

[49] Li K.-C. and Duan N.(1989). Regression analysis under link violation. *Annals of Statistics* **17**, 1009-1052.

[50] Li L.(2007). Sparse sufficient dimension reduction. *Biometrika* **94**, 603-613.

[51] Li L., Cook R. D. and Nachtsheim C. J.(2005). Model-free variable selection. *Journal of the Royal Statistical Society Series B* **67**, 285-299.

[52] Li L., Cook R. D. and Tsai C. L.(2007). Partial inverse regression. *Biometrika* **94**, 615-625.

[53] Li L. and Li H.(2004). Dimension reduction methods for microarrays with application to censored survival data. *Bioinformatics* **20**, 3406-3412.

[54] Li L. and Lu W.(2008). Sufficient dimension reduction with missing predictors. *Journal of the American Statistical Association* **103**, 822-831.

[55] Li L. and Nachtsheim C. J.(2006). Sparse sliced inverse regression. *Technometrics* **48**, 503-510.

[56] Li L. and Yin X.(2008a). Sliced inverse regression with regularizations. *Biometrics* **64**, 124-131.

[57] Li L. and Yin X.(2008b). Rejoinder to ' A note to sliced inverse regression with regulations' by Bernard-Michel, Gardes and Giard. *Biometrics* **64**, 984-986.

[58] Li L. and Yin X.(2009). Longitudinal data analysis using sufficient dimension reduction method. *Computational Statistics and Data Analysis* **53**, 4106-4115.

[59] Li Y. and Zhu L.(2007). Asymptotics for sliced average variance estimation. *Annals of Statistics* **35**, 41-69.

[60] Ling Y., Yin X. and Bhandarkar S.(2003). Sirface vs Fisherface: recognition using class specific linear projection. *Proceeding of the IEEE international conference on Image Processing ICIP*, 885-888. Barcelona, Spain.

[61] Ling Y., Bhandarkar S., Yin X. and Lu Q.(2005). Saveface and sirface: Appearance-based recognition of faces and facial expressions. *Proceeding of the IEEE international conference on Image Processing ICIP*,11-14. Genova, Italy.

[62] Naik P. and Tsai C.-L.(2000). Partial least squares estimator for single-index models. *Journal of Royal Statistical Society* **62**, 763-771.

[63] Ni L., Cook R. D. and Tsai C.-L.(2005). A note on shrinkage sliced inverse regression. *Biometrika* **92**, 242-247.

[64] Samarov A. M.(1993). Exploring regression structure using nonparametric functional estimation. *Journal of the American Statistical Association* **88**, 836-847.

[65] Schott J. R.(1994). Determining the dimensionality in sliced inverse regression. *Journal of the American Statistical Association* **89**, 141-148.

[66] Schwarz G.(1978). Estimating the dimension of a model. *The Annals of Statistics* **6**, 461-464.

[67] Shao Y., Cook R. D. and Weisberg S.(2007). Marginal tests with sliced average variance estimation. *Biometrika* **94**, 285-296.

[68] Shi P. and Tsai C.-L.(2002). Regression model selection – a residual likelihood approach. *Journal of the Royal Statistical Society B* **64**, 237-252.

[69] Tibshirani R.(1996). Regression shrinkage and selection via the lasso. *Journal of Royal Statistical Society B* **58**, 267-288.

[70] Velilla S.(1998). Assessing the number of linear components in a general regression problem. *Journal of the American Statistical Association* **93**, 1088-1098.

[71] Wang H., Ni L. and Tsai C.-L.(2008). Improving dimension reduction via contour-projection. *Statistica Sinica* **18**, 299-311.

[72] Wang H. and Xia Y.(2008). Sliced regression for dimension reduction. *Journal of the American Statistical Association* **103**, 811-821.

[73] Wang Q. and Yin X.(2008). A nonlinear multi-dimensional variable selection method for high dimensional data: Sparse MAVE. *Computational Statistics and Data Analysis* **52**, 4512-4520.

[74] Xia Y.(2007). A constructive approach to the estimation of dimension reduction directions. *The Annals of Statistics* **35**, 2654-2690

[75] Xia Y.(2008). A multiple-index model and dimension reduction. *Journal of the American Statistical Association* **103**, 1631-1640.

[76] Xia Y., Tong H., Li W. K. and Zhu L.(2002). An adaptive estimation of dimension reduction space (with discussion). *Journal of the Royal Statistical Society Series B* **64**, 363-410.

[77] Ye Z. and Weiss R. E.(2003). Using the bootstrap to select one of a new class of dimension reduction methods. *Journal of American Statistical Association* **98**, 968-979.

[78] Yin X. and Cook R. D.(2002). Dimension reduction for the conditional k-th moment in regression. *Journal of the Royal Statistical Society B* **64**, 159-175.

[79] Yin X. and Cook R. D.(2003). Estimating central subspace via inverse third moment. *Biometrika* **90**, 113-125.

[80] Yin X. and Cook R. D.(2004). Dimension reduction via marginal fourth moment in regression. *Journal of Computational and Graphical Statistics* **13**, 554-570.

[81] Yin X. and Cook R. D.(2005). Direction estimation in single-index regressions. *Biometrika* **92**, 371-384.

[82] Yin X., Li B. and Cook R. D.(2008). Successive direction extraction for estimating the central subspace in a multiple-index regression. *Journal of Multivariate Analysis* **99**, 1733-1757.

[83] Yin X. and Seymour L.(2005). Asymptotic distributions for dimension reduction in the SIR-II method. *Statistica Sinica* **15**, 1069-1079.

[84] Yu Z., Zhu L.-P.and Zhu L. X.(2009). On splines approximation for sliced average variance estimation. *Journal of Statistical Planning and Inference* **139**, 1493-1505.

[85] Zeng P.(2008). Determining the dimension of the central subspace and central mean subspace. *Biometrika* **95**, 469-479.

[86] Zhong W., Zeng P., Ma P., Liu J. S. and Zhu Y.(2005). RSIR: regularized sliced inverse regression for motif discovery. *Bioinformatics* **21**, 4169-4175.

[87] Zhou J. and He X.(2008). Dimension reduction based on constrained canonical correlation and variable filtering. *The Annals of Statistics* **36**, 1649-1668.

[88] Zhu L.-P. and Zhu L. X. (2006). Asymptotics for kernel estimation of slici-

ng average third-moment estimation. *Acta Mathematicae Applicatae Sinica (English Series)* **22**, 103-114.

[89] Zhu L-P. and Zhu L. X.(2007). On Kernel Method for Sliced Average Variance Estimation. *Journal of Multivariate Analysis* **98**, 970-991.

[90] Zhu L-P. and Zhu L. X.(2009a). Dimension reduction for conditional variance in regressions. *Statistica Sinica* **19**, 869-883.

[91] Zhu L-P. and Zhu L. X.(2009b). On distribution-weighted partial least squares with diverging number of highly correlated predictors. *Journal of Royal Statistical Society B* **71**, 525-548.

[92] Zhu L. X. and Fang K. T.(1996). Asymptotics for kernel estimates of sliced inverse regression. *The Annals of Statistics* **24**, 1053-1068.

[93] Zhu L. X., Miao B. Q. and Peng H.(2006). On sliced inverse regression with high-dimensional covariates. *Journal of the American Statistical Association* **101**, 630-643.

[94] Zhu L. X. and Ng K. W.(1995). Asymptotics of sliced inverse regression. *Statistica Sinica* **5**, 727-736.

[95] Zhu L. , Ohtaki M. and Li Y. X.(2007). On hybrid methods of inverse regression based algorithms. *Computational Statistics and Data Analysis* **51**, 2621-2635.

[96] Zhu Y. and Zeng P.(2006). Fourier methods for estimating the central subspace and the central mean subspace in regression. *Journal of the American Statistical Association* **101**, 1638-1651.

Chapter 10

Combining Statistical Procedures

Lihua Chen *and Yuhong Yang [†]

Abstract

There are two general directions of combining statistical procedures: combining for adaptation and combining for improvement. The former tries to have a final procedure that behaves like the best among the procedures that are considered, and the latter intends to improve over all of the candidate procedures. We review tools available for achieving those goals, present their properties, and discuss related issues. It turns out that one must pay a higher price for combining for improvement, but fortunately, if one combines the candidate procedures in a multi-directional fashion, the combined procedure can be both aggressive and conservative whichever is better.

Keywords: Aggregation of estimators, combining statistical procedures, model averaging, model selection.

1 Introduction

Model selection is perhaps involved in almost all statistical applications. By a model, in this work, we mean a family (typically finite-dimensional) of probability distributions that are intended to describe the observations in a statistical experiment. Given a finite dimensional model, parametric estimators are commonly used to estimate unknown parameters associated with the family of distributions. Since usually a number of viable families can be naturally considered, a statistician needs to choose one of the candidates for his/her goals of data analysis. Alternatively, one can somehow assign probability weights on the candidate models and then average estimates of the same quantity of interest from the different models. This is often called model averaging in the literature. Model averaging, or more generally, model combining, has become increasingly important for analyzing high-dimensional data.

In this chapter, we focus on the statistical problem of estimating (or predicting) quantities of interest. By a statistical procedure, we mean an estimation/prediction method that yields estimate/prediction based on data at each sample size. The candidate statistical procedures for estimating a target quantity that

---

*James Madison University, chen3lx@jmu.edu

[†]University of Minnesota, yyang@stat.umn.edu

we consider in this work may be parametric, semi-parametric, nonparametric, or of other types. The procedures are to be combined, i.e., a final estimate/prediction is to be built based upon the individual procedures and the available data. This task is called *combining statistical procedures.* In the process of combining procedures data-splitting is often used, training the procedures with part of the data and assessing their performance with the rest to determine how to integrate the estimates into a final estimate.

In the literature, aggregation of estimates typically means one starts with a given collection of fixed estimates and construct an aggregated estimate from these based on *an additional sample.* Thus aggregating estimates is a step in combining procedures when data splitting is employed. Various approaches to integrating candidate procedures or estimates may be taken. The approaches include selection of a candidate based on a criterion, convex combinations of the estimates, linear combinations of the estimates, and non-global selection/combination of the candidates for which weights depend on the location where estimation/prediction is to be made.

Combining procedures is not limited to weighting the candidates, although weighting is a natural and popular way to proceed. When probability weights are used, they may either be constructed directly for each candidate model (as in Bayesian model averaging, where posterior probabilities of the models are obtained) or the weights may be specifically for the target quantity to be estimated. The former, we call model averaging or mixing. Combining statistical procedures does not have to follow convex weightings, and thus has a broader meaning. In particular, estimators of the target quantity from finite-dimensional candidate models and nonparametric estimators such as kernel, smoothing spline, and series expansion based estimators may well be considered at the same time. For various reasons that we will discuss, the weights might not be restricted to being convex, and even if they are, they may be constructed in a way that is closely tied to the estimation problem and thus they may not necessarily be meaningful as weights on the models.

The purpose of this chapter is to review the literature on combining statistical procedures. In the remainder of the introduction, we will briefly present the motivations for combining procedures, the main approaches, and the organization of the chapter.

1.1 Why combine procedures?

Statistical applications including regression, density estimation, forecasting, and quantile regression provide numerous reasons to consider model/procedure combination. We give an overview here and go into more details in the next section.

1. *Handling model selection uncertainty.* When facing high model selection uncertainty, mixing properly can reduce the possibly large influence of randomness in selection. The aim of mixing is to obtain an estimate more robust than those of winner-takes-all procedures.

2. *Reducing variance in estimation.* When model selection uncertainty is high, mixing can significantly reduce variability in estimation and consequently

improve estimation accuracy. Substantial evidence exists in the literature in the form of simulations and real examples showing mixing reduces instability and improves over model selection.

3. *Combining to pool information.* In the context of forecasting, multiple commercial entities or research institutes may issue forecasts of a quantity (e.g., number of new home sales). The different parties may not share the information available to them. Combining the different forecasts is a way to pool the information to some extent.

4. *Combining for improvement.* Here the goal is to improve over the individual procedures. The rationale is that if all the procedures have some "missing components", then a proper combination by weighting or other methods may yield a procedure that remedies the shortcomings of the individuals, and thus has a smaller risk than the best candidate.

The above items are related. Model averaging to deal with model selection uncertainty often reduces variance in estimation. Combining to pool information often leads to performance improvement over the candidates, but combining for improvement can also be done when all the procedures are based on the same data.

1.2   Approaches to combining procedures

The following are among the combining procedures that have been proposed:

1. Model averaging. A number of probabilistic models are considered and probability weights are assigned to each based on Bayesian reasoning or ad hoc methods. This includes BMA and information criterion based weighting methods.

2. Regression-type combining. The estimates from the candidate procedures are treated as predictors and some type of regression is run (e.g., least squares linear regression, minimizing variance of linear or convex combination based on estimated covariance matrix of the individual estimators, cross validation based weighting). Stacking is a typical regression-type combining method.

3. Ensemble methods. Multiple estimates of a quantity of interest are produced by applying the same method to modified versions of the original data set. A typical technique for modifying data is obtaining bootstrap samples of the original data, as in Bagging (Breiman, 1996a), random forest (Breiman, 2001) and boosting (Freund and Schapire, 1996). Friedman et al. (2000) have interpreted boosting as fitting an additive model. Here the covariates and the residuals from the previous fitting can be taken as the modified data.

4. Mixing on the product space (MPS). This approach is rooted in information theory, particularly universal coding. Each candidate procedure is associated with a coding strategy, which corresponds to a probability distribution on the product space of the observations. These joint probability distributions can be mixed to produce convex weights on the candidates. A salient feature of this approach is that oracle inequalities show it performs optimally in terms of matching the best candidate procedure performance with few assumptions

on the candidate procedures. This category includes adaptive estimation and prediction via mixing in various statistical settings developed by Yang (2001b, 2004).

5. Localized model combination (including localized model selection). Instead of globally declaring one candidate the winner by a criterion or weighting the candidates by global weights, regions are specified in which each of the candidate should be used or weights that depend on the predictor values are assigned to the candidates.

When sequential prediction is the goal, one very fruitful line of research focuses on *prediction of individual sequences*, which does not assume any probabilistic nature of the sequence of observations (typically assumed to be bounded) to be forecast sequentially. Interestingly, probabilistic tools are involved in analysis of various prediction strategies and the performance bounds have connections to statistical risk bounds of related methods by the statistical approach. The reader is referred to the book by Cesa-Bianchi and Lugosi (2006) for details. This chapter will mainly focus on introducing statistical approaches.

1.3 Importance of data splitting when combining procedures

On the surface, training the candidate procedures on the full sample size rather than on split data seems more efficient. While losing efficiency to some degree, data-splitting-based combining nonetheless has its advantages. First, the data-splitting-based approach typically requires fewer or far fewer distributional assumptions. Second, the data-splitting-based approach tends to be more robust. This is particularly important when we consider a situation where some of the candidate procedures have already been "fine-tuned" based on the same data. For such a situation, the performance assessment statistics of the models/procedures (such as AIC or BIC values) that are used to compute weights without data splitting are often distorted, possibly to a large extent if care is not taken. Consequently, the resulting weights may poorly reflect the actual performances of the procedures. We give a simple demonstration below.

We generate 200 observations with 50 predictors that are iid standard normal. The response Y is standard normal as well, independent of the predictors. Thus, none of the predictors should be used for estimating the true regression function.

We use Efroymson's stepwise method (Miller, 1990) to select a subset model. We also consider two alternative subset models, the null model with no predictor and the full model. We want to compare combining techniques with or without data splitting. The data-splitting-based combining method is ARM (Yang, 2003) with equal size splitting (see Section 2 for more details). We use ARM to combine the null model, the full model, and the model selected by Efroymson's method. The non-data-splitting method is to use convex weight proportional to $\exp(-AIC)$ over the three models, where AIC is the AIC value of the model.

Simulation shows the global squared L_2 risks of the two combining methods are drastically different: 0.012 (standard error 0.001) for ARM and 0.074 (standard error 0.009) for AIC weighting. A main reason for the poor performance of the non-splitting-based estimator is that the AIC value of the selected model is simply too optimistic due to the searching process (the issue can be even more serious if the stepwise selection is done directly in terms of AIC value). Data splitting, in contrast, can wash out such biases.

In conclusion, when some of the candidate procedures are obtained based on a preliminary study of the same data (which is usually the case), data-splitting-based combining is more robust and reliable. For complicated modeling, the cost of a slight reduction in estimation efficiency is usually much smaller than the potential bias of non-data-splitting methods.

1.4   Organization of the chapter

In the above-mentioned approaches, ensemble methods mix by manipulating the data and the other methods mix over intrinsically different estimates. We will focus on the latter. These combining approaches can be categorized in two main directions: One is combining for adaptation (i.e., with the goal of matching the performance of the best candidate model), whereas the other is improving over the performance of the candidate models.

The rest of the chapter is organized as follows. Sections 2 and 3 will cover some influential combining procedures in the two above-mentioned directions: combining for adaptation, combining for improvement. Some remarks and conclusions will be given in the last section.

2   Combining for adaptation

Combining for adaptation methods mainly take the form of convex combination of different procedures. Suppose a quantity Δ which is of common interest for the J procedures under consideration is to be estimated. This Δ can be a future response value, a class label, a density function or a survival function. Each procedure produces an estimate $\hat{\Delta}_j, j = 1, \ldots, J$. Through combining, Δ can be estimated as $\hat{\Delta} = \sum_{j=1}^{J} w_j \hat{\Delta}_j$ where w_j are weights assigned to each procedure.

For instance, in the regression setting, the target is to estimate the regression function. Let (X_i, Y_i), $i = 1, ..., n$ be iid observations with X_i taking values in $\mathcal{X}$, a measurable set in R^d. Let $Y_i = f(X_i) + \varepsilon_i$, where f is the regression function and ε_i is normally distributed error with mean zero and variance σ^2. In the linear regression case where subset selection is often a basic part of building a linear model, $f(X_i)$ takes the form $X'_{i(s)}\beta^s$ where $X_{i(s)}$ is a subset of X_i and β^s is the corresponding parameter vector. Different subsets of X_i form different models. When model selection uncertainty exists, $f(X_i)$ can be estimated by the weighted average of the estimates from these different models.

Different combining approaches differ on weighting schemes. In Bayesian Model Averaging (BMA), the weights are naturally taken to be the posterior probabilities of the models in the model list. Some combining approaches assign weights based on the performance of each model. Buckland et al. (1997) used a weighting scheme based on a model selection criterion, such as the AIC or BIC value of a model. Leung and Barron (2006) combined models based on the estimator's risk and obtained tight risk bounds on the combined estimator under normal errors. Yang (2001b, 2004) used a prediction based weighting scheme where the model weight depends on the predictive performance of each model. This weighting scheme has a close connection to information theory which guarantees good theoretical risk bounds on the combined estimator.

2.1 BMA

In BMA, the weights for different models are their posterior probabilities. By Bayes rule, the posterior probability for model M_j is computed as

$$p(M_j|D) = \frac{p(D|M_j)p(M_j)}{\sum_k p(D|M_k)P(M_k)}.$$

Here D stands for data, $p(M_j)$ is the prior probability of model M_j, and $p(D|M_j)$ is the integrated likelihood under model M_j.

The integrated likelihood $p(D|M_j)$ in BMA weighting can in general be hard to compute. Three main approaches exist to address the computational difficulty.

The first approach is to reduce the model list and average over a subset of the models. Madigan and Raftery (1994) use the Occam's window method to trim the model list. Compared to the model with the highest posterior probability, models with posterior probabilities under a certain threshold value are excluded. This threshold value can be determined by the data analyst. Models whose simpler models have higher posterior probabilities are also excluded. This can greatly reduce the number of models in the list.

The second approach is Markov Chain Monte Carlo Model Composition (MC^3) (Madigan and York, 1995). It uses a Markov Chain Monte Carlo method to approximate the posterior distribution of a quantity of interest. It generates a Markov chain that moves through the model space. MC^3 offers great flexibility, however, as with other $MCMC$ methods, convergence can be a problematic issue.

The final method is based on the Laplace method for approximating the integrated likelihood. Raftery (1995) developed a simplified formula for weighting models

$$p(M_j|D) = \frac{\exp(-\frac{1}{2}BIC_j)}{\sum_k \exp(-\frac{1}{2}BIC_k)},$$

where BIC_j is the BIC value for model M_j.

BMA has been widely used in different statistical settings to account for model selection uncertainty, including averaging over linear models, generalized linear models, discrete graphical models, loglinear models, and survivor models. Interested readers can refer to the BMA homepage www.research.att.com/ volinsky/bma.html for related publications, manuals and software.

2.2 Model selection criterion based combining

A common alternative approach is to weight models based on a model selection criterion. Models with better performance as measured by a model selection criterion receive higher weights.

Buckland et al. (1997) used weights

$$w_j = \frac{\exp(-\frac{1}{2}I_j)}{\sum_k \exp(-\frac{1}{2}I_k)},$$

where I_j is an information criterion, such as the AIC or BIC value of model j. When BIC is chosen, it coincides with Raftery's simplified BMA weighting formula. Note that when two models have the same dimension, the ratio between their weights is just the likelihood ratio. For this reason, Buckland et al. (1997) call the ratio between the weights of two models "the relative penalized likelihood factor".

In the linear regression setting under normal errors where the goal is to estimate the conditional mean, Leung and Barron (2006) weighted models by the estimate of the risk of the estimators

$$w_j \propto \exp[-\beta \frac{\hat{r}_j}{2\sigma^2}],$$

where $\hat{r}_j$ is the risk estimate of the estimator under model j, and σ^2 is the variance of the error assumed to be known for simplicity. When $\beta = 1$, this becomes the weighting scheme of Buckland et al. (1997).

Leung and Barron (2006) obtained the risk estimate of the combined estimator using Stein's formula, and demonstrated that the risk of the combined estimator is bounded by the risk of the best estimator plus a constant penalty term. They also showed that the sharpest bound was achieved when choosing $\beta = \frac{1}{2}$. This is equivalent to weighting models by $w_j \propto \exp(-\frac{1}{4}AIC_j)$.

In the direction of obtaining optimal model assessment, Shen and Huang (2006) developed a general framework for model combining based on the concept of generalized degree of freedom. This framework enabled the computation of the risk function of the combined estimator via a data perturbation technique and consequently enabled the computation of the weights through minimizing the risk. They also considered using the information criterion approach to weight models. Their information criterion, $l(\lambda, j)$ for model j, is the negative loglikelihood plus a more general penalty term, λK, where K is the dimension of the model. This criterion includes AIC, AICc, BIC, and RIC when assigning different values to λ. In their approach λ is adaptive in the sense that it is estimated through minimizing the loss of the estimator. This yields the proposed weights

$$w_j = \frac{\exp(l(\hat{\lambda}_j, j))}{\sum_k \exp(\hat{\lambda}_k, k)}.$$

In a similar direction, Hjort and Claeskens (2003) built a general large sample likelihood apparatus for obtaining the limiting distributions of the risk estimate of the combined estimator. Again, model weights can be obtained as the minimizer of this risk. In using the information criterion for weighting, they followed the weighting scheme of Buckland et al. (1997), and in particular promoted the use of FIC (focused information criterion) which aims at finding the best candidate model for a given focus parameter. Their proposed model weights are

$$W_j \propto \exp\left(-\frac{1}{2}\kappa\frac{FIC_j}{\omega^t K\omega}\right),$$

where κ is an algorithmic parameter and $\omega^t K\omega$ is the constant risk of the minimax estimator. A special case of this weighting formula agrees with weighting using AIC. Some empirical studies have demonstrated the advantage of using FIC over AIC for weighting.

2.3   Mixing on the product space

Yang (2001b, 2004) developed a weighting scheme via mixing on the product space where the weights are determined by the predictive performance of a model. He used a data splitting technique with one part of the data used for estimation and the second part for model evaluation. This weighting scheme has a risk comparable to that of the best candidate model.

2.3.1 Statistical settings

Yang and co-authors explored this combining technique in several statistical settings in addition to the linear regression setting mentioned above. Some results are also obtained by Catoni (2004).

1. Density estimation. Let X_1, X_2, ..., X_n be i.i.d. observations with density $f(x), x \in \mathcal{X}$ with respect to a measure μ. Here the space $\mathcal{X}$ is general and could be any dimensional. The goal is to estimate the unknown density f based on the data.

2. ANOVA modeling. Suppose there are Φ ($\Phi \geqslant 2$) factors with levels $I_1, \ldots, I_\Phi$ ($I_1, \cdots, I_\Phi \geqslant 2$) respectively. Consider a balanced factorial design with J ($J \geqslant 2$) replicates. Let $Y_{i_1 \ldots i_\Phi, j} = \mu_{i_1 \ldots i_\Phi} + \epsilon_{i_1 \ldots i_\Phi, j}$, where $Y_{i_1 \ldots i_\Phi, j}$ is the jth observation in cell $i_1 \cdots i_\Phi$, $\mu_{i_1 \ldots i_\Phi}$ is the mean response at that cell and $\epsilon_{i_1 \ldots i_\Phi, j}$ are independent Gaussian errors with mean 0 and unknown variance σ^2. ANOVA concerns how the cell means $\mu_{i_1 \ldots i_\Phi}$ depend on the factors and also the estimation of the main factor and interaction effects.

3. Times series forecasting. The goal is to forecast or predict a real-valued continuous random variable Y. Let Y_1, Y_2, ... be the true values at time 1, 2, ... respectively. At each time $i \geqslant 1$, (possibly) related qualitative and/or quantitative information, denoted jointly by X_i (which thus could be multi-dimensional) is observed prior to the occurrence of Y_i. The task is to predict Y_i based on X_i and the earlier data $Z^{i-1} = (X_l, Y_l)_{l=1}^{i-1}$.

4. Quantile regression. We observe (Y_i, X_i), $i = 1, \cdots, n$, where $X_i = (X_{i1}, \cdots, X_{ip})$ is a p-dimensional predictor. The target is to estimate $q_T(x)$, the con-

ditional quantile function of the response variable Y given the input X at a given probability level T ($T \in (0,1)$).

5. Classification or pattern recognition. The response variable Y is categorical. For simplicity, consider the two-class case with class labels $Y \in \{0,1\}$. One observes $Z_i = (X_i, Y_i)$, $i = 1, ..., n$, which are independent copies of the random pair $Z = (X, Y)$. Let $f(x) = P\{Y = 1 | X = x)$ be the function of the conditional probability of Y taking label 1 given the feature variable $X = x \in \mathcal{X}$. The goal here is to estimate f.

2.3.2 Performance assessment

To assess the performance of the estimators, Yang mainly used the Kullback-Leibler loss and the squared L_2 loss for density and probability estimation and the squared L_2 loss for other settings. The L_2 distance between functions g_1 and g_2 with respect to a measure μ is $\| g_1 - g_2 \|_2 = \left(\int |g_1 - g_2|^2 d\mu \right)^{1/2}$. The Kullback-Leibler (K-L) divergence between two densities f and g with respect to a measure μ is defined as $D(f \| g) = \int f \log (f/g) \, d\mu$.

2.3.3 Main idea

A key insight can best be illustrated in the density estimation setting where X_1, X_2, ..., X_n are i.i.d. observations with density $f(x)$. The goal is to estimate the unknown density f based on the data.

The idea is to mix on the product space of the observations (see Barron (1987), Yang and Barron (1999)). Suppose a family of densities $q_j, j = 1, \dots$, are considered. Mix the joint densities $q_j^n(x_1, ..., x_n) = \Pi_{i=1}^n q_j(x_i)$ to get a centroid, $q^n = \sum_j \pi_j q_j^n$, $\sum_j \pi_j = 1$. The insight is that in terms of the K-L divergence, the distance between the true joint density f^n and the centroid is upper bounded by the minimum distance between f^n and the members plus a penalty term, $D(f^n, q^n) \leqslant \inf_j \left(D(f^n, q_j^n) + \frac{1}{n} \log \frac{1}{\pi_j} \right)$.

Based on this insight, we can construct a mixing strategy in a sequential way. Assume data arrive one at a time. At sample size i, for each model/strategy j, obtain the predictive density for observation X_{i+1} conditional on available data $(X_1, \dots, X_i)$. Mix these predictive densities from different procedures to get the combined predictive density for X_{i+1}. The weight for each procedure is proportional to the product of the predictive densities for $X_1, \dots, X_i$. This is similar to weighting by the prequential likelihood of the data in Dawid (1984). Dawid advocated using prequential likelihood to assess model performance as it is based on "genuine observables". By the chain rule, the cumulative K-L loss between the true density f and the combined predictive density is just the K-L distance between f^n and the centroid which is the mixture of the joint predictive densities for $X_1, \dots, X_{n+1}$. It follows that, in terms of the cumulative K-L risk, up to an additive penalty term of order $\frac{1}{n}$, the combined density estimator performs as well as the best estimator, the oracle. Similar conclusions can be obtained for squared error risk bounds by translating the K-L loss to the squared error risk under some mild technical conditions. Some oracle inequalities follow next.

2.3.4 Key oracle inequalities

Density estimation: A density estimation procedure δ refers to a sequence of estimators $\{\hat{f}_{\delta,n}(x; X^n), n \geq 1\}$ based on observation(s) X^n, $n \geq 1$ respectively. Let

$$R(f; n; \delta) = ED(f \parallel \hat{f}_{\delta,n})$$

denote the K-L risk.

Let $\{\delta_j, j \geq 1\}$ be a given general collection of density estimation procedures. Procedure δ_j produces density estimators $\{\hat{f}_{j,n}(x; X^n), n \geq 1\}$ based on observation(s) X^n, $n \geq 1$ respectively.

Yang (2000b) gives the following result on combining different density estimation procedures. Let $\underline{\pi} = \{\pi_j, j \geq 1\}$ be a set of positive numbers satisfying $\sum_{j \geq 1} \pi_j = 1$. Consider a sequence $\{N_n, n \geq 1\}$ with $1 \leq N_n \leq n$. The procedure producing $\{\hat{f}_{\delta^*,n}, n \geq 1\}$ will be called a combined procedure denoted by δ^*.

Theorem 1 *For any given countable collection of estimation procedures $\{\delta_j, j \geq 1\}$ and $\underline{\pi}$, we can construct a single estimation procedure δ^* such that for any underlying density f,*

$$R(f; n; \delta^*) \leq \inf_{j \geq 1} \left(\frac{1}{N_n} \log \frac{1}{\pi_j} + \frac{1}{N_n} \sum_{i=n-N_n+1}^{n} R(f; i; \delta_j) \right). \qquad (2.1)$$

Risk bounds under L_2 loss can also be obtained.

Regression: Consider the regression model

$$Y_i = f(X_i) + \sigma \cdot \varepsilon_i, \ i = 1, \ldots n,$$

where $(X_i, Y_i)_{i=1}^n$ are i.i.d. copies from the joint distribution of (X, Y) with $Y = f(X) + \sigma \cdot \varepsilon$. The (possibly high-dimensional) explanatory variable X has an unknown distribution P_X. The variance parameter $\sigma^2 > 0$ is unknown and the random variable ε is assumed to have a known density function $h(x)$ (with respect to Lebesgue or a general measure μ) with mean 0 and variance 1. The goal is to estimate the regression function f based on the data $Z^n = (X_i, Y_i)_{i=1}^n$.

Let δ be a regression estimation procedure producing estimator $\hat{f}_i(x) = \hat{f}_i(x; Z^i)$ for each sample size $i \geq 1$. Let $\| \cdot \|$ denote the L_2 norm with respect to the distribution of X. Let $R(f; n; \delta) = E\|f - \hat{f}_n\|^2$ denote the risk of the procedure δ at the sample size n under squared L_2 loss.

Let $\Delta = \{\delta_j, j \geq 1\}$ be a collection of regression procedures, and let $\hat{f}_{j,i}(x) = \hat{f}_{j,i}(x; Z^i)$ denote the estimator of f based on δ_j given the observations Z^i for $i \geq 1$. The index set $\{j \geq 1\}$ is allowed to degenerate to a finite set. Let π_j be positive numbers summing up to one, i.e., $\sum_{j=1}^{\infty} \pi_j = 1$. The algorithm ARM (adaptive regression by mixing) can be used to combine candidate procedures for adaptation in Yang (2001b). The following conditions are needed.

A1. The regression function and the estimators are uniformly bounded: there exists a constant $A > 0$ such that $\| f \|_\infty \leqslant A$ and $\| \widehat{f}_{j,i} \|_\infty \leqslant A$ with probability one for all i, j.

A2. The variance parameter σ is bounded above and below by known positive constants $\overline{\sigma} < \infty$ and $\underline{\sigma} > 0$.

A3. The known error distribution h has a finite fourth moment and satisfies that for each pair $0 < s_0 < 1$ and $T > 0$, there exists a constant $B_{s_0, T}$ (depending on s_0 and T) such that

$$\int h(x) \log \frac{h(x)}{\frac{1}{s} h(\frac{x-t}{s})} dx \leqslant B_{s_0, T} \left((1-s)^2 + t^2 \right)$$

for all $s_0 \leqslant s \leqslant s_0^{-1}$ and $-T < t < T$.

Theorem 2 *Assume Conditions A1-A3 hold. Then a convexly combined estimator $\widetilde{f}_n$ by ARM satisfies*

$$E\|f - \widetilde{f}_n\|^2 \leqslant C \inf_j \left(\frac{\sigma^2}{n} \left(1 + \log \frac{1}{\pi_j} \right) + E\|f - \widehat{f}_{j,n/2}\|^2 \right),$$

where the constant C depends only on A, $\overline{\sigma}$, $\underline{\sigma}$, and h. In particular, if there are M_n procedures to be combined with uniform weight, then

$$E\|f - \widetilde{f}_n\|^2 \leqslant C \left(\frac{\sigma^2 \log M_n}{n} + \inf_j E\|f - \widehat{f}_{j,n/2}\|^2 \right).$$

2.3.5 Implications: minimax-rate adaptation

Adaptive function estimation is an important topic in nonparametric function estimation. Under various performance measures, many results have been obtained on minimax-rate adaptation (even with the right constant in some cases) for specific function classes such as Sobolev and Besov classes. Results include Efroimovich and Pinsker (1984), Efroimovich (1985), Härdle et al. (1985), Lepski (1991), Golubev and Mussbaum (1992), Donoho and Johnstone (1994), Delyon and Juditsky (1994), Mammen and van de Geer (1997), Tsybakov (1995), Goldenshluger and Nemirovski (1997), Lepski et al. (1997), Devroye and Lugosi (1997) and many others. General schemes have been proposed for the construction of adaptive estimators in Barron and Cover (1991) based on minimum description length (MDL) criterion using ϵ-nets. Other adaptation schemes and adaptation bounds by model selection have been developed later including very general penalized contrast criteria in Birgé and Massart (1997), and Barron et al. (1999) with many interesting applications on adaptive estimation; penalized maximum likelihood or least squares criteria in Yang and Barron (1998) and Yang (1999); and complexity penalized criteria based on V-C theory (e.g., Lugosi and Nobel (1999)). For estimating functionals or pointwise performance assessment, see e.g., Lepski (1991), Cai and Low (2005) and Cai (2008) for references.

A fundamental question on adaptive function estimation is: Given a general collection of function classes, is it possible to achieve minimax-rate adaptation over them? It turns out that the results on combining procedures for adaptation readily give positive answers.

Regression: Yang (2000b) demonstrated that by the recipe to combine density estimators, one can easily construct minimax rate adaptive estimators over general function classes.

Let $\mathcal{U}$ be a class of regression functions. Consider the minimax risk for estimating a regression in $\mathcal{U}$:

$$R(\mathcal{U};n) = \min_{\hat{u}_n} \max_{u \in \mathcal{U}, \sigma \leqslant \bar{\sigma}} E \parallel u - \hat{u}_n \parallel^2,$$

where $\hat{u}_n$ is over all estimators based on $Z^n = (X_i, Y_i)_{i=1}^n$ and the expectation is taken under u and σ. Let $\{\mathcal{U}_j, j \geqslant 1\}$ be a collection of classes of regression functions uniformly bounded between $-A$ and A. Assume the true function u is in (at least) one of the classes, i.e., $u \in \cup_{j \geqslant 1} \mathcal{U}_j$. The question to be addressed is: Without knowing which class contains u, can we have a single estimator such that it converges automatically at the minimax optimal rate of the class that contains u? If such an estimator exists, we call it a minimax-rate adaptive estimator with respect to the classes $\{\mathcal{U}_j, j \geqslant 1\}$.

Definition: If the minimax risk sequence satisfies that $R(\mathcal{U}; \lfloor n/2 \rfloor)$ and $R(\mathcal{U}; n)$ are of the same order, then the minimax risk of the class $\mathcal{U}$ is said to be *rate-regular*.

Yang (2000b) gives the following theoretical result.

Theorem 3 *Let $\{\mathcal{U}_j, j \geqslant 1\}$ be any collection of uniformly bounded function classes. Assume further that the minimax risk of each of the classes is rate-regular. Then we can construct a minimax-rate adaptive procedure such that it automatically achieves the optimal rate of convergence for any class in the collection with a nonparametric regular-rate and it is within a logarithmic factor of the optimal for a parametric rate.*

For a demonstration, consider estimating a regression function on $[0,1]^d$. Applying the above result, we have an estimator that has the desired properties: it is consistent for all functions that are bounded between $-C$ and C for a known positive constant C, it adapts to smoothness and interaction order of Besov classes, and it converges at a rate $o(n^{-1/2})$ if the true regression function has a neural net representation.

Density estimation: Let $\{\mathcal{F}_\theta : \theta \in \Theta\}$ be a collection of density classes indexed by a hyper-parameter $\theta \in \Theta$. Here θ could be multi-dimensional with possibly continuous components (e.g., Besov classes). For simplicity, consider the case that Θ is a subset in a finite-dimensional Euclidean space R^m, $1 \leqslant m < \infty$. Let $\parallel \theta \parallel_2 = \sqrt{\theta_1^2 + ... + \theta_m^2}$ denote the Euclidean norm of $\theta = (\theta_1, ..., \theta_m)$ on R^m. Assume that there is a partial order on the hyper-parameter space Θ and that the order of the hyper-parameters is in accordance with the order of the corresponding density classes: if $\theta_1 \preceq \theta_2$ then $\mathcal{F}_{\theta_1} \subset \mathcal{F}_{\theta_2}$.

If for every choice of N_n with $N_n = o(n)$, we have

$$(1/N_n) \sum_{i=n-N_n+1}^{n} R(\mathcal{F}; i) \sim R(\mathcal{F}; n),$$

then we say the class has *a regular risk*. If $R(\mathcal{F}; n)$ converges essentially more slowly than the parametric rate, i.e., $R(\mathcal{F}; n) \geqslant cn^{-\theta}$ for some $0 < \theta < 1$ and a constant $c > 0$, we say the class has *a nonparametric risk*. Let $M(n; \theta)$ be an upper bound on the minimax risk being considered. A certain continuity condition on $M(n; \theta)$ is assumed. Yang (2001a) gives the following result.

Theorem 4 *Consider either the K-L risk or the squared L_2 risk. Let $\{\mathcal{F}_\theta, \theta \in \Theta\}$ be any collection of density classes each with a nonparametric rate-regular risk under the loss being considered. Under some continuity conditions on the minimax risks of the classes, we have:*

1. There exists a minimax-rate adaptive procedure under the K-L loss. If in fact, each class in the collection has a regular risk with $M(n; \theta)$ satisfying $\lim_{n \to \infty} R(\mathcal{F}_\theta; n)/M(n; \theta) = 1$ for each θ, then there exists an adaptive estimator asymptotically achieving the minimax risk under the K-L loss for every class $\mathcal{F}_\theta$, $\theta \in \Theta$.

2. Assume that $\{\mathcal{F}_\theta, \theta \in \Theta\}$ is uniformly bounded with $\sup_{\theta \in \Theta} \sup_{f \in \mathcal{F}_\theta} \|f\|_\infty \leqslant A < \infty$. If, in addition, either each $\mathcal{F}_j$ is convex including the uniform density 1 or the classes are uniformly bounded away from zero, then there exists a minimax-rate adaptive estimator over the classes under the squared L_2 loss.

For example, we can have an adaptive estimator for estimating a density on $[0, 1]^d$ with respect to Lebesgue measure μ under K-L loss such that 1) it is consistent for all densities with finite entropy; 2) it is simultaneously minimax-rate optimal for piecewise Besov classes with different interaction orders and continuous smoothness parameters.

Adaptive estimation by combining procedures has been considered for other statistical problems. We mention a few below.

Nemirovski (2000) proposes the combination of a large number of estimators for adaptive estimation for Barron's classes. He utilizes the aggregation bound for linear combination with constraint (details will be given in the next section). Gaïffas and Lecué (2007) give optimal rates and adaptation in the single-index model using aggregation. Chesneau and Lecué (2009) obtains minimax-rate optimal adaptation over Besov classes by using a convex combination of weighted term-by-term thresholded wavelet estimators.

2.3.6 Algorithms

In developing practically feasible algorithms, a simplification can be made to ease the computational burden of evaluating at every sample size. A general algorithm can be summarized as follows:

1. Data splitting: randomly split the data into two parts.
2. Estimation: Use the first part of the data to fit each candidate model.

3. Evaluation: For each model, obtain the likelihood of the second part of data E_j.

4. Weighting: Assign weight w_j to model j by $w_j = \frac{E_j}{\sum_j E_j}$.

5. Smoothing: Repeat the above procedure multiple times to average out data splitting variability and use the averaged weight over these repetitions.

Following this general recipe, Yang and co-authors developed feasible algorithms in different statistical settings, including Adaptive Regression by Mix (ARM) for regression and ANOVA models (Yang, 2001b) (Chen et al., 2007), Aggregated Forecasting Through Exponential Re-weighting (AFTER) for times series forecasting (Zou and Yang, 2004), Adaptive Classification by Mixing (ACM) (Yang, 2000a) for classification, and Adaptive Quantile Regression by Mixing (AQRM) (Shan and Yang, 2009) for quantile regression. Both theoretical risk bounds and extensive numerical results were demonstrated in these settings.

2.4   More recent developments

A number of interesting results on combining procedures for adaptation have appeared in recent years. Audibert (2006) gives fast learning rates in statistical inference through aggregation. Birgé (2006) develops optimal risk bounds for combining procedures based on a general idea of overcoming weaknesses of maximum likelihood estimation and other methods through many simultaneous tests between probability balls in a suitable metric space to attain robust optimal performance. Lecué (2007) derives classifiers that adapt to margin and complexity parameters by combining classifiers that work well non-adaptively. Dalalyan and Tsybakov (2007) study the problem of aggregation for regression with deterministic design. Györfi and Ottucsák (2007) study sequential prediction of unbounded stationary time series by a combination of many simple predictors. Giraud (2008) handles the problem of combining least squares estimators when the noise variance is unknown. Goldenshluger (2009) constructs general combining methods that work for a variety of global risk measures.

3   Combining procedures for improvement

The results given so far deal with combining statistical procedures for adaptation, i.e., matching the best performing candidate procedure. A more aggressive goal can be to do better than the best candidate procedure. This is most relevant when none of the candidate procedures are optimal, leaving the possibility of a combination doing better.

Some ways to combine candidate procedures for improvement include constrained or unconstrained linear combination and localized combination. More flexibility in combining the candidates allows greater potential to reach a combined procedure that is better than the individuals, but at the same time, a penalty is expected from searching for the best combined procedure in a larger class of allowed combinations. Results in these directions will be reviewed next, all on regression.

3.1 Global combining

3.1.1 Regression-type combining

A regression type combining for improvement originated with "stacked generalization" of Wolpert (1992). Breiman (1996a) developed this idea in the regression setting and called it "stacked regression" in which he formed linear combinations of predictors to improve prediction accuracy. Denote K different regression predictors by $c_k(x), k = 1, \ldots, K$. Let $c_k^{(-i)}(x_i)$ denote the leave-one-out cross-validated fit for $c^k(x)$, evaluated at $x = x_i$. The stacking method minimizes

$$\sum_{i=1}^{n}[y_i - \sum_{k=1}^{K} \beta_k c_k^{(-i)}(x_i)]^2,$$

producing $\hat{\beta}_k, k = 1, \ldots, K$. The final predictor is $v(x) = \sum \hat{\beta}_k c_k(x)$.

Breiman found that minimizing under the constraint $\beta_k \geqslant 0$ often produced a lower prediction error than the single predictor with the minimum cross validation error.

Instead of bias-correcting each individual predictor through cross validation, LebBlanc and Tibshirani (1996) used bootstrap techniques to estimate the bias of the empirical error $\sum_i (y_i - \sum_k \beta_k c_k(x_i))^2$, and minimize the sum of the empirical error and the estimate of the bias to produce β. Through approximating the cross-validation estimates, they also derived an analytical estimate of β. Their empirical studies demonstrated the good performance of nonnegativity constraints on β.

Clarke (2003) compared BMA with stacking in a variety of simulations and concluded that stacking is more robust than BMA in the most important settings.

These findings motivate the theoretical characterization that follows.

3.1.2 What is the tradeoff in different ways of combining for improvement?

Let (X_i, Y_i), $i = 1, ..., n$ be iid observations with X_i taking values in $\mathcal{X}$, a measurable set in R^d. Let $Y_i = f(X_i) + \varepsilon_i$, where f is the regression function and ε_i is normally distributed error with mean zero and variance σ^2. Unless stated otherwise, ε_i is assumed to be independent of X_i.

Now let $\Delta = \{\delta_1, \delta_2, ..., \delta_{M_n}\}$ denote a finite collection of candidate procedures for estimating the regression function f. For each $\delta_j, j \in J$, at a given sample size n, δ_j produces a regression estimator $\hat{f}_{j,n}(x)$. The number of procedures, M_n, changes according to the sample size n. In particular, we will consider the case when M_n is of order n^T for some $0 \leqslant T < \infty$. A larger sample size allows more candidate procedures (possibly more and more complicated) to be considered.

Let $\mathbf{F}_{n,C} = \{\sum_{1 \leqslant j \leqslant M_n} \theta_j \hat{f}_{j,n}(x) : \sum_{1 \leqslant j \leqslant M_n} |\theta_j| \leqslant 1\}$ be the collection of linear combinations of the original estimators in Δ with coefficients summing to no more than 1 in absolute values. Let $\| \cdot \|_1^{M_n}$ denote the l_1 norm on R^{M_n}, i.e.,

$\| \theta \|_1^{M_n} = \sum_{1 \leqslant j \leqslant M_n} |\theta_j|$. Define

$$R_C^* (f; n; \Delta) = \inf_{\|\theta\|_1^{M_n} \leqslant 1} E \| f - \sum_{1 \leqslant j \leqslant M_n} \theta_j \widehat{f}_{j,n} \|^2.$$

It is the smallest risk over all the estimators in the linear aggregation class $\mathbf{F}_{n,C}$. Without the constraint on the linear coefficients, we define $\mathbf{F}_n = \{\sum_{1 \leqslant j \leqslant M_n} \theta_j \widehat{f}_{j,n} ($ $\theta \in R^{M_n}\}$ be the collection of all linear combinations of the original estimators in Δ, and define

$$R^* (f; n; \Delta) = \inf_{\theta \in R^{M_n}} E \| f - \sum_{1 \leqslant j \leqslant M_n} \theta_j \widehat{f}_{j,n} \|^2.$$

Obviously, $R^* (f; n; \Delta) \leqslant R_C^* (f; n; \Delta) \leqslant \inf_{1 \leqslant j \leqslant M_n} R(f; n; \delta_j)$, which indicates more flexible combinations cannot hurt potential gain of accuracy.

Theorem 5 (Upper bounds) *Assume that f and all the regression estimates $\widehat{f}_{j,n}$ are uniformly upper and lower bounded. For any given collection of estimation procedures $\Delta = \{\delta_j, 1 \leqslant j \leqslant M_n\}$,*

1. *we can construct a combined procedure δ^* such that*

$$R(f; n; \delta^*) \leqslant \begin{cases} C_1 R_C^* \left(f; \frac{n}{2}; \Delta\right) + C_2 \frac{M_n}{n} & \text{when } M_n < \sqrt{n} \\ C_3 R_C^* \left(f; \frac{n}{4}; \Delta\right) + C_4 \sqrt{\frac{1}{n} \log \left(\frac{M_n}{\sqrt{n}} + 1\right)} & \text{when } M_n \geqslant \sqrt{n} \end{cases} .$$

In particular, if $M_n \leqslant C_0 n^T$ for some $T > 0$ and $C_0 > 0$, then

$$R(f; n; \delta^*) \leqslant C' \begin{cases} R^* \left(f; \frac{n}{2}; \Delta\right) + \frac{1}{n^{1-T}} & \text{when } 0 \leqslant T \leqslant 1/2 \\ R^* \left(f; \frac{n}{4}; \Delta\right) + \left(\frac{T \log n}{n}\right)^{1/2} & \text{when } 1/2 < T < \infty; \end{cases} \quad (3.1)$$

2. *we can construct a combined procedure δ^* such that $R(f; n; \delta^*) \leqslant R^*(f; \frac{n}{2}; \Delta)$* $+ C \frac{\min(M_n, n)}{n};$

where the constants C's do not depend on n.

Note that for both parametric and nonparametric regression, for a good procedure, $R(f; n; \delta)$ and $R(f; n/2; \delta)$ are usually of the same order. Thus $R^*(f; n; \Delta)$ and $R^*\left(f; \frac{n}{2}; \Delta\right)$ typically converge at the same rate. From the result, when $T \geqslant 1/2$, the penalty term for pursuing the best linear combination of n^T procedures is of order $((\log n)/n)^{1/2}$ (independent of T). When $T < 1/2$, the penalty is smaller in order, resulting in a possibly much faster rate of convergence. For an extreme example, when M_n is fixed, the price we pay is only of order $\log n/n$.

How good are the upper bounds derived here? They turn out to be optimal in some sense.

Theorem 6 (Lower bounds) *Consider normal errors with mean zero and variance 1.*

1. *There exist M_n procedures $\Delta_{M_n} = \{\delta_j, 1 \leqslant j \leqslant M_n\}$ such that for any aggregated procedure $\delta^{(n)}$ based on Δ_{M_n}, one can find a regression function f with $\| f \|_\infty$ uniformly upper bounded satisfying*

$$R\left(f; n; \delta^{(n)}\right) - R_C^*\left(f; n; \Delta_{M_n}\right) \geqslant C \begin{cases} \dfrac{M_n}{n} & \text{when } M_n < \sqrt{n} \\[2mm] \sqrt{\dfrac{1}{n} \log\left(\dfrac{M_n}{\sqrt{n}} + 1\right)} & \text{when } M_n \geqslant \sqrt{n} \end{cases}$$

   where the constant C does not depend on n.

2. *There exist M_n procedures $\Delta_{M_n} = \{\delta_j, 1 \leqslant j \leqslant M_n\}$ such that for any aggregated procedure $\delta^{(n)}$ based on Δ_{M_n}, one can find a regression function f with $\| f \|_\infty$ uniformly upper bounded satisfying*

$$R\left(f; n; \delta^{(n)}\right) - R^*\left(f; n; \Delta_{M_n}\right) \geqslant C \min\left(\frac{M_n}{n}, 1\right),$$

   where the constant C does not depend on n.

3. *There exist M_n procedures $\Delta_{M_n} = \{\delta_j, 1 \leqslant j \leqslant M_n\}$ such that for any aggregated procedure $\delta^{(n)}$ based on Δ_{M_n}, one can find a regression function f with $\| f \|_\infty$ uniformly upper bounded satisfying*

$$R\left(f; n; \delta^{(n)}\right) - \inf_{1 \leqslant j \leqslant M_n} R^*\left(f; n; \delta_j\right) \geqslant C \min\left(\frac{\log(M_n)}{n}, 1\right),$$

   where the constant C does not depend on n.

REMARKS:

1. Nemirovski (2000) describes the three different directions of combining estimates.

2. Both Juditsky and Nemirovski (2000) and Tsybakov (2003) focus on the aggregation stage (i.e., the regression estimates to be combined are treated as fixed) and give methods that have the above risk bounds with the multipliers $C_1 = C_2 = 1$. The multiplying constant being 1 is important in the aggregation problem for its mathematical elegance. When the game is to combine the candidate procedures, aggregating the estimates is only part of the whole combining process. Since $R^*\left(f; \frac{n}{2}; \Delta\right)$ is the risk at the sample size $n/2$ instead of n, which is larger than but typically of the same order as $R^*\left(f; n; \Delta\right)$, the risk bounds with $C_1 = C_2 = 1$ do not seem to have general significant theoretical or practical gain.

3. For linear combining under constraint, Juditsky and Nemirovski (2000) and Yang (2004) give partial or near optimal results (not necessarily requiring normality of error distribution) for upper and lower bounds. Tsybakov (2003) later provides a complete solution for the aggregation stage under the normal error assumption, and also give the answer for linear combination without constraint.

4. In combining for adaptation, some upper bounds are given in the previous section for both density estimation and regression (see also Catoni (2004)).

5. The sample size $n/2$ involved in the upper bound is due to a specific data splitting choice. Similar results hold with other data splitting choices.

Combining the two theorems, without any additional conditions, we somewhat disappointingly still cannot determine the optimal rate of convergence for the combining problems because $R^*\left(f;\frac{n}{2};\Delta\right)$ in the upper bound and $R^*\left(f;n;\Delta\right)$ in the lower bound may or may not converge at the same order. If they do, which fortunately is indeed the case for familiar parametric and nonparametric situations, then uniformly over bounded f, 1), for linear combining with constraint, the optimal rate of convergence of $R\left(f;n;\delta\right)$, in a proper sense, is

$R_C^*\left(f;n;\Delta_{M_n}\right)+\frac{M_n}{n}$ for $M_n<\sqrt{n}$ and $R_C^*\left(f;n;\Delta_{M_n}\right)+\sqrt{\frac{1}{n}\log\left(\frac{M_n}{\sqrt{n}}+1\right)}$ for $M_n\geqslant\sqrt{n}$; 2), For linear combining, the optimal rate of convergence of $R\left(f;n;\delta^*\right)$ is $R^*\left(f;n;\Delta_{M_n}\right)+\min\left(\frac{M_n}{n},1\right)$; 3), For combining for adaptation, the optimal rate of convergence of $R\left(f;n;\delta^*\right)$ is $\inf_{1\leqslant j\leqslant M_n}R^*\left(f;n;\delta_j\right)+\min\left(\frac{\log(M_n)}{n},1\right)$.

It is worth noting how the price (in rate) for constrained linear combining changes according to M_n. In the beginning, it increases linearly in M_n, but after M_n reaches $\sqrt{n}$, it increases much more slowly in a logarithmic fashion at rate $\left(\frac{\log n}{n}\right)^{1/2}$ as long as M_n increases polynomially in n.

3.1.3 Multi-directional combining

Observing the dramatic difference between the different combination scenarios, we naturally face the question: Should we combine for adaptation or for improvement and in which way? If one of the original procedures happen to behave the best (or close to the best) among all the linear combinations, or at least one of the original procedures converges at a rate faster than $n^{-(1-T)}$ (for $0\leqslant T<1/2$) or $\sqrt{\log n/n}$ (for $T\geqslant 1/2$), if one linearly combines the procedures with or without the constraint, one could be unfortunately paying too high a price for nothing but hurting the convergence rate in estimating f. In terms of rate of convergence, linear combining with constraint is worth the effort for sure only if $R^*\left(f;n/2;\Delta\right)$ plus the price of $n^{-(1-T)}$ (for $0\leqslant T<1/2$) or $\sqrt{\log n/n}$ (for $T\geqslant 1/2$) is of a smaller order than $(\log M_n)/n+\inf_j R(f;n/2;\delta_j)$. Since the risks are obviously unknown, in applications, one cannot know in advance whether to combine for adaptation or combine for improvement.

Multi-directional combining is considered in Yang (2004). It is shown that when combining the procedures carefully, one can have the potential of obtaining a large gain in estimation accuracy yet without losing much when there happens to be no advantage considering sophisticated linear combinations. Specifically, by applying combining at two levels, the first level to achieve the best performance of a class of combinations and the second to combine these combined strategies, the overall estimation method intelligently behaves like the best way of combining being considered. He shows that the multi-directional combining simultaneously serves three directions in terms of rate of convergence. First of all, the final estimator converges as fast as any original procedure. Secondly, when linear combinations of the first L_n procedures (for some $L_n>1$) can improve estimation accuracy dramatically, one pays the price at most of order $\frac{L_n\log(1+n/L_n)}{n}$ when $1\leqslant L_n<\sqrt{n}$ and $\frac{\log L_n}{\sqrt{n}\log n}$ when $L_n\geqslant\sqrt{n}$ for the best performance of linear combinations with the summability constraint. When L_n is small, the gain is substantial. When certain linear combinations of a small number of procedures perform well, the final estimator can also take advantage of that. In summary, the final estimator can behave both aggressively (combining for improvement) and conservatively (combining for adaptation) which ever is better.

Bunea et al. (2007) in the same spirit constructs a two-level combining method with the second level based either on BIC type or data-dependent l_1-type penalty, and show that the final estimator nearly achieves the best performance among combining for adaptation, linear combining with or without constraint and sparse combining when only a subset of candidate procedures is important.

3.2   Localized combining/selection

In the above-mentioned combining procedures, each model/procedure is assessed and weighted on its global performance, and the weight a model receives is constant across the domain of the input x. However, in some complicated cases, the relative performance of different models may vary across the domain of x, and it is desirable to consider localized combining where the weight of a model depends on x. For example, in the regression setting, the combined estimator may have the form $\widehat{f}(x) = \sum_j w_j(x; \theta_j)\widehat{f}_j(x)$ where $\theta_j \in \Theta$ is a parameter vector.

3.2.1 Localized combining

To obtain the input dependent weights, Pan et al. (2006) used a regression scheme based on the training data. The idea is to create a new response variable $z_i, i = 1, \ldots, n$ which measures the empirical or predictive performance under each model, and regress z on x to get the parameter estimate $\hat{\theta}$. The weight can be computed as $w_j(x) \propto g(x'\hat{\theta}_j)$ where $g(z)$ is a one-dimensional, nonnegative function such as the logistic function. Pan et al. (2006) demonstrated the advantage of input dependent weighting through some empirical studies.

Let the combined estimator be of the form

$$\widehat{f}(x) = \sum_j w_j(x; \theta)\widehat{f}_{j,n_1}(x) : \theta \in \Theta,$$

where $\widehat{f}_{j,n_1}$ are based on the first part of data of sample size $n_1 = n - n_2$. Let $w_j(x; \theta^*)$ be the minimizer of the expectation of $L(\theta; n_1) = \| \sum_j w_j(; \theta)\widehat{f}_{j,n_1} - f \|^2$. Use least squares method to estimate θ^* and let $\widehat{\theta}$ be the minimizer of the average squared prediction error

$$\min_{\theta \in \Theta} \frac{1}{n_2} \sum_{i=n_1+1}^{n} \left(Y_i - \left(\sum_j w_j(X_i; \theta)\widehat{f}_{1,n_1}(X_i) \right) \right)^2.$$

From Yang and Barron (1998) and Huang (2004), if the weight set $\{w_j(x; \theta) : \theta \in \Theta\}$ has a finite metric dimension (see Assumption 7 of Huang (2004)), then

$$EL(\widehat{\theta}; n_1) = O\left(EL(\theta^*; n_1) + \frac{1}{n_2} \right).$$

Thus for linear localized combining, with weighting functions from a class with a finite metric dimension, the least squares method typically achieves the optimal convergence rate.

3.2.2 Localized selection

Yang (2008) considered a special form of localized combining, localized selection. For simplicity of illustration, consider two procedures to combine:

$$\widehat{f}(x) = \widehat{f}_1(x)I_{\{x \in A\}} + \widehat{f}_2(x)I_{\{x \in A^c\}}$$

for some A. Here A is assumed to be in a given collection of sets $\mathcal{A}_n$.

The weighting reduces to estimate A^*, the area at which the first procedure performs no worse than the second. Yang proposed two directions to estimate A^*.

Local cross validation method. When it is difficult to describe A^* by simple sets, instead of directly estimating A^*, at each given x_0, try to find the candidate that performs better in the local neighborhood around x_0. For a given x_0, consider the ball centered at x_0 with radius r_n for some $r_n > 0$. Randomly split the data into two parts, and then use the training set to do estimation and only the test set in the defined neighborhood to do evaluation. Identify the procedure with the smaller average squared prediction error. This process is repeated with a number of random splittings of the observations to avoid the splitting bias. The procedure that wins more frequently among the permutations is the final winner. For this method, Yang (2008) proved that as long as the neighborhood size and the data splitting ratio are chosen properly, the locally best estimator will be selected with probability tending to 1.

Optimal region selection. The goal is to directly estimate the preference region. In general, the optimal set can be very complicated and hard to estimate. Yang (2008) considered a collection of sets of manageable complexity and used empirical minimization or cross validation to select a set in the class. Yang (2008) proved that when the complexity of the collection of the candidate sets is properly controlled, the combined procedure based on the empirically selected optimal-region behaved well in squared error risk.

4 Concluding remarks

The theoretical and empirical researches on model combining have clearly shown its power as an important tool for statistical estimation and prediction. In the modern era of rich and complex data, we expect that the approach of combining statistical procedures will prove to be essential for sharing strengths of different modeling ideas and tools.

Is model combining always better than model selection? If not, when is combining better than selection and how do we know it? Yang and his co-authors (Yuan and Yang, 2005) (Chen et al., 2007) tried to answer these questions through studying the relationship between model selection instability and the relative performance of model combining. Following Breiman (1996b), they proposed bootstrap instability, perturbation instability and sequential instability in selection and perturbation instability in estimation to measure selection instability. Their empirical studies showed that model combining performed better with higher selection instability and selection worked better with low selection instability. This suggests a data analyst should be aware of the uncertainty of model selection as well as the pitfall of blind combining. Just assigning a set of weights to models does not automatically solve the problem of model selection uncertainty. The measures proposed by Yang and his co-authors can serve as a useful guide for deciding between selection and combining for adaptation. More generally, when other goals of combining procedures are involved, much is to be learned about how to statistically characterize relative performance of and relationships between the candidate procedures in ways that lead to insight on choosing the right direction and method of combining.

Combining confidence intervals is a difficult problem. When there is non-negligible uncertainty about the "true model", clearly confidence intervals based on the selected model are not trust-worthy. One naturally hopes that a suitable combination of the intervals from the candidate models can much help the situation. However, theoretical developments seem very difficult and we are not aware of developments in this direction.

Localized combining presents another challenge. Identifying local or non-global relative performance patterns in a computationally feasible way is generally difficult, especially for high-dimensional data. Here non-asymptotic results are sought.

A major criticism of the approach of combining procedures is that it lacks interpretation and does not offer much understanding by giving a set of most relevant predictors, which is a practically important issue in many applications. Chen et al. (2007), in the context of factorial data analysis, made an effort to use combined estimates of cell means to infer which factors were likely to be most significant. The more accurate combined regression estimate may be better for identifying important predictors than the estimate based on model selection. More work in this area may prove to be fruitful.

References

[1] Audibert J.-Y.(2006). A randomized online learning algorithm for better variance control. *Lecture Notes in Computer Science (including subseries Lecture Notes in Artificial Intelligence and Lecture Notes in Bioinformatics)*, 4005 LNAI, 392-407.

[2] Barron A. R.(1987). Are Bayes rules consistent in information? In T. M. Cover and B. Gopinath, editors, *Open Problems in Communications and Computation*, 85-91. Springer-Verlag, New York.

[3] Barron A. R. and Cover T. M.(1991). Minimum complexity density estimation. *IEEE Transactions on Information Theory* **37**, 1034-1054.

[4] Barron A. R., Birgé L. and Massart P.(1999). *Probability Theory and Related Fields*, Chapter Risk bounds for model selection via penalization, 301-413. Springer, Berlin.

[5] Birgé L.(2006). Model selection via testing: An alternative to (penalized)

maximum likelihood estimators. *Annales de l'institut Henri Poincare (B) Probability and Statistics* **42** (3), 273-325.

[6] Birgé L. and Massart P.(1997). *From model selection to adaptive estimation*, pages 55–87. Festschrift for Lucien Le Cam: Research Papers in Probability and Statistics. Springer-Verlag, New York.

[7] Breiman L.(1996a). Bagging predictors. *Machine Learning* **24**, 123-140.

[8] Breiman L.(1996b). Heuristics of instability and stabilization in model selection. *Annals of Statistics* **24**, 2350-2383.

[9] Breiman L.(2001). Random forest. *Machine Learning* **45**,5-32.

[10] Buckland S. T., Burnham K. P. and Augustin N. H.(1997). Model selection: An integral part of inference. *Biometrics* **53**, 603-618.

[11] Bunea F., Tsybakov A. B. and Wegkamp M. H.(2007). Aggregation for gussian regression. *Annals of Statistics* **35**(4), 1674-1697.

[12] Cai T. T.(2008). On information pooling, adaptability and superefficiency in nonparametric function estimation. *Journal of Multivariate Analysis* **99**, 421-436.

[13] Cai T. T. and Low M. G.(2005). On adaptive estimation of linear functionals. *Annals of Statistics* **33**, 2311-2343.

[14] Catoni O.(2004). *Statistical Learning Theory and Stochastic Optimization*, volume 1851 of *Lecture Notes in Mathematics*. Springer, New York.

[15] Cesa-Bianchi and Lugosi G.(2006). *Prediction, Learning and Games*. Cambridge University Press, Cambridge.

[16] Chen L., Giannakouros P., and Yang Y.(2007). Model combining in factorial data analysis. *Journal of Statistical Planning and Inference* **137**(9), 2920-2934.

[17] Chesneau C. and Lecué G.(2009). Adapting to unknown smoothness by aggregation of thresholded wavelet estimators. *Statistica Sinica* **19**, 1407-1417.

[18] Clarke B.(2003). Comparing Bayes model averaging and stacking when model approximation error cannot be ignored. *Journal of Machine Learning Research* **4**, 683-712.

[19] Dalalyan A. S. and Tsybakov A. B.(2007). Aggregation by exponential weighting and sharp oracle inequalities. *Lecture Notes in Computer Science (including subseries Lecture Notes in Artificial Intelligence and Lecture Notes in Bioinformatics)*, **4539 LNAI**, 97-111.

[20] Delyon B. and Juditsky A.(1994). Wavelet estimators, global error measures revisited. *Technical report*, IRISA.

[21] Devroye L. P. and Lugosi G.(1997). Nonparametric universal smoothing factors, kernel complexity, and Yatracos classes. *Annals of Statistics* **25**, 2626-2637.

[22] Donoho D. L. and Johnstone I. M.(1994). Ideal denoising in an orthonormal basis chosen from a library of bases. *C. R. Acad. Sci. Paris*. **319**, 1317-1322.

[23] Efroimovich S. Yu.(1985). Nonparametric estimation of a density of unknown smoothness. *Theory of Probability and its Applications*.

[24] Efroimovich S. Yu. and Pinsker M. S.(1984). A self-educating nonparametric filtration algorithm. *Automation and Remote Control.*

[25] Freund Y. and Schapire R.(1996). Experiments with a new boosting al-

gorithm. In *Machine Learning: Proceedings of the Thirteenth International Conference,* 148-156.

[26] Friedman J., Hastie T. and Tibshirani R.(2000). Additive logistic regression: a statistical view of boosting. *Annals of Statistics* **28**(2), 337-407.

[27] Gaïffas S. and Lecué G.(2007). Optimal rates and adaptation in the single-index model using aggregation. *Electronic Journal of Statistics* **1**, 538-573.

[28] Giraud C.(2008). Mixing least-squares estimators when the variance is unknown. *Bernoulli* **14**(4), 1089-1107.

[29] Goldenshluger A.(2009). A universal procedure for aggregating estimators. *Annals of Statistics* **37**(1), 542-568.

[30] Goldenshluger A. and Nemirovski A.(1997). Adaptive de-noising of signals satisfying differential inequalities. *IEEE Transaction on Information Theory* **43**, 872-889.

[31] Golubev G. K. and Mussbaum M.(1992). Adaptive spline estimates for nonparametric regression models. *Theory of Probability and its Applications* **37**, 521-529.

[32] Györfi L. and Ottucsá k.(2007). Sequential prediction of unbounded stationary time series. *IEEE Transactions on Information Theory* **53**(5), 1866-1872.

[33] Härdle W. and Marron J. S.(1985). Optimal bandwidth selection in nonparametric regression models. *Annals of Statistics* **13**, 1465-1481.

[34] Hjort N. L. and Claeskens G.(2003). Frequentist model average estimators. *Journal of the American Statistical Association* **98**(464), 879-899.

[35] Huang T.(2004). Convergence rates for posterior distributions and adaptive estimation. *Annals of Statistics* **32**, 1556-1593.

[36] Juditsky A. and Nemirovski A.(2000). Functional aggregation for nonparametric estimation. *Annals of Statistics* **28**, 681-712.

[37] LebBlanc M. and Tibshirani Rob.(1996). Combining estimates in regression and classification. *Journal of American Statistical Association* **91**, 1641-1650.

[38] Lecué G.(2007). Simultaneous adaptation to the margin and to complexity in classification. *Annals of Statistics* **35** (4), 1698-1721.

[39] Lepski O. V.(1991). Asymptotically minimax adaptive estimation i: Upper bounds. optimally adaptive estimates. *Theory of Probability and its Applications* **36**, 682-697.

[40] Lepski O. V., Mammen E. and Spokoiny V. G.(1997). Ideal spatial adaptation to inhomogeneous smoothness: an approach based on kernel estimates with variable bandwidth selection. *Annals of Statistics* **25**, 929-947.

[41] Leung R. and Barron A. R.(2006). Information theory and mixing least-squares regressions. *IEEE Transactions on Information Theory* **52**(8), 3396-3410.

[42] Lugosi G. and Nobel A.(1999). Adaptive model selection using empirical complexities. *Annals of Statistics* **27**, 1830-1864.

[43] Madigan D. and Raftery A.(1994). Model selection and accounting for model uncertainty in graphical models using Occam's Window. *Journal of the American Statistical Association* **89**, 1535-1546.

[44] Madigan D. and York J.(1985). Bayesian graphical models for discrete data. *Internation Statistical Review* **63**, 215-232.

[45] Mammen E. and van de Geer S.(1997). Locally adaptive regression splines. *Annals of Statistics* **25**, 387-413.

[46] Miller A. J.(1990). *Subset Selection in Regression.* Chapman and Hall.

[47] Nemirovski A.(2000). *Topics in Non-parametric Statistics*, volume 1738 of *Lecture Notes in Mathematics.* Springer, New York.

[48] Pan W., Xiao G. and Huang X.(2006). Using input dependent weights for model combination and model selection with multiple sources of data. *Statistica Sinica* **16**(2), 523-540.

[49] Raftery A.(1985). Bayesian model selection in social research. *Sociological Methodology* **25**, 111-163.

[50] Shan K. and Yang Y.(2009). Combining quantile regression estimators. *Statistica Sinica.*

[51] Shen X. and Huang H.(2006). Optimal model assessment, selection, and combination. *Journal of the American Statistical Association* **101**(474), 554-568.

[52] Tsybakov A.(2003). Optimal rates of aggregation. *Computational Learning Theory and Kernel Machines*, volume 2777 of *Lecture Notes in Artificial Intelligence.* Springer, Heidelberg.

[53] Tsybakov A. B.(1995). Pointwise and sup-norm adaptive signal estimation on the sobolev classes. Preprint.

[54] Wolpert D. H.(1992). Stacked generalization. *Neural Networks* **5**, 241-259.

[55] Yang Y.(1999). Model selection for nonparametric regression. *Statistica Sinica* **9**, 475-499.

[56] Yang Y.(2000a). Adaptive estimation in pattern recognition by combining different procedures. *Statistica Sinica* **10**, 1069-1089.

[57] Yang Y.(2000b). Combining different procedures for adaptive regression. *Journal of Multivariate Analysis* **74**, 135-161.

[58] Yang Y.(2001a). Minimax rate adaptive estimation over continuous hyperparameters. *IEEE Transactions on Information Theory* **47**(5), 2081-2085.

[59] Yang Y.(2001b). Adaptive regression by mixing. *Journal of the American Statistical Association* **96**, 574-588.

[60] Yang Y.(2003). Regression with multiple candidate models: selecting or mixing? *Statistica Sinica* **13**, 783-809.

[61] Yang Y.(2004). Combining forecasting procedures: Some theoretical results. *Econometric Theory* **20** (1), 176-222.

[62] Yang Y.(2008). Localized model selection for regression. *Econometric Theory* **24**, 472-492.

[63] Yang Y. and Barron A. R.(1998). An asympotic property of model selection criterion. *IEEE Transactions on Information Theory* **44**, 95-116.

[64] Yang Y. and Barron A. R.(1999). Information theoretic determination of minimax rate of convergence. *Annals of Statistics* **27**, 1564-1599.

[65] Yuan Z. and Yang Y.(2005). Combining linear regression models: When and how? *Journal of the American Statistical Association* **100** (472), 1202-1214.

[66] Zou H. and Yang Y.(2004). Combining time series models for forecasting. *International Journal of Forecasting* **20** (1), 69-84.

Subject Index

L_1-norm penalty 25, 33, 39, 47, 60, 129, 226, 248, 249

A

accelerated failure time model 212, 224, 247, 251

accelerated life model 196, 201, 202, 211-213, 215

adaptive Lasso 61, 134, 142, 170-172, 187, 227, 231, 232, 234, 245, 250

B

Bayes classifier 5-7, 14, 15

Binary Markov random field (BMRF) 148, 149, 151

C

Censoring 177, 196-198, 201, 208, 212, 213, 224, 225, 227

central k-th moment subspace 259

central mean subspace 259, 261, 263, 264

central subspace 258-267

classification error rates 3

combining statistical procedures 275-277, 288

Cox model 174, 182, 224, 226, 228, 229, 232, 233, 247

D

dimension reduction subspace 258, 259

distanced-based classifier 18, 20

double censoring 196-198

E

elastic net 134, 142, 171, 181, 247-249

eQTLs 173, 182, 183, 185, 188

F

failure time 204, 211, 212, 213, 223-225, 234, 247

false discovery rate 30, 75, 77, 87, 88, 93, 97, 105, 183

feature selection 3, 4, 13, 14, 19-21, 23, 24, 27, 29

Fisher consistency 25, 32, 33, 39, 52, 56

G

Gaussian Markov random field (GMRF) 151

gene network 148, 150, 151, 157

grouped hypotheses 79, 93, 95, 96

grouped variables 231, 232, 238

H

hierarchical penalty 233, 234, 242

I

impact of dimensionality 8, 9, 11, 12, 16, 19

independence learning 28

independence rule 3, 4, 9-11, 15, 16, 20, 21

information criteria 119, 264

intensive longitudinal data 195, 196, 199, 200, 215, 220

interval censoring 195-198, 201, 217

L

large margin 39-41, 43, 45-47, 49, 51, 53, 55, 57, 59, 61-63, 65, 66, 67, 69-71

large-scale multiple testing 75, 76, 78, 83, 85, 102, 110

Author Index

A

Abecasis G., 189
Agresti A., 66
Ahlquist P., 114
Akaike H., 106, 123, 207, 250, 265
Alon U., 189
An L. T. H., 33, 56
Andersen P. K., 198, 224, 230, 231, 232, 241, 242
Anderson P., 238-240, 243
Anderson T., 120
Antoniadis A., 233, 251
Apanasovich T., 218
Audibert J.-Y., 288
Augustin N. H., 296

B

Babu D., 217
Bach F. R., 263
Bae K., 148
Bair E., 13, 34, 65, 69
Banerjee M., 198
Bannerjee S., 185
Barbacioru C., 192
Barron A. R., 279-282, 284, 286, 292, 294, 296, 297
Barski A., 188
Bartlett P., 25
Bartlett P. L., 56
Baum L., 105, 106, 112, 284, 296
Becker C., 262
Beer D. G., 253
Bellman R. E., 257
Benjamini Y., 77, 80, 93, 103, 112
Bennett K., 33, 53
Bennett K. P., 43
Berger J., 125
Bernstein C. N., 190
Besag J., 148, 151, 163

Best N., 145
Bhandarkar S., 268
Bickel P., 106, 112, 172, 189
Bickel P. J., 8-10, 15-17, 34, 224, 227
Bik E. M., 190
Binder H., 158
Birgé L., 285
Bishop C. M., 40
Blair C. B., 216
Bodeau J., 192
Boldrick J., 112
Bondell H. D., 265
Borgan O., 198
Boser B., 25
Boulesteix A. L., 13
Bowden D., 164
Bradfield J., 113
Bradley P., 47, 59
Bredensteiner E., 33, 53
Breheny P., 171
Breiman L., 26, 34, 40, 46, 67, 125, 129, 142, 143, 242, 251, 276, 287, 288, 293, 295
Brem R. B., 184
Breslow N. E., 198, 246
Brown E. R., 196, 205
Brown M., 114
Brown P. J., 125
Buchholz B. A., 220
Buckland S. T., 220
Buhlmann B., 170, 172
Buhlmann P., 191
Bumgarner R., 115
Bunea F., 227, 292
Bura E., 13, 34, 262, 264, 268
Burdick J. T., 189
Burge C. B., 36, 37, 70, 192
Burnham K. P., 296
Bycott P., 196, 203, 204

Bayes boundary

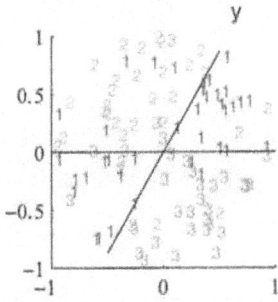

y

Boundary using loss function H_1

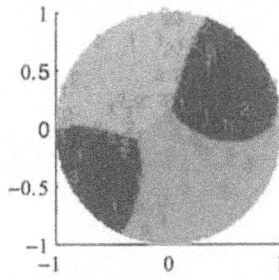

Boundary using loss function T_0

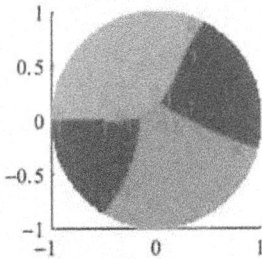

Boundary using loss function $T_{-0.5}$

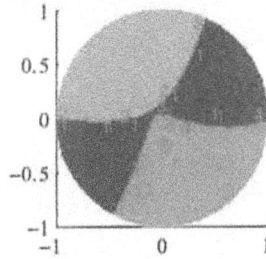